The Institution of
StructuralEngineers

Computational engineering

Author

P Debney   BEng(Hons) DipComp(Open) CEng FIStructE (Arup)

Reviewers

D Brohn   CEng FIStructE (New Paradigm Solutions Ltd)
G Evans   BSc(Hons) PhD CEng FICE FIStructE MBCS (Constructex) *Technical Products Panel*
R Feigin   BEng(Hons) BSc MIStructE CEng MIEAust (AECOM)
C Hickey   MMath(Hons) (Arup)
P Jeffries   MEng(Hons) CEng MIED (Ramboll)
R Kannan   PhD (Arup)
J Leach   MEng CEng FIStructE MICE (AECOM)
T Q Li   PhD MIStructE CEng (Arup)
I A MacLeod   BSc PhD FIStructE FICE (Prof. Em., University of Strathclyde)
S Melville   BEng MSc DIC CEng MIStructE FRSA (Format Engineers)
P Shepherd   MA(Cantab) PhD PGCAPP CEng CMath CSci MICE FIMA SFHEA (University of Bath)
W Wild   MEng(Hons) MIStructE CEng (Arup)

Publishing

L Baldwin   BA(Hons) DipPub (The Institution of Structural Engineers)

Published by The Institution of Structural Engineers
International HQ, 47–58 Bastwick Street, London EC1V 3PS, United Kingdom
T: +44(0)20 7235 4535
E: mail@istructe.org
W: www.istructe.org

First published (version 1.0) October 2020
Version 1.1 (published February 2021) corrects a small error in an equation in Appendix G
Version 1.2 (published July 2021) corrects a small error in a matrix on p.87 and contains minor textual
improvements throughout
This version (1.3 – published July 2022) corrects errors in Figures 8.1 and 10.6

978-1-906335-44-1 (print)
978-1-906335-45-8 (pdf)

Contents

Foreword

An alien looking down at the occupants of planet Earth during the COVID-19 pandemic might have marvelled at the adaptability and resourcefulness of humankind… changing in the space of a few days from the traditional, physical face-to-face relationships of our hunter-gather origins to a new world heavily dependent on virtual remote-working, sharing information, friendships, culture, humour, shopping and news online, with no real physical equivalent. I fear for my fast-disappearing legs. This foreword is written and transmitted, from my house, in the midst of a lockdown, in a way that would have been impossible until the development of sophisticated and publicly available computational tools. I've checked the facts, remotely; I've read the original source material; looked at conflicting arguments; checked a couple of formulae and run some numerical variations; and I'm about to have a meeting with my design team all without leaving my seat. This has only become possible in the last decade or two of our planet's 13.5 billion year history, and probably there's no going back: we are at a watershed and it signals an end to the "old way of working".

Four decades ago, when I began as a practising engineer, I was told that "the mark of a good engineer was to be able to produce 10 pages of neat, accurate calculations per day". Slowly, the first finite element programmes churned away, with the DEC 10, a machine the size of a small building churning out reams of 16 digit numbers that you would hunt through to try and spot the one critical condition… No previews, and a mistake took 24 hours to re-run. Even in 2005 the idea of being able to shift an entire workforce out of its expensive offices to work at home without major disruption would have been fanciful… yet our office has made the transition seamlessly over a weekend. And many have said they like commuting virtually, and have no plans to go back to physical travel.

Such is the power and convenience, and the adaptability of our digital tools. Eric Schmidt, former CEO of Google, once remarked we are now in a partnership between humans and machines in which humans should continue to do what they do best. On Artificial Intelligence, Jason Collins, a behavioural and data scientist from PwC Australia, inverts it more pithily in a down-under-ish fashion: "Before humans become the standard way in which we make decisions," he writes, "we need to consider the risks and ensure implementation of human decision-making systems does not cause widespread harm."

Of course we enjoy the digital world, which brings us possibilities that we have never before contemplated as a species. But the Law of Unintended Consequences still applies: There is a school of thought, and a thousand Hollywood movies heavily CGI-d, that we have never made contact with aliens because, in advanced technological civilisations, the ability for intelligent life to control its tools is eventually surpassed by the capability of those tools to act autonomously. That once a civilisation has the power to destroy itself, it is only a matter of time before that power is accidentally or deliberately used. Just as they were able to talk to us, those aliens blew themselves up, and perhaps that version of the omniverse explains why there's been no sign of aliens in the 80 years since we invented radio receivers and have been able to detect them.

While humans are not the only creatures who use tools, we have over the past 10,000 years taken them to levels exponentially beyond those of our planet's fellow species. The relationship between ourselves and our computers large and small, local or dispersed, is at the heart of our generation's success. Yes, if we abuse that relationship the digital era may bite us, let's hope nothing worse, no asteroid to wipe us out. But used intelligently and in partnership, it offers the key to our future development. The speed and force of the change in that partnership turns it into a moment in the evolution of our species of Darwinian significance to us all.

Chris Wise
April 2020

Foreword

I joined the structural engineering profession forty years ago when computer analysis was just taking hold as a mainstream design activity. My whole career has been built upon trying to use new digital techniques as and when they become available. In general, my observation is that their use is more limited by our lack of imagination and curiosity, combined with a natural risk aversion common among engineers, than it has been by the technology itself.

Over that period, structural analysis, or performance simulation, has not changed very much but our attitude towards it has. We used to use it to prove ideas and designs that we had already fully conceived. It simply checked the sums and made sure that we had not made a numerical mistake.

Nowadays we use simulation to explore conceptual possibilities: what if? How might this structure compare with that one? Which is more carbon efficient (cost efficient, material efficient…) while being easier to build and providing the best human experience? These are examples of the sort of questions that we increasingly use digital simulations, building information models, visualisations and virtual reality to help answer.

At all times, simulation models have the capability of being double-edged swords. We still need to use our engineering judgement and experience to decide what to model and how to represent it. When I was first taught about finite element analysis (FEA), I was told that such an analysis would converge on a good solution only as you continuously refined the mesh. From that, I have learned to always build lots of different models; simple ones and complicated ones; two dimensional and three dimensional; of components, assemblies and of the whole structure. I do not agree with those that assert that you need to be able to check the model by hand, but I do agree that all models need to be checked by making lots of other models — ideally by different people using different software.

In the immediate future we can start harnessing Machine Learning (ML) and Artificial Intelligence (AI) to better assess the outcomes of these simulations. Once we have processed a whole set of parametric simulations in the cloud, we can unleash ML/AI to sift and sort the results, and extrapolate towards more fertile possibilities.

We can use 3D printing to test the forthcoming ideas at model scale and additive (and subtractive) manufacturing to test them at full scale; we can simulate all aspects of their real-world performance, such as auralisation, before committing to a physical reality. In the truly virtual world, we can simulate millions or billions of possibilities and not care at all if 99.999% of them fail, provided we find the single solution that best meets the project needs — such profligacy does not always sit comfortably with engineers.

These tools and techniques will result in a more delightful and efficient built environment. But what are the opportunities for then taking a further step and harnessing "internet of things" devices to monitor real world performance? We could reduce our design loads and material factors of safety if we can reliably monitor real-world loads and real-material performance.

We are also likely to see other branches of technology, for example biotechnology and nanotechnology, help us in our search for more carbon-friendly materials that are also robust and durable.

Right now, we have the ability to halve our usage of new materials, to reuse more existing structures, to increase recycling and to minimise waste, while improving the world around us and regenerating ecosystems — just by using the computational techniques currently available and well described by Peter Debney in this book.

Tristram Carfrae
May 2020

Foreword

Every generation of structural engineers goes through a combination of evolutionary and revolutionary change throughout their career. We are always learning; through formal education, our career experience, sharing knowledge and our observations through life. But we often become very set in our ways, with fixed ideas of what is right and wrong.

The potential of computational design and the rapidly expanding role of digital technology throughout our lives and workplaces still invokes fear in many, and it is essential that we do not blindly use the powerful tools we now have at our disposal without understanding the fundamental engineering principles behind them.

Not all of us will be software developers or coding specialists, but when we use these tools it is essential that we at least understand what is going on inside the "black box". Using Peter's eloquent analogy in this book, knowing the mechanics of an engine does not make you an experienced driver, and *vice versa*. But as the driver it is essential to understand how to control the vehicle and safely harness its power, so you can trust the relationship between human and machine.

Our understanding of structural materials, new and old, and of simulating and modelling structural behaviour continues to expand. However, in terms of the first principles of mechanics and mathematics, the algorithms and calculations we are working with day to day do not fundamentally change. Computers can do more, faster, but they are still only tools and as engineers we are still the craftspeople. Tools can speed up common and repetitive tasks, but in the right hands they can also create beauty and innovation.

So, by embracing a more balanced partnership between the engineer and digital technology, we can use computational design not only to automate the mundane tasks but to unleash our creativity. Using generative design and multi-objective optimisation algorithms, we can explore and objectively assess a myriad of efficient design options that our experience-biased minds might not have considered in the first place. And using Machine Learning to make sense of the huge amounts of data at our fingertips we can place ourselves in a much more central and active role in the design process — something we can all aspire to as professionals.

This book provides an enjoyable and holistic introduction to many of these concepts — serving as a stepping stone for those embarking on the early part of their engineering career, but also as a very useful point of reference for those of us more set in our ways, making us critically assess our philosophy as designers, our imperfect relationship with technology, and our understanding of why things sometimes still go wrong!

Jon Leach
April 2020

Peter Debney

Arup/Oasys Software

Since gaining my degree in Civil Engineering at Surrey University over 30 years ago I have designed sewage treatment plants and petrochemical refineries, houses and shopping centres, portal frames and reservoirs. I have worked as a CAD technician and as an engineer, creating drawings, calculations and macros. I have also worked directly in the engineering software industry, helping develop the CAD programs *Xsteel* (*Tekla Structure*) and *3D+*, and the finite element analysis and design program *GSA*.

Computing has dominated my life and career; from trying (and failing) to write games for the Sinclair ZX81, via using analysis and CAD programs to design buildings, to designing and supporting those programs for others to use, I have seen every side of the life of high-tech structural design. I have also been teaching those programs and the engineering behind them for more than 20 years.

I am also conscious that, though it began over 50 years ago, we are still at the start of the engineering computing revolution. We can now achieve things that were not possible only a decade ago, but are also in danger of forgetting important lessons of the past. Most importantly, the climate crisis means that we must do more than design structures that work: we must do more with less. Optioneering and optimisation are now key tasks for the modern engineer, and the computer is the tool we need to help us fulfil them.

Acknowledgements

Permission to reproduce the following has been obtained, courtesy of these individuals/organisations:

Cover *The Lattice Ceiling in Kings Cross Station* © chrisdorney — stock.adobe.com
Author image © Beatrice Debney
Figures 1.1 and 6.17 © Beatrice Debney
Figure 1.2 © Tim Ibell
Figure 2.1 © Daniel Imade_Arup
Figures 2.2, 2.5, 3.12, 3.13 and 8.9 © Arup
Figure 2.4 © Seagate Mass Timber Inc./Pollux Chung
Figure 2.7 © Benh LIEU SONG (CC BY-SA 3.0)
Figure 2.8a © George Gastin (CC BY-SA 3.0)
Figure 2.8b © Simon Johnston (CC BY-SA 2.0)
Figures 3.1, 3.23, 9.12, 9.16 and 10.14 © xkcd https://creativecommons.org/licenses/by-nc/2.5/
Figure 3.3 © Paul Carstairs
Figure 3.4 © Nigel Whale_Arup
Figure 3.5 © Rasmus Hjortshoj_COAST
Figure 3.9 © Grant Smith_VIEW
Figure 3.11 © Hufton+Crow
Figure 3.18 © Robert Leighton
Figures 3.19–3.22 © The Institution of Structural Engineers
Figure 6.22 derived/adapted from Figure 3.1 of BS EN 1998-1 (BSI)
Figure 7.1 Jerry Williams. Copyright © 2020. Hartford Courant. Used with permission
Figure 7.2a © Bair175 (CC BY-SA 3.0)
Figure 7.2b © Bluemoose (GNU free documentation licence)
Figure 7.3 © Mark Mitchell, The New Zealand Herald
Figure 7.32 © Transport Scotland
Figure 7.41 © Vincent Ramet
Figure 8.5 Courtesy of © SOM
Figure 8.6 © MAXXI Museo nazionale delle arti del XXI secolo, Rome. MAXXI Architettura Collection (F5506)
Figure 8.8 © David de Jong_Arup
Figure 8.10 (top) © Emerald Publishing Limited
Figure 8.30 © ETH-Bibliothek Zurich, Bildarchiv (CC-BY-SA 4.0)
Figure 8.32 ETH Zurich, Block Research Group © Iwan Baan
Figure 9.11 derived/adapted from Dobbyn *et al.*[146]
Figure 9.13 © Johann Dréo (GNU free documentation licence)
Figure I1 © David Brohn

Others:
Figures 3.16, 7.31 and 10.15 public domain
Figure 3.17 US Army (public domain)
Figures 4.7, 6.32, 6.36, 6.40, 6.45 and 6.46 derived/adapted from Hellen and Becker[53]
Figure 4.14 derived/adapted from *Guidelines for the use of computers for engineering calculations* (IStructE Ltd)
Figure 5.4 US Government (public domain)
Figure 6.44 derived/adapted from *Train Test Crash 1984 — Nuclear Flask Test*[81]
Figures 6.47 and 8.33 derived/adapted from Oasys GSA training material
Figure 7.20 derived/adapted from Martin and Delatte[85]
Figures 7.21, 7.22 and 7.23 derived/adapted from Schlaich and Reineck[88]
Figure 7.40 NZ Government (Crown Copyright)
Figure 8.2 derived/adapted from Michell[108]
Figure 8.10 (bottom) derived/adapted from Fairclough *et al.*[114]
Figure 8.16 derived/adapted from Allen and Zalewski[116]
Figure 10.1 Eadweard Muybridge (public domain)
Figures 10.16 and 10.17 © M.Bongard/Macmillan Publishers
Figure 10.18 derived/adapted from 'Fibonacci' (Wikipedia) (CC BY-SA 3.0)
Figure 11.5 derived/adapted from *Understanding Quantum Computers*[230]

Permission to reproduce extracts from British Standards is granted by BSI Standards Limited (BSI). No other use of this material is permitted. British Standards can be obtained in PDF or hard copy formats from the BSI online shop:
www.bsigroup/Shop

Preface

"A lot [of structural engineers] are acting as human calculators and that's not engineering. If that is what people are doing, they will soon be replaced by computers, and that's a good thing. Because then they will be free to do what humans do best: complex problem solving; dealing with new phenomena; and being human."
Chris Wise[1]

The engineer and the computer

In my first year at university I dutifully attended the maths classes. Although I had been top of my school in maths at 16, I disliked A-level calculus and despaired to learn that I would need to do a lot more of it, both at university and in my career. It was a relief then, when I went into the subsequent structures class and was told that we could ignore calculus; instead there were plenty of standard formulas that we would use to calculate the results. This, I thought, was more like it. Finally, I attended the computing class, where we were told that we did not even need to know the standard formulae but could just put the structure into the computer and let it do all the calculations. This was the answer I was looking for!

Of course, as I subsequently learned, these statements were not entirely correct. Rather, like so much in engineering, they were 'near enough'. I have subsequently found occasional uses for calculus, and standard formulas can be incredibly useful. Where would we be without $M = wL^2/8$ for example: it is the structural engineer's equivalent to $E = mc^2$! Such a simple formula yet it allows us to rapidly analyse not only beams, but also trusses, arches and catenaries, if you know how to apply it.

But it was to the computer that I kept returning, whether for creating computer-aided drawings or finite element analysis models, many of which would be extremely difficult, if not impossible, to design with paper and pencil.

I also discovered that computers are not the answer to every problem and, like the oracles of old, their answers need the most careful scrutiny and thought. They will, usually, give you an answer, but it is for you to ensure that you asked the right question and received the right answer. The more experienced you are as an engineer the better you become at judging the results of computer calculations. "The purpose of computing is insight, not numbers" said Richard Hamming back in 1962 and he was, in many respects, correct, though numbers are still particularly important to the engineer[2]. I have learned a lot from building computer models of structures to see how they behave, but usually the models are of structures that I already understand, so I use them to extract the value of the forces and moments.

As a graduate I focused a lot on engineering computing, so it came as quite a shock when I started preparation for the IStructE Chartered Membership exam: computers were not allowed. But as my preparations continued, I realised that this was quite correct. To be a chartered structural engineer you must know the answers before putting pen to paper, or mouse to computer. Also, you cannot analyse a structure until you've designed it, and the design needs to be decided by you. Computers are an excellent tool for engineers to fine-tune their designs: check exactly what section sizes are required, and perhaps tweak the geometry to maximise efficiency. They also help when it comes to revising the design when the architect has, invariably, changed their mind again.

Computer programs are not universal panaceas: they will help a good engineer produce better designs and a bad engineer produce worse ones. They are only tools to magnify our abilities. There is no substitute for engineering skill, but the best engineers also make the most of computers.

Workshop manual or driving handbook?

Computers have enabled the modern engineer to create more efficient, more elegant, and more sustainable structures faster than they ever could have done using the methods of previous generations. While we may know less now about advanced mathematical and empirical methods than our forebears, who had no choice in the matter, we instead need to know more about the electronic computer, the programs that run on them, the methods they use, and how to use them all to best advantage. Among everything else that project work demands, today's graduate engineer must also be knowledgeable in computational engineering. So how is this knowledge acquired?

One of the primary tools of the modern computational engineer is finite element analysis (FEA), which allows us to analyse almost any structure in multiple ways. Most universities and textbooks approach FEA by teaching about its inner workings in detail but then ignore how to use it. This is like teaching someone how the internal combustion engine works, then after passing a written exam on the advantages of fuel injection over a carburettor and calculating torque from a diesel motor, awarding them a driving licence. We would not let anyone drive a car unless they had proved their competence. Yet junior engineers are often just as inexperienced in 'driving' engineering software.

So, what does a driver need to know? Do not put diesel into a petrol engine and *vice versa*; keep the oil and water topped up; and get the car serviced regularly. You also need to know which side of the road to drive on, how to behave at junctions, and how to stop in an emergency. Driving an engineering computer program is just the same. You need to know how to build good models and avoid bad ones, to let the computer deal with the repetitive and mundane tasks, and to use the computer to explore efficiencies and possibilities in design that were just not possible a generation ago. As practicing engineers, we need to have a working knowledge of FEA even if we are not experts, just as we need to know something about materials, construction methods, fire, architecture, and other problems that we must deal with on a typical design project.

There are many excellent textbooks available on the detailed workings of FEA, on optimisation methods, on the detail of the Industry Foundation Classes that allow communication between engineering programs, and so on. These are all invaluable if you are going to write your own programs or want to deepen your understanding of how those programs work. If the chapters in this book ignite your interest, then do seek them out. Likewise, talk to the highly skilled engineers and mathematicians who teach FEA modules and write programs — they've spent years studying engineering computing in detail. Like all engineering experts they are a great resource for those occasions where we need to dig deeper into a project problem. Meanwhile this guidance will introduce you to the many aspects of computing for structural engineers, bringing them together to look at the whole design process, with a minimum of maths but a maximum of explanation and context.

Computational engineering

The structural engineering profession is on the verge of a new generation of technology and innovation: the fourth industrial revolution. The 1960s and 70s gave us the mainframe and the first finite element analysis; in the 1980s CAD became mainstream and personal computers (or PCs) replaced the behemoths; mobile phones in the 1990s lead to the mobile technology revolution and smart phones of the new century; and more recently the industry (finally) woke up to the communication and workflow possibilities afforded by BIM. Now we are beginning to see artificial intelligence (AI) and machine learning (ML) applied to everyday engineering tasks.

Artificial intelligence has been an ambition of computing since its very beginnings with Alan Turing's 1950 paper *Computing Machinery and Intelligence*[3]. AI has had its setbacks, and while it has not achieved the original goal of creating a general intelligence, certain aspects have been phenomenally successful. Computing power, especially the parallelisation possible with distributed cloud computing, is now making optimisation a working proposition for everyday engineering. 3D printing, robotics, and zero-carbon construction means that very soon material usage will dominate over labour costs, necessitating the application of optimisation techniques that our mechanical engineering colleagues have been using for some years. And the computational power of modern engineering computer programs means that the traditional tasks of sizing steel members and rebar quantities are now regularly automated.

This all means that the time has come for a fresh look at computing for structural engineers and what our role is in this age of machine learning and automation. What does the graduate of today actually need to know and how can they compete? The engineers of today must embrace computers as digital engineering assistants if they are not to replace us. We must use their strengths and know their limitations, let them free us up to do that what computers can (probably) never do — understand the client's needs — then use imagination and ingenuity to solve their problems.

So, what do engineers need to know about these computer programs? As always it is best to start at your goals and work backwards. Think of it like those puzzles we had as children, where there are several staring points and some end points, joined by lines resembling a plate of spaghetti. I realised quite young that the way to solve those puzzles was not to try the start points, but to instead start at the end and work back. Sometimes trial and error can be a useful thing though, as we will explore in the chapter on optimisation: answers can sometimes depend on where you start from. There is rarely only one 'correct' answer.

Apart from computing, this book is about design, but that also raises the question: 'What is design?' If you ask any civil engineering student, they might tell you that design is determining the size of the steel beam or the reinforcement needed in a concrete beam using a design code. But these are both trivial problems. You put the forces and moments into the formula, turn the handle, and the answer drops out. In the case of a steel beam you may then need to choose a section that meets the requirements from a table, but little more than that. It is just a mechanical process with a single answer. Or is it that simple?

It is easy to find the lightest steel section, but is it the cheapest or the one with the lowest environmental impact? What happens when we take the connections into account, is it still the best, or are we better-off using a larger section and make the connections simpler? Should we make the section the same as some others to increase manufacturing efficiency? But clearly making all the beams on the project the same depth, whether they are spanning 1m or 10m, is not going to be efficient — so what is the answer?

If you are just focusing on optimising the section sizes you are missing the bigger picture: should the member even be there, or should it be somewhere else? You may tune the chord sizes on a truss but varying the depth of the truss will have a far greater impact on the end result; but did you test that or just use a standard span-to-depth ratio and make it work? What if you change the number of diagonals in the truss, will that make it better or worse? Does the bottom or top of the truss really need to be straight, or would a different shape be better? Should it be a Warren, a Fink, a Prat, or even a Michell truss? We have not seen many Michell trusses so far, but I expect that they will be far more popular in the future. What if the columns were further apart, or closer together? What if the beams spanned the other way? What if we change the roof slope or even invert it? Should we use steel, concrete, timber, or consider an alternative material?

The lightest design is not the cheapest design is not the design with the lowest environmental impact is not the design with the shortest time on site is not the design with the simplest construction. Usually. Design codes do not tell us how to design, despite their name, but just give us the basics: strengths, formulas, minimum requirements, and so on. But how do we find the best design?

> *"Engineering problems are under-defined, there are many solutions, good, bad and indifferent. The art is to arrive at a good solution. This is a creative activity, involving imagination, intuition and deliberate choice."*
> Ove Arup

Despite our training in maths and physics, where we are taught that there is one answer, design is not so simple as we may have been led to believe. As some of my colleagues would put it: it is "non-trivial". Personally, I would say that the challenge makes design more interesting. How do we produce good designs when there are so many variables and no single answer? It takes a long time for us to check a single design. We cannot check them all, but this is where computers can help us.

One advantage of computers is that they are incredibly good at doing the maths for us — the clue is in the name. Another is that they do the maths very quickly. And a third is that they do not tire or complain about doing the same thing over and over again. While we, if working with pencil and paper, might choose just one design, we can tell the computer to test 100, then take the best and from those make 100 more. Or we might ask it to keep adjusting the position of a connection until it has found the best location. Or we might ask it to examine all the designs that we have done in the past and suggest what the best answer is likely to be this time.

As a young structural engineering graduate, I started in the industry in the early days of the computer analysis and drafting revolution. Now we see the same happening to design, just at the time when design, especially the environmental impact of our designs, is becoming so important. We have the responsibility and we have the tools at hand to make this world a better place, if we know how to use the tools wisely. Let's get started.

1 Introduction

"I think there is a world market for maybe five computers."
Thomas Watson, president of IBM, 1943[4]

"There is no reason anyone would want a computer in their home."
Ken Olsen, founder of Digital Equipment Corporation, 1977[4]

1.1 Computation and the structural engineer

On the morning of 23 August 1991, the newly constructed Sleipner A gravity base slipped into the water at Gandsfjord. Its destination: the Sleipner East gas fields of the North Sea. It was designed to withstand the pressures of 82m depth of water, but during the preparations for towing, the structure imploded and sank. The impact registered 3 on the Richter scale and $700 million on Statoil's accounts. The cause was a poorly detailed finite element analysis model, even poorer detailed reinforced concrete, with insufficient checks made on both.

2 August 2012 saw the first track cycling event at the London 2012 Games. Rather than a more conventional frame the engineers chose to make the Velodrome roof a hyperbolic paraboloid constructed of cables. Careful computer analysis to find the best form of the roof, along with a stiff substructure to provide the prestress, enabled them to reduce the weight to one tenth that of the adjacent venues[5]. In recognition of its accomplishment, the Velodrome won the IStructE's Supreme Award for Engineering Excellence in 2011.

The common theme between these two projects is the application of computerised design: one bad and one good.

1.2 The computer as Engineering Assistant

Since the 1970s, structural engineers have relied increasingly on the calculation speed and power of electronic computers. In times 'BC' (before computers) engineers needed specific formulas to analyse particular structural shapes. Now 'AD' (after digitalisation) electronic computers enable us to analyse any shape of structure, take second-order and nonlinear behaviours into account, and automate the detailed design.

Computers have also revolutionised other parts of the design office: where before there may have been five drafters for every engineer, there are now five engineers for every 3D CAD technician. Engineering design teams have become more productive and used that power to create more efficient and complex structures. Without doubt engineering is changing, construction is changing, the environment is changing, new materials and technologies are changing our industry and we need to understand how best to make use of these new methods and options.

What is the role of an engineer in the modern day? Where Artificial Intelligence programs use Machine Learning to produce a structure, FEA programs analyse them, optimisation programs hone them to the ideal, and knowledge-based systems check them against the regulations, why do we still need humans if everything can be done better and faster by machine? The answer is that machines can do some things well, such as maths or checking hundreds of options to see which is best, but they are terrible at other tasks. Computers do what they are told, but they do not understand. They can explore possibilities, but they cannot invent. They can check for things that have been problems in the past, but not for things that have not been problems until now, and it will be a long time before a computer can attend a site meeting, discuss the issues with the contractor, and suggest a solution!

1.3 The purpose of design and analysis

On my first day at university, we were addressed by the head of the department with the following 'joke':

*"What is the difference between a doctor and a civil engineer?
A doctor can only kill people one at a time!"*

This left a marked impression on me, as it rightly highlighted the responsibility that an engineer has. But it is also a negative view on what civil engineers do, so I would like to propose a revision:

"What is the difference between a doctor and a civil engineer?
A doctor can only save lives one at a time!"

Civil engineering has the power to improve the quality (and therefore length) of people's lives. The work of engineering hero Joseph Bazalgette saved the lives of countless thousands with the creation of the London sewers in the 19th century. Water engineers today continue that vital work, whether supplying clean and safe drinking water or taking away the foul. Transport engineers check that station platforms do not become overcrowded (the only way out is onto the tracks). Fire engineers ensure that people can escape from a burning building in time. And so on.

With structures the main work is indeed ensuring that it does not collapse and that it functions in useful ways, but there is more that we do to positively affect the world. Bridges and tunnels allow people to cross wide rivers; giving access to food, medicine, business and society. Buildings provide shelter from cold, heat, rain and wind — and provide safe places to live, learn and work.

"Architects do not make buildings, they make drawings of buildings."
Robin Evans[6]

Likewise, engineers do not make structures, we make models of structures. These models can be in many forms: they can (and should) be mental models of how the building, bridge etc. will behave, physical models[†] and, of course, computer models of the structural behaviour and physical detail.

Unlike manufacturing, construction is very inefficient. Some of this is because of unknowns that are out of our control, such as the weather. Some unknowns are hard to resolve in advance, such as ground conditions — boreholes can only tell us a little about the site. And some is the difficulty in automation on yet another unique project.

While engineering is an art, it is a technological one. It has, at its core, a tool called mathematics. Mathematical ability is viewed as an important attribute of any engineer, but modern computers are far, far better at maths than we are. They are faster, more accurate, and tireless. They are also stupid to the point that they cannot do anything at all without being told to do so first. It is true that computers can now learn, but only because we tell them how. They cannot think, they cannot imagine, but they can explore possibilities far faster than we can.

When Big Blue beat Garry Kasparov, the world chess champion, it did so by testing millions of possible moves to choose the most promising one. Kasparov, on the other hand, always considered just one move — the right one[7]. Kasparov won the first round, but he lost the second: the age of Artificial Intelligence had finally arrived. While AI then went on to conquer the game of 'Go'[8], which is several orders of magnitude more complex than chess, it is not all-conquering. Although the best AI beats the best humans at games, they are both trounced by centaurs: humans and AI working together in partnership.

Arguably, structural engineering design is more complex than games like Go; it is certainly less structured[††]. Engineering is an art, a science, an invention, and an application of technology. It has the rigour of the mathematician and the inspiration of the sculptor. And structural engineering is at its best when human and machine work together, using the talents of each to bring out their maximum potential.

[†]   Though not so much today. Physical models were useful in the past, especially for form finding, and are making a comeback thanks to 3D printing.
[††]   Ironically.

1.4 Garbage in, Garbage out

"On two occasions I have been asked, — 'Pray, Mr. Babbage, if you put into the machine wrong figures, will the right answers come out?'... I am not able rightly to apprehend the kind of confusion of ideas that could provoke such a question."
Charles Babbage[9]

One of the big problems facing engineers is that we can solve only the simplest of structures by hand because the mathematical process rapidly becomes too complex. This means that we need techniques such as FEA working on computers to solve them for us. Our job is to produce good answers, and we do that by putting good data into appropriate software. But how do we know what is good data and good software? This is where 'validation' and 'verification' come in. These terms have several definitions, mostly depending on what you are trying to do.

Validation asks whether the software does the right thing. For developers of the software this means whether it does what it is supposed to do; this often means whether it gives the right answers to standard questions. For engineers, the question is more whether the software is appropriate for this particular problem: Is it geared to give me the answers that I need, or should I use something else?

Verification asks whether the output matches reality: are the answers correct? This is tricky as we are using the program to find the answers! You need to have a good idea of the approximate answer and use the computer program to find the numbers. For a structural problem, you should know what the deflected shape and bending moments, etc. should look like in advance.

Of course, there will be times when you want to use the FEA program as a sandbox: to run experiments. When I was a graduate, I had a debate with my boss about how reciprocal structures worked[†]. We eventually resolved it when I created an FEA model of an example that we were happy with. Experiments like this can be particularly useful, but are to be used with care, especially if they are on a live project rather than just a theoretical exercise.

So, good answers are ones you expect, and bad answers are often ones you do not. Answers that do not match your expectations are usually because:

1. You have modelled it wrong.
2. The answers are correct, but your expectations are wrong.
3. The software is wrong.

And it is rarely number 3!

Modelling errors are the most common source of bad answers. They might come from the wrong sign on loads, the wrong units being chosen, errors on calculating the input, typos, and many other factors. Errors in FEA models have led to structural collapses, so we will look deeper into this important topic later.

Having wrong expectations is an interesting problem: you think that you know how things work but you are still learning. Well get used to that, because to work in engineering is to experience a lifetime of learning. For example, I once worked on an asymmetric portal frame that always swayed the same way despite the wind direction. Despite my expectation that a structure always sways with the wind, this one defied that thinking because of its unusual roof geometry. Likewise, there are examples of multi-storey structures published in *The Structural Engineer* where a push on the first level resulted in a deflection of the upper levels in the opposite direction[10,11]. If in doubt talk to your more experienced colleagues.

[†] Reciprocal structures can lack robustness, so are not suitable for all situations, but can be very elegant.

As for bad software, anyone who has programmed knows it is easy to make mistakes, but professional commercial software writers also know that and put their programs through banks of tests. They will have validated and verified their software to the best of their ability, but it is possible that you might have found a bug that slipped through. It is most likely to be your mistake but do speak to them if you have exhausted possibilities 1 and 2, as professional programmers are always striving to perfect their programs.

In summary: assume that you are wrong until proven correct, and if in doubt, ask.

1.5 The engineer of today and tomorrow

"Tools are mediators, and much of human progress has come about because someone invented a better and more powerful tool. Physical tools speed up work and rescue people from hard labour. The plough and the wheel, the crane and the bulldozer, amplify the physical abilities of the people who use them. Information tools are symbolic mediators. They amplify intellect rather than muscle."
Bill Gates[12]

The maths involved in engineering design has seen several 'revolutions' over its time (Figure 1.1). First, slide rules[†] meant that we did not have to look up logarithms in tables. They did not stop engineers from working, quite the reverse. Instead they automated the drudgery of multiplying large numbers — allowing the engineer to concentrate more on the engineering and less on the maths.

I still have my father's slide rules and they are beautiful and tactile things, though I find calculators considerably faster and more adaptable. These quickly replaced slide rules, though there were grumbles from traditionalists more familiar with mechanical calculations. True that there was no scope to "bend the wire to get the right answer" as one senior colleague told me, but also there was less scope for getting the order of magnitude of the answer wrong.

When I was at school we were constantly told of the importance of mental arithmetic as we "would not have a calculator in our pockets". With the advent of the smart phone, that is no longer true.

Figure 1.1: Several 'revolutionary' engineering tools

[†]   Analogue mechanical computers, sometimes known as 'slipsticks'.

That said, the ability to solve approximate sums in your head is still incredibly useful, whether in the supermarket or the design office, at least as a check.

Pocket calculators replaced slide rules because they could do more and do it faster and more accurately. And programmable computers do more tasks faster than calculators. At base level, they are both doing the same thing: adding up zeros and ones very quickly (and not even that if you speak to electronic engineers) but computers can add layers of control and meaning to the numbers to make them easier for us to work with. The numbers can represent a load, a beam, a floor, or a bridge; more numbers can show us on the screen what the structure looks like or how it will behave.

Computers are tools to take some of the drudgery out of engineering tasks and allow us to concentrate more on the engineering. Of course, the tools do require some attention as well, but a good tool delivers more savings and efficiency than it takes. A saw might need sharpening occasionally, but it will cut a lot quicker and easier if you invest that time (or money getting someone else to sharpen it for you). Likewise, computer tools need some maintenance, whether the hardware or the software, commercial or self-written.

1.5.1 To code or not to code
As long ago as 2003, *New Civil Engineer* magazine ran a survey asking working civil engineers if they thought that it was still essential to have a maths A level. The conclusion was mixed, but all agreed that maths is extremely useful[13]. Today's engineer might not need to be a maths whizz in the same way as our predecessors, because computers are better and faster mathematicians than we will ever be; the maths that you need in practice is not normally that difficult. Consider what tasks you will need to undertake on a typical project (Figure 1.2).

Figure 1.2: Time spectrum on a project for a typical structural engineering designer[14]

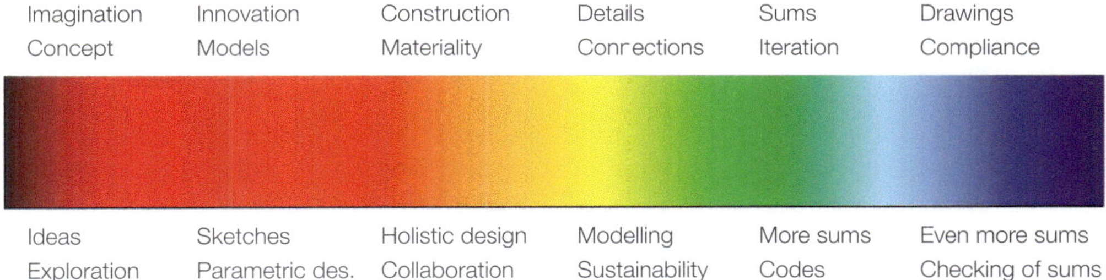

Some tasks involve calculation, many do not. For those that do, much will involve building models in commercial software, some creating programs to work with or within commercial software, and (only occasionally) require writing a standalone program. In general, computers are involved more as the project progresses, whether for calculation or communication.

Modelling with commercial software requires little maths skill, other than that required to ensure you have the right inputs. If you want to program computers to do engineering tasks, say setting up a parametric model, then that requires good understanding of the principles involved. Programming tasks, such as writing FEA solvers or creating optimisation macros, does require a deep understanding of mathematics.

The maths that you need to program a computer is often simple but needs breaking down into tiny steps. Computers do exactly what you tell them, but even the Artificial Intelligence programs have no knowledge or understanding. We can take shortcuts and remember answers (such as: "What is 5×5?"). We know the answer and can repeat it as many times as we are asked, but a computer will calculate it every single time no matter how often it is asked that question.

In the early days, if you wanted to run an analysis program you had to write it yourself first. Then came the multi-purpose commercial programs, and the 'engineer-programmer' became a rarity. Today the tide is turning again as the commercial programs become platforms for your scripts and code. You might not write a FEA solver, but you are likely to write a program to generate the geometry that the FEA program will analyse. The commercial programs may have section-sizing tools, but your macro may work with it to optimise the section choice over the whole building. Many consulting engineers, and some contractors, expect their graduates to have some programming skills.

Besides, programming is not only a useful skill, it also teaches you about sequencing and creating instructions such that anyone can follow them, even a computer.

We are now also seeing Machine Learning and Artificial Intelligence programs being used for structural engineering. While these too are not going to replace engineers, or indeed our existing analysis programs, they are going to make our work more efficient. ML techniques are providing new opportunities for engineers to focus more on the important things by prefiltering information and making recommendations. AI programs might not be trusted to make life-safety decisions, they are too vulnerable to bad data, but they may act as assistants.

Might a computer program one day be able to pass our own Turing Test and take the IStructE Chartered Membership exam? I believe it's possible, even if not in the short term, but such a program would still not be able to attend a client or site meeting. Engineers are creative problem-solvers, computers are lightning-fast calculators; each has strengths and weaknesses. The best engineers recognise both and use the computer to find the optimum answers, explore options, react to changes, and minimise rework faster than they could by themselves.

1.5.2 The need for better design

The environmental impact of civil engineering is becoming increasingly important. Not only must we provide cost-effective designs for our clients, we must also consider the cost to the rest of the world. The manufacturing of structural steel and concrete consumes a lot of energy, often derived from polluting fossil fuels, as well causing damage through the excavation of the raw materials. It is our responsibility as engineers to produce designs that are as efficient as possible. We have a major impact on the world, and we need to minimise that.

CO_2 is a greenhouse gas that is contributing to global warming. 40% of all global energy-related emissions are due to buildings and construction[15]. About half of a reinforced concrete structure is there to support its self-weight, yet we usually just pick a beam or column size and make it work. Simply choosing the best section size could make a significant difference to the environment; reducing CO_2 emissions, sand and gravel extraction for aggregate, limestone for the cement, timber for the shuttering, and so on.

It was once thought that 75% of the energy use of a building came from its operation, but this is no longer the case. Operational energy usage has plummeted thanks to better insulation, energy-efficient heating, renewable energy production, and so on. Now it is the other way around, with nearly 70% of the energy used embodied in the structure[16]. This means that it is even more important today to optimise our structures, and not just pick a size and make it work.

While we must never under-design, over-design is being increasingly seen as an industry-wide problem. Studies have found that structural engineers regularly over-design, with much of a structure working at about 50% utilisation[17,18] and most structural material is only at about 10% utilisation[19]. There is much to be done in the overall design of structures to make them more efficient, in choosing the correct section size for the load, and ensuring that the section shape is also the most efficient. If we can design the structures better, then together we can have a significant impact on climate change.

1.6 The computational engineer

"To engineer is human."
Henry Petroski[20]

So, what are the roles of the engineer and the computer in this digital age?

Computers are fast, powerful, and will give the right answers if given the right information and algorithms; and will give the wrong answers just as fast if the information they're given is incorrect. The engineer's role is to choose the data and the method, then to judge the answer. All engineers must know the answers, at least approximately, before they run any analysis or design.

But how do you spot errors and how do you predict the answers when you need the computer to calculate them? There are techniques and heuristics to create good models, predict the answers, and to track down errors.

Knowledge of the inner workings of the program might not be as important as understanding structures, but it can help.

How might you create a perfect model then? Alas, perfection is not possible as:

> *"... all models are wrong; some, though, are more useful than others..."*
> McCullagh & Nelder[21],[†]

You cannot model all the possible imperfections and tolerances, the exact loadings, and the actual materials used that may be in the final structure. The aim of the structural analysis is not to work out how the structure will stand up, but to ensure that it will not fail or otherwise misbehave, while still being economical and efficient. You need to work out all the loads that the structure is likely to experience and ensure that there is sufficient strength to cope, plus a bit extra as we can never be entirely sure. But there are also some risks that are so unlikely/insignificant that they are not worth designing for. In the UK for example, we do not design for earthquakes[††]. Similarly, it is generally only government, military, and some industrial buildings that are designed for explosions.

While there may be a correct answer to an analysis, design often has no definitive answers, only better and worse ones. In these situations, the computer can quickly explore different alternatives and quantify some of the advantages and disadvantages of the options. Engineering is the art of getting the answer near enough, with failure not being an option. The definition of failure can vary though. A structure might be said to have not failed during an earthquake because everyone was able to escape safely, even though it needs demolition afterwards; for another structure, such as a hospital, non-failure means that it can be used immediately after an earthquake. Cars are designed to protect occupants in the event of an accident by absorbing the damage itself. All models are wrong, but is your model near enough?

[†] The origin of this quote is in some dispute — although it is commonly attributed to George Pelman Box[22].
[††] In the UK, seismic design is mostly confined to nuclear power stations.

2 Design

"As in art, [engineering's] problems are undefined, there are many solutions, good, bad, and indifferent. The art is, by a synthesis of ends and means, to arrive at a good solution."
Ove Arup[23]

2.1 Introduction

'Design' is a curious word in engineering and the source of much confusion due to its dual definitions. If you read the design codes, and you will need to, then design is mostly concerned with the sizing of sections, the provision of rebar, and the ensuring of strength, stability and longevity. This is mostly what I would call 'detailed design'. Detailed design is essentially a mechanical process with few options; as long as you know the inputs then you can calculate the output. As such, it is a common task for both graduates and computer programs, as well as a skill needed to pass your university exams.

The other form of design is more appropriately called 'scheme' or 'concept' design, and is a far more complex, challenging and interesting problem. While detailed design deals with 'how much', scheme design deals with 'what' and 'where'. It is a skill needed to pass the IStructE's Chartered Membership exam.

Creativity is necessary in scheme design. It is also a talent that some find natural, but mostly it is something that can be both learned and honed. Firstly, let us put one myth to rest: creativity is not best with a completely blank piece of paper. It is best when you start with the following:

- **Requirements**
- **Constraints**
- **Options**

Requirements

Remember art classes, or on rainy holidays, when an enthusiastic teacher or desperate parent would say: "Draw me a picture". That is a requirement. In our case the requirement might come from a client saying: "Design me a crossing for this river", or an architect saying: "Let's design a building together".

As leading artists might sometimes work to a commission, and at other times create work because they themselves find the need, so it can be with engineering. While we usually work to a commission, there are times when we see the need and convince others that it needs to be built. Isambard Kingdom Brunel saw the economic benefits of a train line between London and Bristol — shortening the shipping times of overseas goods and people into the capital city. Joseph Bazalgette knew that a sewer system in London would save thousands of lives from cholera, while Parliament and much of the medical profession was still stuck with the medieval idea that bad smells caused disease[†]. Frank Whittle saw the need for faster aeroplanes and solved it with the jet engine. James Dyson knew that vacuum cleaner bags were a problem and replaced them with a vortex. Creative engineers through the ages have seen the requirements (and solutions) that others have either missed, or considered a part of life to be endured.

You will also note that these examples were innovations with existing technology. The technique for vortex separation was already in use in the paint industry; trains had been used in mines for centuries and in other parts of the UK for years; the Romans invented sewers. They all drew from existing ideas and did something new with them.

*"What has been will be again,
what has been done will be done again;
there is nothing new under the sun."*
Book of Ecclesiastes 1:9

[†]   The Miasma theory, which lingered for far longer than it should have.

Sir Issac Newton said: "If I have seen further it is because I have stood on the shoulders of giants". He took existing ideas and built upon them. The website *Everything is a Remix*[24] shows to great visual effect, that while many Hollywood films are simply derivative, the most innovative films drew wholesale from earlier works. *Star Wars* for example, had an early draft made from film clips taken from classics like *633 Squadron*, *The Battle of Britain*, and *The Hidden Fortress*, plus inspiration from books like *The Hero with a Thousand Faces*[25].

> ***"If all you have is a hammer, then everything will look like a nail."***
> anon

Know. Combine. Do something new…

> ***"Good artists borrow,***
> ***Great artists steal."***
> Pablo Picasso (attributed)

As all structures are built on firm foundations, so structural designs are built on the design ideas of others. Review, renew, recycle, recombine, create something new.

Immerse yourself in the works of other engineers. Good engineers are well read, and not just within engineering text books, but across other disciplines and sciences. Read biology books, as nature has been designing efficient structures for millions of years. For example, how does a horse carry the load of its rider? Is it like a beam? Unlikely, as beams are inefficient, and nature abhors inefficiency[26].

Constraints

While constraints are an important part of creativity, you must start by pushing them to the background[†]. You must instead start with what might be possible. This opens your mind to what could work: to options. Once you know the range of things that you might do, you can filter down to what you will do. Information inspires imagination and innovation.

Conversely, the worst place to start is to look at what is not possible. There are some people who excel at telling you what will not work. They are deadly to any scheme design and must be kept out of the way at this stage. Later, when you reach the detailed design stage, they are invaluable. Negative thinkers are great at finding the flaws in a design, which can be a lifesaver when you are considering the nuts and bolts and welds, the construction sequence, the accuracy of the results, and so on. As the saying goes: "the devil is in the detail". But at scheme stage: no, keep them away.

If you read the IStructE Chartered Membership examiners' comments, you will see a depressing trend. One of the most common complaints is that candidates, while required to produce two workable schemes, present only one. The options are almost limitless, and you will be a better engineer when you realise that. The best engineers are those who can quickly filter down to the best options.

But what makes a good structure? There are as many answers to this question as there are engineers and architects. Simplicity is usually seen as a good thing[27], as is efficiency. But some beautiful structures are both complex and sculptural. Expression of structure is often welcome, but that can lead to thermal inefficiencies and cold-bridging if you put the structure outside the cladding.

Consider the diagrid of 30 St Mary Axe, London a.k.a The Gherkin (Figure 2.1) The structure that resists gravity and that which stiffens against the wind are combined, then expressed in the building facade. Its regularity also makes it easy to define using just a few rules.

[†]   It is usually the same when optimising a structure, but more of that later.

Figure 2.1: The Gherkin, London, UK

Stadium tensegrity roofs are another example of a form that is both expressed and suitable for parametrising (Figure 2.2). Unlike the circular curves of the Gherkin, here the cables need to be parabolic (Figure 2.3). The entire layout is driven by the outer and inner stadium circle or ellipse, as well the roof depth and cable spacing. Light and robust, but not a structural form that is taught in many undergraduate courses.

Figure 2.2: Ken Rosewall Arena, Sydney, Australia

Figure 2.3: Tensegrity roof section

The structure must give shape to the form: a steel structure is different to a concrete structure, and if you build in masonry then it's not going to be the same if you build in timber. But there are also similarities. Masonry works best with compressive forms like walls, arches and buttresses. Concrete is also compressive, but reinforced it becomes beams and slabs. Steel is equally good in tension and compression, but unlike concrete is slender not stocky, and in pure tension can span stadia, rivers and canyons. Timber is weaker than steel but is sustainable and a good all-rounder; coming from trees it is happiest when in a frame or truss; long, slender and flexible, with stiffness coming from triangulation. But you can also make walls from steel, timber (Figure 2.4) or concrete catenaries (Figure 2.5), reinforced brick beams, and multi-storey timber. You are not limited to the standard forms, though some will work better than others.

Figure 2.4: Grandview Heights Aquatic Centre, BC, Canada

Fractals are self-similar shapes, regardless of how close you look. Familiar examples are clouds and coastlines — if you see a picture of one with no other reference, you cannot tell how large it is or how far away[†]. Fractal structures, such as Sierpiński triangles work by repeatedly removing the middle area of a triangle or square (Figure 2.6). If you continue forever, the result is a form that has zero area or volume but an infinite perimeter or surface.

[†] Coastlines also have the strange property that their length depends on the length of your ruler: the shorter the ruler, the longer the perimeter.

Figure 2.5: Portuguese National Pavilion, Lisbon, Portugal

Figure 2.6: Sierpiński triangle

Fractals are theoretical forms, but do have their application in real structures, particularly historical ones when large elements were difficult to manufacture. Iconic structures like the Eiffel Tower (Figure 2.7) and the Forth Rail Bridge (Figure 2.8) use fractal principles, though they were not recognised as such at the time, as Benoît Mandelbrot did not coin the term until 1975[28]. The tower and cantilevers are broken down into trusses, then each member in the truss is itself a truss (Figure 2.9). Each level of detail is driven by rules and the forces that every element and member must carry.

Figure 2.7: Eiffel Tower, Paris, France

Figure 2.8a: Forth Bridge, Edinburgh, Scotland

Figure 2.8b: Forth Bridge detail

Figure 2.9: Fractal truss

Options

There are all the standard building forms of course: beams and slabs, columns and braces, portal frames and trusses and so on. While this is not going to be an exploration of structural form, an important thing to remember is that there is never just one answer to a design problem, or even just one *satisfactory* answer — there are many. Another is that there are certainly more options and alternatives than you realise. When I was preparing for my IStructE Chartered Membership exam, I decided to draw up a list of all the possible ways of either spanning a gap or supporting a load at height. While there may not be an infinite number, there always seemed to be one more.

2.1.1 Learning to be creative
Creativity is not just a talent reserved for artists; good engineers are also creative, and you can learn to improve your own skills in this area.

In *Hegarty on Creativity: There Are No Rules*[29], the author says that the biggest misconception regarding creativity is that it means originality. In fact, nothing can be truly original and thus a fresh new idea actually "draws its inspiration from the world around and reinterprets in a way we haven't seen before". The creative process is made up of four steps. The first three were formulated by Hermann Von Helmholtz in the 19th century, with a fourth added by Henri Poincaré at the start of the 20th:

1. Saturation
2. Incubation
3. Illumination
4. Verification

Saturation means that you must immerse yourself in as many ways of solving engineering and design problems as possible. If you only know one way of doing something, then you will struggle to find alternatives. Study hard, read engineering and architecture books, plus anything else that interests you. I found this list in a book called *The Quark and the Jaguar*[30] by Nobel-winning physicist, Murray Gell-Mann (it is well worth the read) — so try not to stick to the obvious examples.

Incubation comes by thinking about the problem, pondering, and searching. This stage can take time.

Illumination can happen at the strangest of times. Ove Arup said that many of his ideas came to him in his sleep: he went to bed worrying about a problem and awoke with the solution. Murray Gell-Mann got his quantum mechanics Nobel prize by making a mistake in a lecture, then recognising its truth. This shows the importance of saturation as well, as he would not have recognised the significance of his mistake if he had not been prepared.

Finally, **verification**. This stage is a vital one and natural to engineers. You may have had a promising idea, but now model it to prove that it will work, then finalise the details.

2.1.2 Structural art

> *"Robert Maillart, the Swiss bridge designer, developed in 1923 a limited theory for one of his arched bridge types which violated in principle the general mathematical theory of structures... he developed the theory to suit the form, not the form to suit the theory. In the United States, by contrast, some of our best engineers understood the general theory well, but not understanding Maillart's specific ideas, they failed to see how new designs could arise. They were trapped in a view of engineering which was so complex it obscured new design possibilities. Today the undue reliance on complex computer analysis can have the same limiting effect on design."*
> David Billington[31]

As Professor Billington says, just because you can analyse it in a computer and make it work does not make it the best structure. You need to understand how forces flow; to channel them in the most efficient and effective way. Architects say that form follows function, but engineers know that form also follows force.

"I think that there used to be an inherent criticism of technology in that it was seen as the exact opposite of creativity, in a way. And that it was a sort of damper on creativity. I think now, with everybody's connection to the computer age and so on, almost the reverse has happened. In a lot of ways people think that you can't be creative without understanding technology. ...There is a strong feeling now that technology is there to be used to help you be creative."
Nicholas Grimshaw[32]

Does information technology restrict or release our engineering imagination? As usual, with these binary questions, the answer is not always definitive. Programs can free us to create in ways that are in line with how that program works but can otherwise stifle us. We need to be masters of multiple tools, aware of their strengths and weaknesses; to know when to work with the grain or against it.

Complete freedom is the blank page where anything can happen, but usually does not because there are too many possibilities; complete restriction means that nothing can happen. Life occurs not in the total order of a crystal nor the total chaos of the heart of a sun, but in the complexity of the "Goldilocks zone" where everything is *just right*. It is the same with creative engineering: 'structural art' if you will. We need requirements and constraints as well as options to choose from. As Ove Arup said: "there are so many possible designs, some good, some bad, some indifferent. Our quest is to avoid the bad, improve the indifferent, and seek the [common] good."

2.2 The design process

In 2004 the UK's Design Council launched their framework for design: the Double Diamond[33] (Figure 2.10). Each diamond represents the concept of first thinking widely and deeply about a particular problem, then focusing on the preferred solution. The first diamond is where you seek to understand exactly what the problem is, so that you can define the problem and produce a solution-neutral design brief. The second diamond explores the options then focuses on the solution you'll deliver.

Figure 2.10: Double Diamond design framework

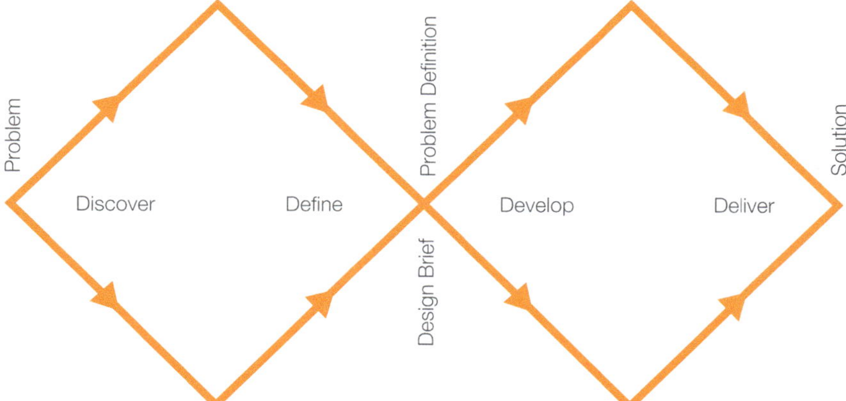

- **Discover** — make sure that you and the team understand the problem. Answering the wrong question will never give the right answer
- **Define** — take what you have discovered and define the challenge in a way that can be addressed
- **Develop** — explore different answers to the defined problem
- **Deliver** — test the different solutions, reject those that don't work, combine and improve those that do, until you have a solution to deliver to the client

In the first half of each diamond you must exercise divergent thinking to explore your options, then practise convergent thinking to exploit what you have discovered. Converging when you should be diverging will lead to a substandard result but diverging when you should be converging will lead to no results at all.

In general, graduate engineers will be most involved to the right of the design process and gradually move left as their experience increases.

Remember that design is an iterative process. There will be false starts and blind alleys, promising avenues, and round trips back to the beginning. To make matters worse, the client may well redefine the initial problem before you have finished.

2.3 Design personalities

"The natural sciences are concerned with how things are.
Design, on the other hand, is concerned with how things ought to be.
Everyone designs who devises courses of action aimed at changing existing situations
into preferred ones."
Herbert A Simon[34]

The best design teams have a mix of personality types and there are a number of personality models available to examine how we might think or work. These include Belbin[35] and Myers Briggs[36], as well as Arup Design School's 'Design personalities' (Table 2.1).

Table 2.1: Design personalities

The Artist
- Pursues interest
- Focuses on ideas
- Finds it easy to generate ideas and is happy to ditch a bad idea in the hope of a better one
- Struggles to explain why one idea is better than another
- Is likely to move onto something more interesting when it surfaces

The Philosopher[†]
- Pursues meaning
- Focuses on problem definition
- Is obsessive about matching answers to questions, and solutions to problems
- Struggles to start because they first need perfection of meaning
- Can, and will, explain precisely, and at length, why one thing is better or worse than the other

The Artisan
- Pursues perfection
- Focuses on improvement
- Draws on precedent and the ideas of others
- Struggles to begin without a pre-existing concept
- Determined and difficult to shift

The Pragmatist
- Pursues solutions
- Focuses on practical considerations
- Likely to find a solution within a limited period, but is not overly concerned with the quality of the solution
- Unless there is a change of inputs, sees no reason to iterate
- Might not inspire, but are useful in a team

Note: Derived/adapted from Arup Design School and 'Engineering is the opposite of science' lecture (C. Wise, 2018 IStructE Academics Conference).

If you compare these personality types to the design Double Diamond you will note that different people will be most comfortable and useful at various stages of the design process. An optimal team will have a good mixture of all these types so that all stages are properly addressed.

But this book is about computational engineering. How do we design using computers?

[†] Fun fact: Ove Arup initially studied philosophy at university before transferring to engineering.

3 Design parametrically

"Algorithmic design is not simply the use of computer[s] to design architecture and objects.
Algorithms allow designers to overcome the limitations of traditional CAD software and 3D modellers."
Arturo Tedeschi[6]

3.1 Introduction

Design is an iterative process and is never exact. There are so many unknowns, whether in material, load, or constructed geometry, we can never be 100% sure what will happen to the structure. As part of the analysis of our designs we need to understand how sensitive they are to the small (but inevitable) variations incurred by the structure. What if the beam is a little bit bigger or smaller? What if the support is not rigid but flexible? What if the wind blows a little harder or the excavator digs a little deeper? We can deal with some of these unknowns in the factors of safety, but others can induce profound changes in the structural response. We need a way to quickly asses the structure in its variability. The first answer is not going to be the final one.

The language of engineering is changing. The core skills of engineers are changing. The core skills of the past are known: maths, physics, calculus and statics. The core skills today are in flux. Some skills remain the same, as do our responsibilities. Maths is still a core skill but in a different form to our predecessors. Then it was pencil, paper and slide rule; now it is computer, programming, and digital workflows. What the key skills will be in the future remain to be seen, but automation is likely to be included.

Engineers have always spoken with the language of geometry, but the way that we speak it is changing. The medieval mason and carpenter worked with span to depth ratios, compass and ruler to build soaring cathedrals. The engineers of the Industrial Revolution built great tunnels and bridges of brick and iron using pen, slide rule and drawing board. The previous century brought domes and skyscrapers of concrete and steel, initially with the same tools as the Victorians, then gradually also with the electronic computer, enabling designs previously only dreamed of. The engineers of the past designed using straight lines and circular arcs, because they could both draw them and build them. Now we can construct more complex shapes, more efficient shapes, and ideally those that are both.

Parametric design is all about automating work. If I need to calculate the moment on a single beam, will I do it in my head or with a calculator?[†]. If I have about a dozen beams or so I might create a quick spreadsheet where the loads and spans are cells. If I have been given a file containing several hundred beams with different spans and loads then I would write a script to read the data, do the calculations, and output a new file with the results. Parametric design is thus first determining what are the parameters that drive the design, then formulating the design model, whether graphical or numeric, to be controlled by those parameters. By adjusting those parameters, you can both explore the options and refine the results, whether manually or with automation.

Automation, like most programming, is time-consuming, so the time saved must outweigh the time spent (Figure 3.1). If you are creating the automation for someone else, then it is another matter: as a programmer you mostly have to spend your time so that others can save theirs.

3.2 How we work with computers

While you may learn how to size a single beam by hand, what will you do when asked to size 2,000? How will you define the structural layout when the architect gives you a SPLINE or NURB surface to work from? How will you work out the most cost-effective option from a vast range of possibilities? What will you do when the requirements change again? The answer is not to break out the paper and pencil, but to program a computer to do it for you.

[†] Usually the calculator…

Figure 3.1: Automation

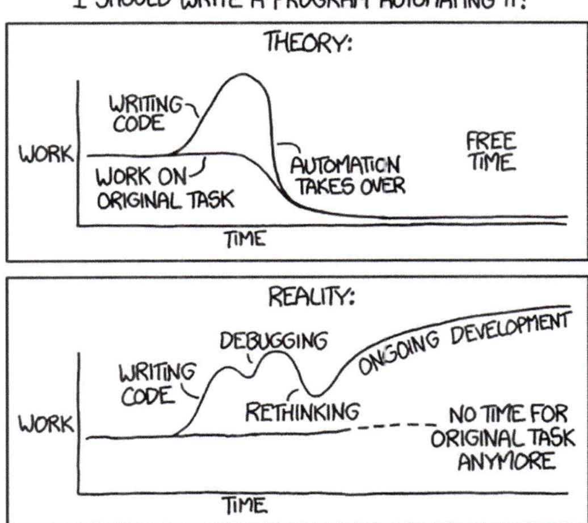

In this age of automation and digital transformation, programming is a differentiator between graduates' CVs. This is especially true for the large companies, who take on huge and complex projects, but is also relevant for many small consultancies who need to repeat certain tasks. One of my first professional programming tasks was as a new graduate, automating the CAD drawing setup at a small consultancy. The management and technicians did not realise that it was even an option, but I saw a need.

c.1990 we had recently introduced CAD to the company to supplement the drawing boards. In those days there were no viewports or drawing sheets, you just placed a drawing border of the right size in the file to the correct scale, set your text sizes, and drew in the plans or details. These were standard operations all geared around a few key variables (mostly the final scale of the drawing). The CAD program had the ability to automate tasks using a programming language called LISP[†]. I created a new menu that allowed the technician to choose the drawing border size and then scale from lists; it would then set the drawing variables, add the border from the library, and change the menu to the standard one to let them start work. As a piece of coding it did not set the world alight but made my colleagues lives that little bit easier, which is the whole point of automation.

3.3 Parametrics

"Communications tools don't get socially interesting until they get technologically boring…
It's when a technology becomes normal, then ubiquitous, and finally so pervasive as to be invisible,
that the really profound changes happen".
Clay Shirky[37]

In the same way that the ubiquitous nature of the mobile phone has transformed society, parametric workflows have been transforming structural design. This is best explained by looking at a few examples.

3.3.1 La Sagrada Família
Antoni Gaudi started the design and construction of The Basílica de la Sagrada Família in Barcelona in 1883[††] and had only completed two facades at the time of his death in 1926. La Sagrada Família was originally designed in the Gothic style but Gaudi knew that the designs were inefficient. Vertical columns need flying buttresses to resist the horizontal thrust from the roof, but he could dispense with the buttresses if he could angle the columns correctly. While he did not have access to electronic computers, he could build an analogue design model. This was a catenary chain model of the structure, with small sand bags to represent the parameters of the permanent and variable actions (Figure 3.2). Careful measurement then gave the geometry for the masons.

[†]   'LISt Processor' — though it was often called *Lost In Stupid Parentheses* as everything had to be nested in brackets.
[††] Another architect originally designed the basilica, in a more traditional style, but he quit after completing the foundations. Gaudi then replaced the superstructure, though still making use of the column layout.

Figure 3.2: La Sagrada Família: design model (a and b) and actual interior c)

(a)

(b)

(c)

Today, computer models have replaced the hanging chains, allowing both analysis and detailed design of the computer-cut prestressed stonework[38].

3.3.2 Selfridges Birmingham

As an example of contemporary parametric design, let's look at Selfridges Birmingham (Figure 3.3). The structural design started with a NURB surface model produced by the architect. The engineer then needed to write some code, which might be text-based or a visual script, that offset this surface by the thickness of the cladding so that they knew where to add the outer surface of the concrete shell. This surface was then offset for the inner surface, and again for primary members and slab edges (Figure 3.4). The results were then exported into an FEA program for analysis and sizing, and a BIM program for documenting, detailing and coordination.

Figure 3.3: Selfridges Birmingham, Birmingham, UK

Figure 3.4: Indicative section through sprayed wall

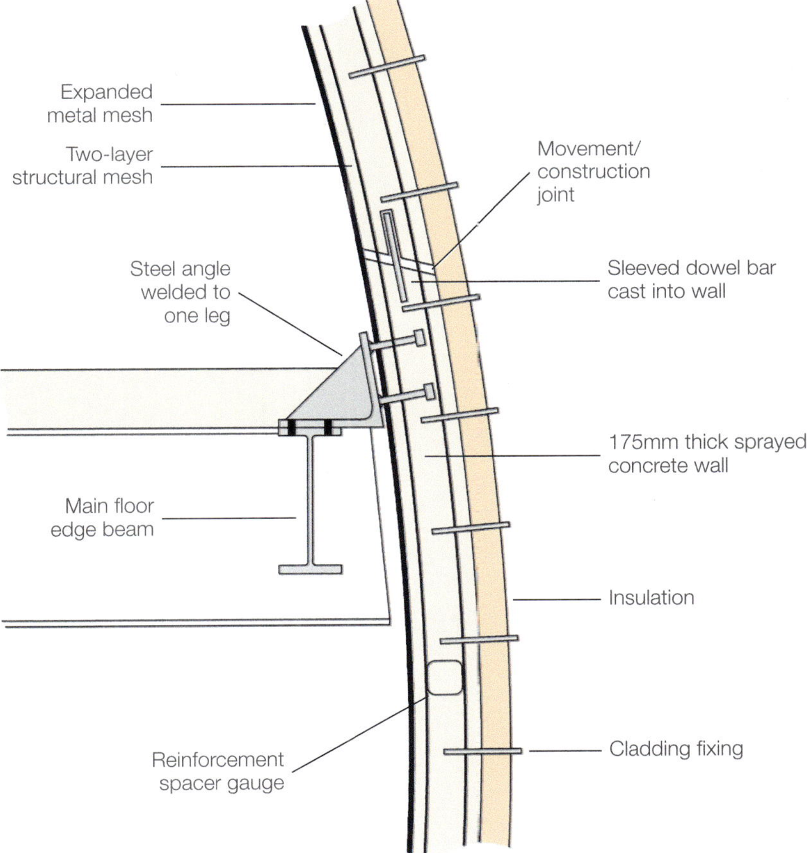

3.3.3 Camp Adventure tower

The Camp Adventure observation tower in Denmark (Figure 3.5) is a hyperboloid structure, which means that while it has a curved profile, it is made up entirely of straight lines: it is a ruled surface. Cylinders are a single-ruled surface (Figure 3.6), which means that there is only one set of straight lines that defines it; hyperboloids are double-ruled, meaning that there are two sets of straight lines on the surface, but with the top line ends rotated round the circumference relative to the bottom (Figure 3.7).

Figure 3.5: Camp Adventure tower, Roennede, Denmark

Figure 3.6: Cylinder

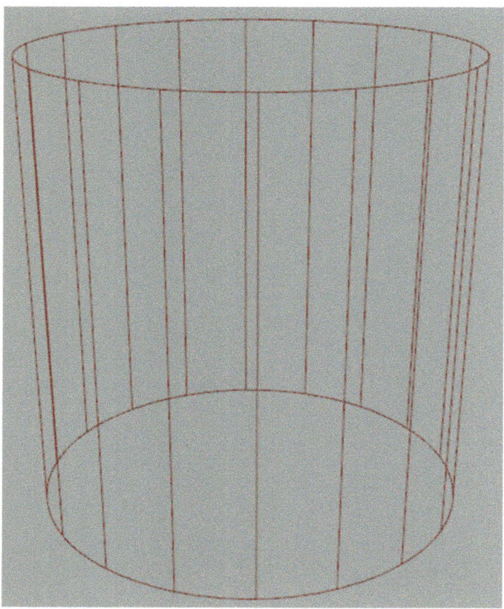

Figure 3.7: Hyperboloid

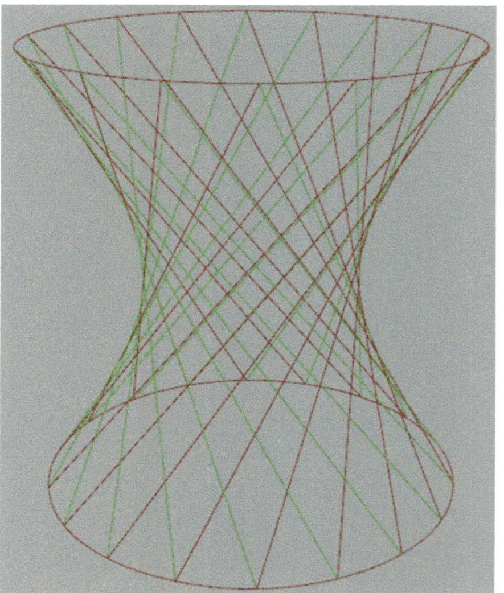

In a recreation of the original design model (Figure 3.8) I have five parameters[†] which I can adjust to get the desired geometry:

1. Height
2. Upper radius
3. Lower radius
4. Number of columns
5. Step shift

Figure 3.8: Full model

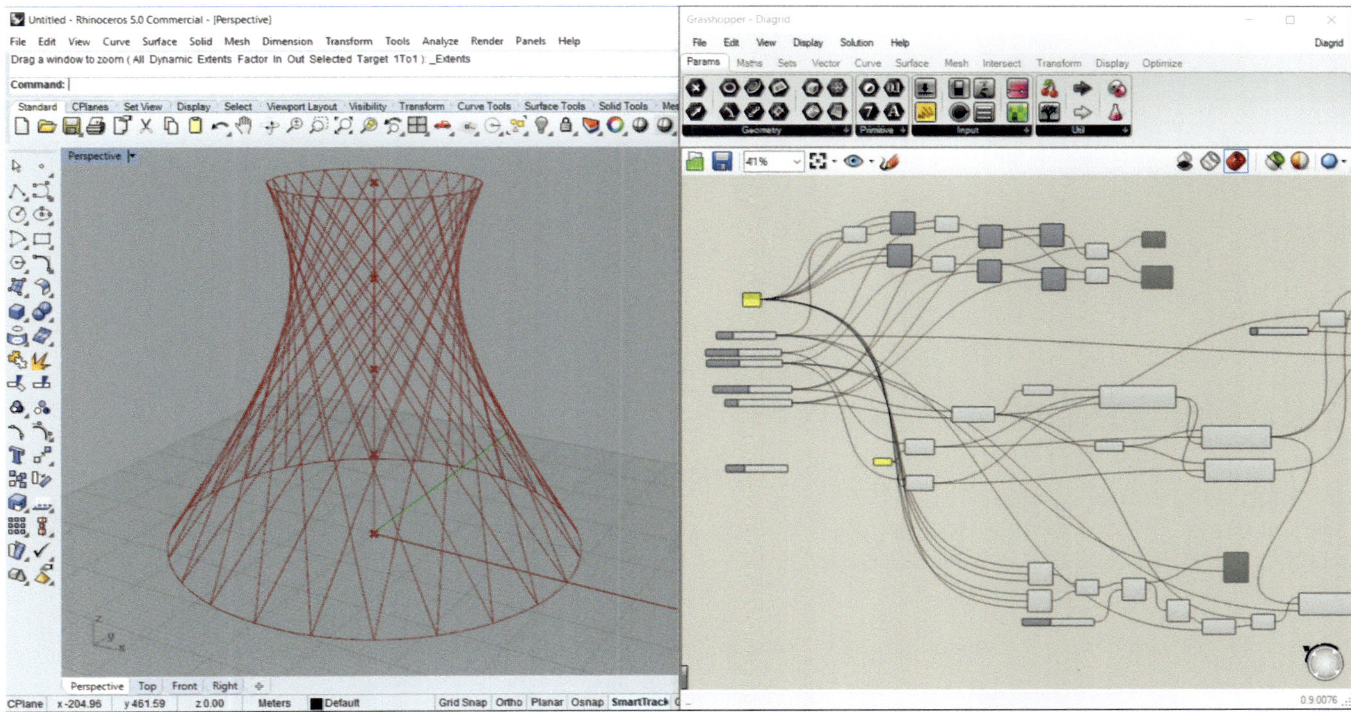

I could have made the upper and lower radius the same, which would have eliminated one parameter. The step shift refers to the rotation of the top ends of the surface columns relative to the bottom. A shift of zero gives the cylinder, and a shift of half the number of columns gives two cones. Here the top points are shifted round six places relative to the bottom, which in this instance equates to 120°. I could have used a rotation angle for the top points but that wouldn't guarantee the columns in each direction would meet on the top ring.

Next in the original design model, the engineers created a constant slope ramp up the inside of the structure[††], then connected this to the columns where they cross with cantilevers. The resulting geometry was then exported for analysis, design, detailing and documenting.

3.3.4 The Gherkin
The Gherkin (Figure 3.9) is another good example of parametric design. It is again a diagrid structure, but the layout comes from a different set of variables: floor-to-floor height, top floor radius, bottom floor radius, and vertical curve radius (though radius of a middle floor would also work). You might also use the number of nodes around each floor as a variable. The analysis model (Figure 3.10) was then generated from the results. Not only is the resulting structure efficient, but also aerodynamic, reducing the wind loads.

[†]   I suspect that the height, radius, and number of column parameters are self-explanatory.
[††] While the ramp slope is constant, the radius is not.

Figure 3.9: The Gherkin, London, UK

Figure 3.10: The Gherkin — GSA model

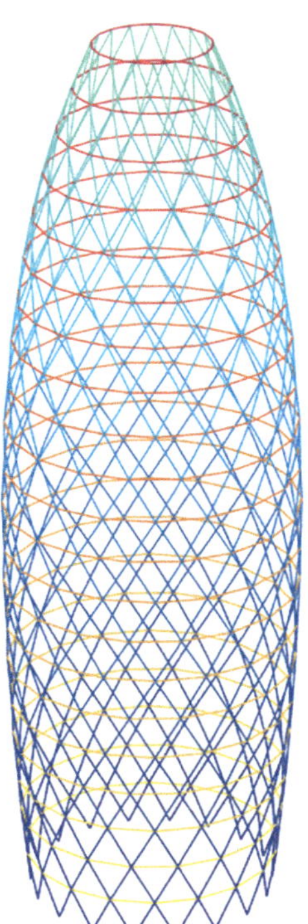

3.3.5 King's Cross concourse

While the King's Cross concourse roof (Figure 3.11) may resemble a form-found structure, its geometry is driven by a number of circular arcs that are subdivided to give the member connection points (Figures 3.12 and 3.13).

Figure 3.11: King's Cross Station concourse, London, UK

Figure 3.12: King's Cross Station concourse — CAD model

Figure 3.13: King's Cross Station concourse — GSA model

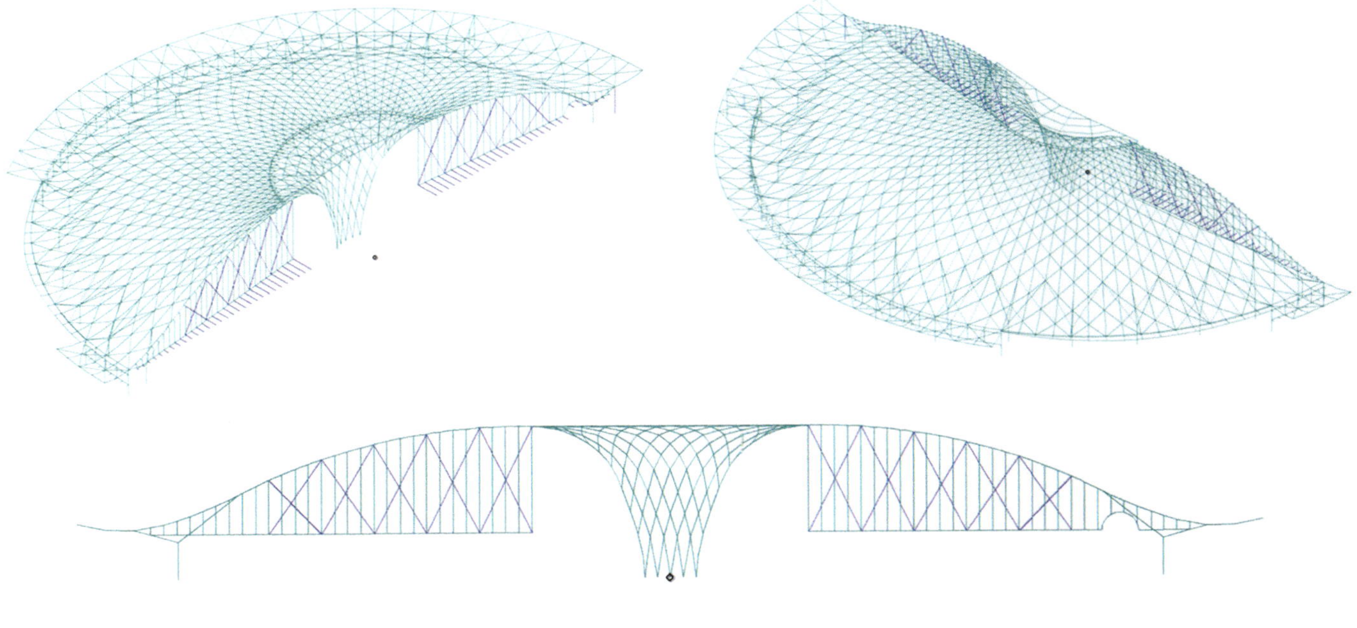

3.3.6 Everyday parametrics

Parametric design is not just for the "weird and wonderful" architectural flights of fantasy, but also down-to-earth projects. Here is an example taken from a project I undertook with students at the University of Bradford to design a swimming pool roof (Figure 3.14).

Figure 3.14: Swimming pool project: design parameters

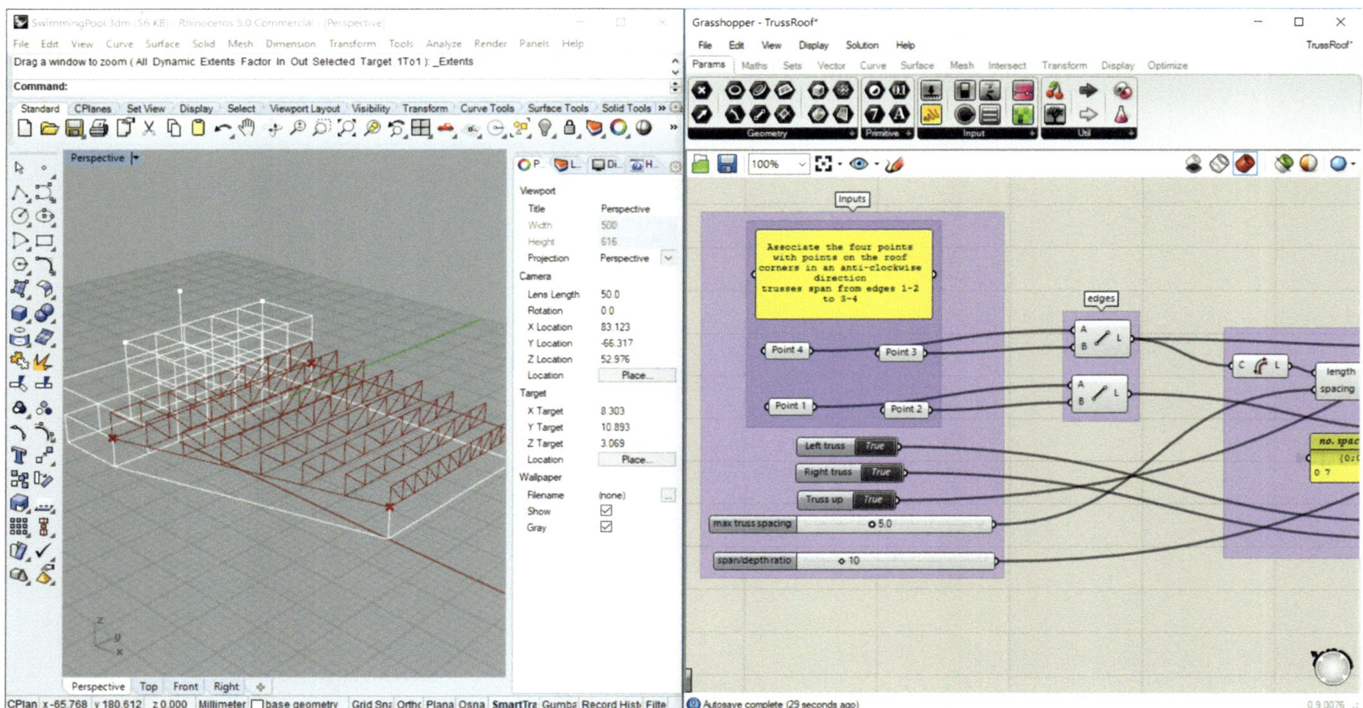

The parameters here are the four corners of the trapezoid pool area, maximum truss spacing, and span-to-depth ratios. Following on from the initial effort of creating the parametric visual script, manipulating the sliders quickly allowed us to explore different layout options (Figure 3.15).

Figure 3.15: Swimming pool project: layout options

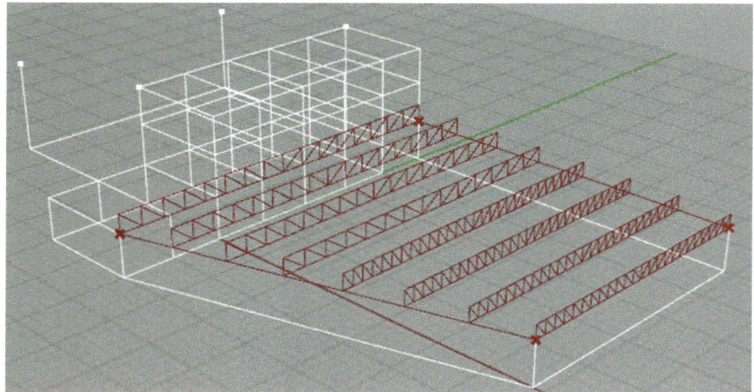 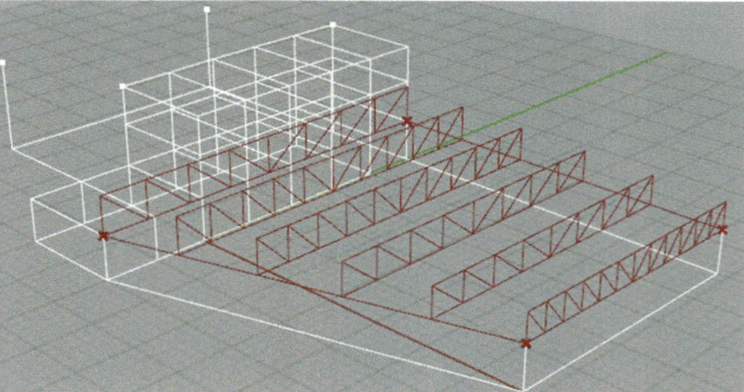

This approach works equally well for portal frame sheds, beam and slab layouts, and so on.

3.3.7 Programming and scripting

> *"Automate the boring stuff."*
> Al Sweigart[39]

Programming, and in particular parametric modelling, is increasingly common for structural engineers. This might be for creating small standalone design programs on a project, macros for Excel spreadsheets, scripts to work with other analysis and design programs, or even creating complete programs that others can use. These scripts might interact parametrically with the CAD or FEA program; automating tasks, data mining and processing, or running design calculations.

All programming is just manipulating numbers. It may not seem like it when you are using a word processor or playing a game, but that is exactly what is happening inside the computer. I once got a glimpse beyond the smoke and mirrors when I decided to change the font in my Word document to Windings[†]. The result was that my document was now a series of random symbols, except that the spell checker still wanted to correct my text. Although I could not read the text on the screen, the program could as the underlying symbols it was using to record the letters were just the same. It translated those symbols via font files into a graphical representation that I could see and understand, but it made no difference to the underlying data.

Working with computers invariably involves layer upon layer of abstraction, hiding the user from the complexities of what is really happening. In fact, the more that I learn about computers, the more amazed I am that they can do anything at all. While the early computer programs required the programmers to specify everything, even down to the physical location of every bit of data, there are now libraries of routines to do this for us. This automation leaves us the space to concentrate on the things that are important.

As users we work with commercially written programs that allow us to model structures, draft reports, communicate with project partners at work, or conquer the world in our leisure time. All these programs are designed to allow us to do the task with the minimum need to understand the underlying workings of the computer.

3.3.7.1 How will the program help?

When writing programs, it's easy to focus on the functions and not on the purpose. There is a story of a power tool company to illustrate this. When the new graduates and executives join the company, they are given their induction training. As part of this they are asked what the customer needs from them. They naturally answer: "power tools", only to be told that they are wrong. What the customers need are holes in walls; the power tools are just a means to an end goal. The executives' job is to determine the best ways to help the customers get what they need.

In the same way, engineering programs must be there to serve the end needs of the user. The analysis and design functions are just tools to get to the desired result. This is much easier if you're writing a program for yourself or your colleagues, but can be much harder when writing for those you have never met and never will.

The principal goal of a structural engineer is to design structures that are: safe, functional, economic, sustainable and elegant. To do this the structure must be strong enough to not break (ultimate limit state) but not too strong (economic and sustainable). The structure must be stiff enough to not deflect too much (serviceability limit state) but too much stiffness can be bad in seismic and blast loading.

Economic and sustainable mean the cost (financial and/or environmental) is minimised. This might be because the beams may be deep, efficient, and cost less themselves, or they are small, maximising the ceiling or minimising the building height. This cost can be to the client and how much they must pay; it might be to your company: how long are you spending designing that beam; and it may be to the world: how much energy and greenhouse gas was involved in the mining, processing, manufacturing and transporting of that beam?

Elegant is, perhaps, the hardest quantity to qualify. Slenderness and efficiency are likely to be factors; so are proportion and aesthetics. Elegance is a value judgement. Note that many of these requirements are in opposition to each other, and an art of engineering is keeping them in balance.

[†] I got where I am today by pushing buttons at random and seeing what happens…

3.3.7.2 A brief history of computer programming

> *"We may say most aptly, that the Analytical Engine weaves algebraical patterns*
> *just as the Jacquard-loom weaves flowers and leaves."*
> Ada Lovelace[40]

One of the earliest computers was the mechanical Difference Engine, designed by Charles Babbage in 1819–22, to produce astronomical and navigation tables[9]. Inspired by the Jacquard looms that used punched cards to weave complex fabrics, and working with Ada Lovelace, he started work in 1833 on the very first programmable computer: the Analytical Engine.

The Analytical and Difference Engines pushed Victorian engineering to its limits, meaning that a fully working version of the Difference Engine was not completed until 1991 and a full Analytical Engine has never been built. Lovelace did create the first manual though[†] and she is regarded as the first computer programmer.

In 1938 the Polish codebreaker and mathematician Marian Rejewski created the electro-mechanical Bombe to decode the German Enigma coded messages. The design was then improved by Alan Turing at Bletchley Park in 1940[41,42]. Like the Difference Engine, the Bombe was also a calculator rather than a programmable (or 'universal') computer. Turing first conceived and wrote of the universal computer in 1936[43], where the input to the machine would contain both the data and the instructions to process that data.

In Germany, Konrad Zuse[††] was working on building the first programmable computer, the Z3, and succeeded in 1941[44], but the German authorities saw little merit in the project. The Z3 and most of the documentation was destroyed in an air raid in 1943. At this time at Bletchley Park, Tommy Flowers created the first programmable electronic computer for the Allies, the Colossus (Figure 3.16). Unlike the programmable computers that followed, it had no memory so could not store the program or data but needed them input by paper tape at runtime.

Figure 3.16: Colossus in operation, 1943

[†] Ada Lovelace translated Luigi Federico Menabrea's description of Babbage's *The Analytical Engine*, then added notes about how to program the Engine that were about twice the length of the original document.
[††] Originally a structural engineer!

The Electronic Numerical Integrator and Computer (ENIAC) followed in 1945 (Figure 3.17)[45], which used punched cards for long-term storage but was programmed by a plugboard and switches. The principal designer was John von Neumann[46], who formulated the various components of the electronic computer and how they work together; this is now known as the 'von Neumann architecture' and is still in use today.

Figure 3.17: ENIAC computer

3.3.7.3 Languages

> *"Everyone in this country should learn to program a computer, because it teaches you to think."*
> Steve Jobs[47]

These initial programs were written in what we would now call machine code, which are the actual instructions for the machine in full detail, including exactly where to store the data. In the 1950s Grace Hopper created the first high-level programming language, COBOL. These programs are written in something close to English and are then converted into machine code by an interpreter (at runtime) or compiler (in advance). Today there are many programming languages and, over the years, I have learnt and used about a dozen of them.

The most common example today is so ubiquitous that it is rarely recognised as programming: the spreadsheet. They were originally created to automate the work of accountants and replaced the bookkeeper's rows and columns of data-filled cells. Other cells would then contain commands to add, subtract, and carry out other mathematical operations. The results of these operations might then be subject to further calculations, and so on. The addition of text into the spreadsheet, plus graphics such as graphs and images, mean that spreadsheets are now extensively used by engineers for reports or calculation sheets.

Spreadsheets do suffer from a few problems that make them risky for anything other than casual use. The first is that they are exceedingly difficult to check. The original accountant's use of adding up rows and columns of numbers is easy to verify, but an engineer's spreadsheet designing a concrete beam might be pulling in values from elsewhere on the sheet, on adjoining sheets, or even completely different files, making it hard to see what data comes from where. Secondly, the formulas in use are all hidden within the cells, making them difficult to read. In addition, the two-dimensional nature of the spreadsheet encourages non-sequential ways of thinking: the values used in a calculation can come from anywhere and the results can show before the data. Finally, their nature means they do not work well with version control methods, which are management systems used to store code and its history. This all means that they are quite easy to create, but the calculations need close scrutiny if they are used for structural design. No matter how nice the formatting, the results must be correct.

Modern spreadsheets that use Macros (text scripts) do go some way to reduce these problems — and they are more useful for everyday calculations and data handling — but while they are a great introduction to programming, they are still a risky option for critical calculations.

Several computer-aided design (CAD[†]) programs have supplemented their text macros with graphical programming options. Instead of writing out all the code, you instead drop components onto the interface, which are then linked to other components — the output of one becoming the input for the next. These components might be as simple as adding two numbers together, or sophisticated programs in their own right that include analysis or links to other programs.

In many respects, the graphical programs hark back to the early electronic computers and the way they were programmed using physical cables. They do suffer from the same problems as those early computers: a complex program soon resembles a plate of spaghetti and is just as difficult to disentangle. Like spreadsheets, they are very efficient at what they are designed to do, which in this instance is create parametric geometries that you can then analyse and take forward to detailed design.

Spreadsheet macros and the CAD graphical programs are examples of what is known as 'application programming', which is writing a program to control the model in another program. If your code is written completely outside the model program then it needs to access it using an Application Programming Interface (API), which is a set of commands and tools that the model will recognise. APIs are a vital component in modern digital workflows as you can use them to extract data from one model and use it again in another.

So, which is the best language to program in? A fun spectator sport is posing that question to a bunch of programmers and watch while they fight it out. There are as many opinions on this question as there are software developers. The truth regarding the best language comes down to a few main factors:

- What do you want to do?
- Where will it run?
- Which are the critical factors when running the program: speed, flexibility, reliability, etc.?
- Are you coding for yourself or will someone else use it?
- Will it be sold to someone else?
- Is coding your full-time job or something you undertake to make your job easier?
- Which coding languages are you familiar with?

Many will start by scripting: creating Excel or CAD macros. With more experience you can start to write more general-purpose programs that other people can use. These are harder to write and maintain as they require user documentation, support, robust error handling, and licensing.

3.4 Design communication

The best design in the world is useless unless it can be made (Figure 3.18). A carpenter or painter may be able to create their work themselves, but as engineers we must document our design so that it can be understood by a variety of individuals across a number of related disciplines. Design communication must be clear, concise and unambiguous. It must tell the architect where the columns will be in their room spaces, the mechanical engineer where the beams and floor voids are for their ducts to pass through or avoid, and the contractor what they must construct.

[†]   As opposed to computer-aided drawing, which is also abbreviated as CAD. Many CAD programs are also CAD programs, which can be confusing.

Figure 3.18: Just because you can draw it, doesn't mean you can build it

Brevity is important as people will tend not read a 200-page specification — just as they tend not to read the terms and conditions on their latest phone operating system upgrade. 20 words are better than 200 if you can pack in what you need to say. Consider this example. Ernest Hemingway is said to have been challenged to write a story in as few words as possible. He allegedly wrote this sad tale: *"Baby shoes for sale, never worn"*. See how much back story and anguish he packs into just six words?

Engineering documents must be concise and unambiguous. It is said that a picture is worth a thousand words, and that is why much of our communication is through drawings. These drawings are not there to look beautiful, though they can do: the colour-washed drawings of the Victorian age can look wonderful in a picture frame. Engineering drawings are there to show others how the design will work.

CAD[†] was first invented in the 1960s but started to become common in the construction industry in the 1980s, and finally supplanted drawing boards in the 1990s. While the first draft of a drawing might take a similar time on a drawing board, changes (and there are always changes) are far quicker with CAD. On drawing boards, we had to erase pencil lines from paper and carefully scratch ink from drawing film with a scalpel, before carefully redrawing the offending item. With CAD you can move, stretch, rotate and edit the images or model with little effort.

CAD drawings were initially just computer versions of the manual drawings, with lines being just lines. In the 1990s 3D modelling packages became popular, where we would add beams and columns as objects and let the computer work out where the lines would show. These were initially most used where 3D was crucial, such as in the petrochemical industry (the pipes can be like a rat's nest) and steelwork fabrication. For much of the 1990s I worked as a 3D structural modeller leading a team designing petrochemical refineries in the UK and Norway. While it was an unconventional experience prior to taking the IStructE Chartered Membership exam, it did teach me how to think about a whole building and not just isolated parts. With 2D drawings it's easy to forget that it must all join up. Even if you don't plan to work full-time as a technician, experience building 3D structural models will not be wasted.

3.4.1 BIM, interoperability and digital workflows
All engineering design needs workflows: information that passes from one stage to the next, whether it is the architect's requirements at scheme stage, or construction documentation for the contractors. As a modern engineer dealing with digital information you need to organise a digital workflow. Building Information Modelling (BIM) is a common expression for this but is also often misunderstood. The original meaning of BIM was the

---

[†] Computer-aided drafting in this instance.

process of sharing data between programs, but the definition has grown to also refer to the models that generate the data[†]. Usually, BIM involves 3D CAD modelling, but not all 3D CAD is BIM and not all data is geometric. BIM as a process used to be known as Electronic Data Interchange (EDI)[48], which is a more accurate description of the process, but it has more syllables and so never achieved the same popularity. 'Interoperability' is a rival term growing in popularity, not least because it is not restricted to buildings, but instead gets more to the heart of the data interchange. 'Digital workflows' is the current popular term, though I wouldn't be surprised if it's called something else in 10 years' time.

At its heart, a digital workflow is about ensuring that information is shared, reused and enriched as it passes through the design process from initial scheme to final construction. Data can also be lost on route through the project[††] as different programs require different representations of the structure, and have different requirements and limitations. For example, the BIM models need to represent the physical nature of the structure, so will have the beams stopping on or before the column faces. The analytical model, on the other hand, must join up at the centroids of the connection. Similarly, the analysis model needs to know about forces and moments, while the BIM model does not. The ideal on a project is to have a single federated model that all programs use and refer to — but it can be difficult to implement unless everyone is using the same software package. While this consistency has some benefits, it can also cause problems, as all programs have strengths and weaknesses. The most suitable program for the architect may be less suitable for engineers.

To explain digital workflows, let's take a simple example: analysis and section design (Figure 3.19). The analysis needs geometry and loads to give you the bending moments, shear forces etc. These analysis results are then passed to the section design, along with other parameters, to give you the required section size or rebar.

Figure 3.19: Workflow analysis to design

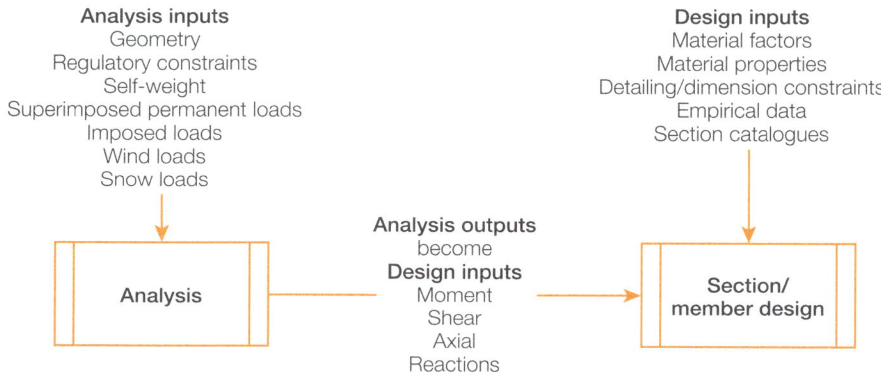

This is ok for when the structure is statically determinate or self-weight is not a particular issue, but for the majority of cases we need to feed information back from the design to the analysis for a second pass (Figure 3.20).

If the updated section sizes change the analysis results sufficiently, the section sizes may need updating again, with a further round of analysis. As you may have realised, automating this workflow will save you a lot of time, especially as the model grows in complexity. This is, of course, only part of the story — the structural layouts need to be created with the correct section sizes added so that the contractors can build what you have designed.

[†]   It is now acceptable to say: "BIM model" (Building Information Modelling model) despite the apparent tautology.
[††] Data relating to security or other private aspects should not be transferred. I have heard of bomb-proof door objects issued with the calculations still embedded, and prison plans online clearly showing CCTV fields of view.

Figure 3.20: Workflow analysis to design and return

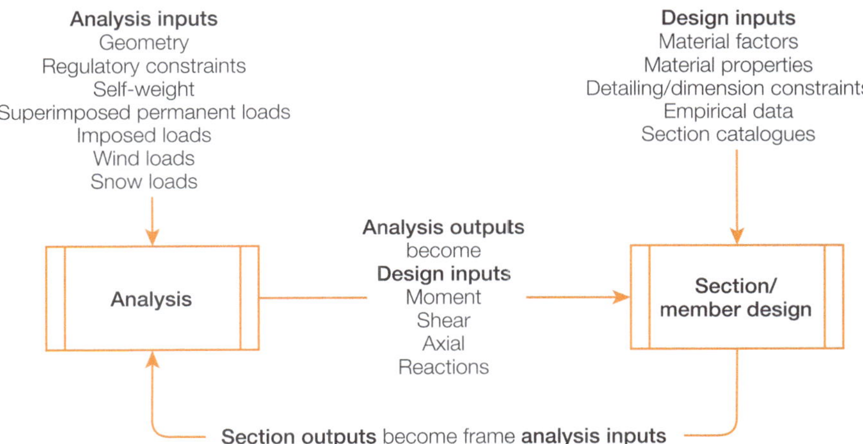

While you can produce both CAD and FEA data from a single source, if you do not keep them connected then any subsequent transfer of data must be carried out manually, with all the risk of error and the time consumption of checking (Figure 3.21).

Figure 3.21: Example of inefficient workflow (as advantages are lost)

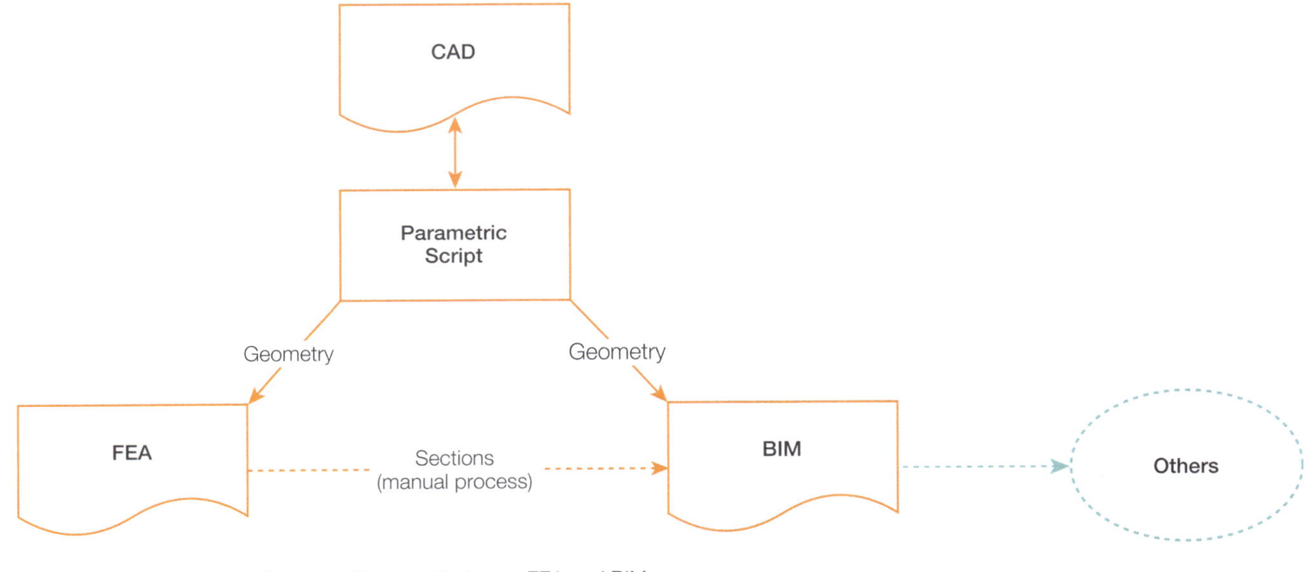

A better workflow might be something like that shown in Figure 3.22.

Figure 3.22: Example digital workflow

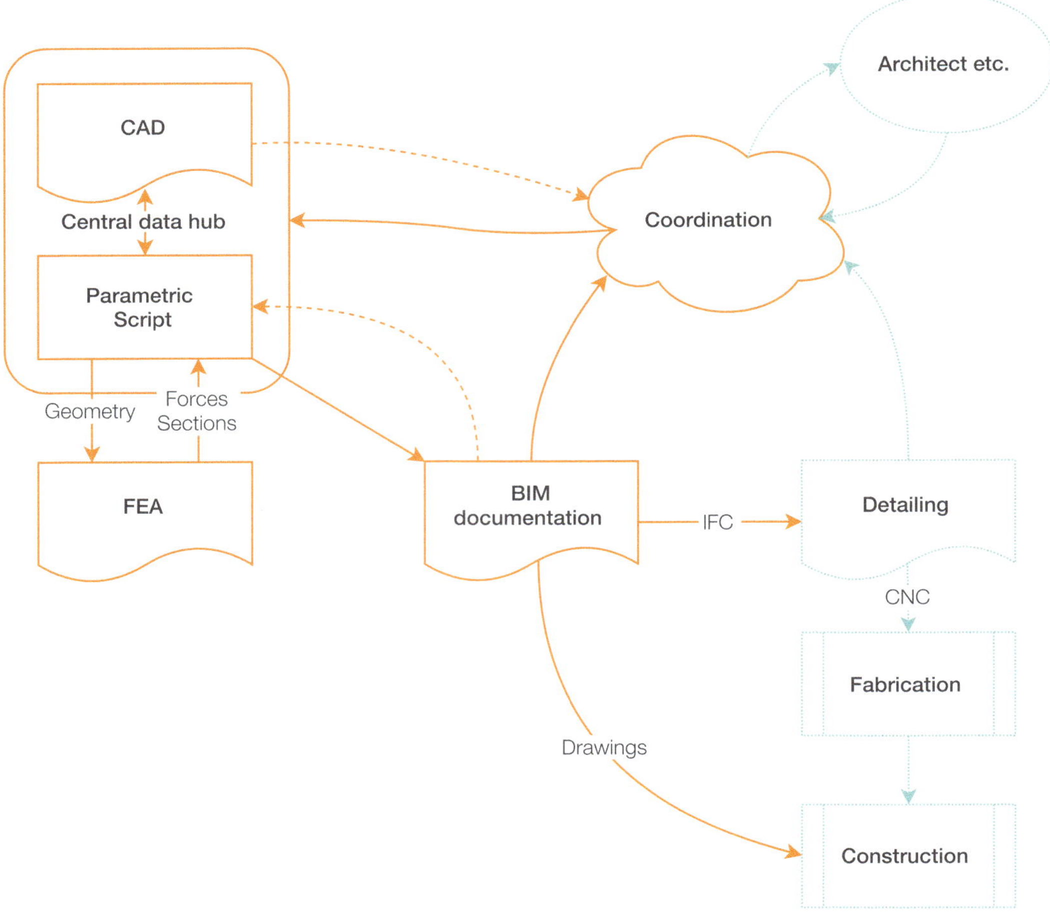

1. The architect creates an initial model and passes it to the engineer.
2. The primary layout of the structure is determined with a parametric CAD program. This then generates:

   a. A detailed BIM model that is capable of documentation (drawing production). This is cross-referenced with the architect's model as well as those from other consultants. Changes are extracted and fed back into the parametric model.
   b. An FEA model — that might be built from either the initial model, or the BIM model. This then analyses the model and sizes the primary members. The section sizes and end reaction etc. are then fed back into the original model.

3. The BIM model is then sent onto a contractor, who either works up detailed models for the steelwork fabrication or reinforcement detailing.
4. The detailed structural parts are then exported to automated fabrication machinery that cuts, bends, and welds the parts together.
5. Humans (currently) then fit the parts together on site, with instructions in the form of drawings from the CAD program or BIM model.

Each step creates a new dataset based on the existing information, plus the skills and knowledge of the engineers. If you can store the information in a consistent manner and automate the flow of that data between the different processes, you will be working very efficiently.

You will note that drawings still feature as part of the workflow. Electronic files might be the best way to transfer data from one program to another, but we need drawings to explain to the humans what is in the model. Drawings are simplified annotated views of the structure, and are there to convey meaning. While electronic checking of models, and indeed drawings, has been around for decades, you still cannot beat going over a printed drawing with a pen and highlighter.

Electronic checks such as clash detection can find what overlaps, but might not spot what is in the wrong place and certainly will not see what's missing. By way of illustration a SCOSS report highlighted a project where the contractor questioned the lack of ground floor columns. It seems that during a transfer between programs, several had gone missing, resulting in a 225mm concrete slab spanning 14m and carrying column loads from above[49]. The consultant blamed BIM, but the real problem was lack of checking. Errors can still occur.

3.4.2 Standards
All modelling programs have their own standards and ways to represent the contained information. To pass information from one program to another you have a choice: you can either create a dedicated program that can translate the standards of all models that you are interested in, or you can use a neutral standard that is supported by all parties. The problems with the first option are obvious: you need to put in a lot of work to connect to every model type, which will need changing if they update their platform. On the other hand, you have complete control over what is written and read.

The standard language would appear to be the best way forward, but the problem is that you are then trying to satisfy everyone — both software providers and users throughout the entire world (Figure 3.23).

Figure 3.23: Standards

This is going to be exceedingly difficult. Specialists have been working on the Industry Foundation Classes (IFC) for over 25 years and have still not reached complete agreement[50]. I myself, played a small part in the community in the late 1990s working on the CIMsteel project, which subsequently evolved to become the structural aspects of IFC.

> *"You can please some of the people all of the time, you can please all of the people some of the time, but you can't please all of the people all of the time."*
> John Lydgate (attributed)

The problem with interchange standards is that a completely universal standard is impossible. While all structural programs, for example, have common concepts such as beams and materials, how they define those concepts and what variables they include will vary from program to program. A drafting program might record only the steel grade name, but an analysis program will need to know the Young's modulus, density, yield stress and so on. Similarly, the analysis program needs only a crude representation of the geometry while the fabrication model needs to know every bolt, notch and weld. This means that as the project data progresses between programs (from original concept to final construction) at every stage some data is enriched, and some is lost.

3.4.3 Beyond 3D CAD: 4D, 5D and 6D

The three dimensions of models are obvious, but what about the higher dimensions?

4D incorporates time aspects into the model, especially construction sequencing. This has been led, naturally, by those who are involved with project management on site, because the exact order of what is built, where and when is crucial: you do not want to install the beams first and add the columns in afterwards if you can help it[†].

Time is also a key factor in the analysis of large bridges and tall buildings, especially if concrete plays a major part. Concrete creeps and shrinks, especially in the initial stages of its life, and the effects can be significant.

5D adds cost to the process over time. As engineers we tend to focus on performance (i.e. the building not falling down), but for the client the cost is a major concern. Cost is, or should be, an integral part of optimisation; unfortunately for the engineer it is also one of the hardest areas as it involves many factors outside our control. There are regularly updated guides to prices from both manufacturers and industry experts, so there is little excuse for engineers to ignore this important aspect. Alternatively, the cost consultant can add this information later in the project, before trying to change your design to cut costs.

6D considers the lifetime of the assets in the building including the structure itself, plus the ongoing performance of the building, especially the heating and cooling requirements. Asset management has long been touted as an important aspect of BIM, allowing the building owner to easily access information regarding the specifications and warranties of all parts of their property. It has not been much used to date for several reasons. One is that it is an extra cost for the consultants and, unlike digital workflows, gives them no benefit. Another is that the client who commissions the building is often not the operator and is thus not interested in paying for the work. Developers build offices and houses to sell as fast as possible and care little what happens to them outside the scope of any warranties.

Asset management comes into its own when the client is also the owner and operator of the facility. An area of growing interest is that of the 'digital twin', where there is a sufficiently detailed model of the facility that has electronic links to the mechanical equipment, room sensors, and so on. This allows the operator to test operational scenarios, predict maintenance tasks etc.

3.5 Conclusion

Parametric design, BIM, interoperability, and digital workflows are all expressions related to the automatic creation, manipulation and translation of project data and designs. No design process is linear, no design project is simple. The key to model project management is to reuse as much information as possible. Anytime you re-enter information by hand there is a chance of making a mistake. Even if you don't make errors you need to check and check again to ensure that you haven't. Then, when a change arrives you will need to do it all over again.

At the risk of repetition: automate the boring stuff and automate the repetitive stuff. Digitise your communications. Check your processes. Control your workflow. Check enough of the output to be confident in the safety of the result. Design parametrically.

[†] I did hear of this being done on a project in India, where the team working on the beams was much quicker than the one doing the columns.

4 Analysis basics

"We actually made a map of the country, on the scale of a mile to the mile!"
"Have you used it much?" I enquired.
"It has never been spread out, yet" said Mein Herr. "The farmers objected: they said it would
cover the whole country, and shut out the sunlight! So we now use the country itself,
as its own map, and I assure you it does nearly as well."
Lewis Carroll[51]

4.1 Analysis — the key things you need to know

There is often a lot of concentration on the 'finite element method' in education, not least because it is objective and easy to test. There are in fact several methods or solvers and you do need to have a basic knowledge of them as different analysis types will treat your model differently and you need to understand why. We will explore the analysis methods in a later chapter. While the software developers create the solvers, you have full control over what element types are used in your model and how they are connected, as well as the analyses that are run. The most important thing is for you to understand them if you are to create a useful model. The problem with modelling is that what constitutes a good model varies from case to case, requiring not just rules but also judgement and experience.

The most important analysis tasks for practicing structural engineers, are to model correctly and to verify the results. This, of course, is easy to say but hard to do as it requires engineering knowledge and model-making practice. Paradoxically, it also requires you to have a good idea of what the answers are before you even build the model. The results will depend on how you build the model, and how you build the model will depend on how you think the structure will behave. If you think that the deflections and stresses will be low then a linear analysis will be sufficient, but if not then you need a nonlinear analysis and that can also require a more sophisticated model. Do you need to look at the dynamic behaviours of the structure? If so, you will need a different model to the static analysis as the structural behaviour can differ between the two cases: pinned connections can act as fixed, and tension-only elements can take reversal of load, but only if the vibrations are small. In later chapters we will look at how to make good models and how to avoid bad ones, but first, what is a model and what goes into it?

4.2 What is a model?

Engineering design is, in many respects, concerned with modelling, either how the world is or how we want it to be. The models may be physical or digital and represent the real or the conceptual nature of what we are modelling. While physical models can play a useful role in both education and design, the two meanings that are particularly relevant here are that a model is a mathematical representation of an aspect of the world in general and an implementation of that theoretical model as a specific piece of that world. This means that we can model how structures work in general, and how structural elements such as a beam or column might perform or be constructed.

Over the course of a project you are likely to build or encounter many models, from the initial geometry to the detailed CAD model, and from the conceptual design to the analysis model. The architect, project manager, and client will each have a different model to yours, reflecting their interests and responsibilities. As a structural engineer we are primarily interested in two types of computer model. One looks at the geometry of the structure: how it is defined, how it will fit together, and how that is communicated and coordinated with others in the project team. The other focuses on how the structure will behave, how it might misbehave, and how we can make it strong, tough, flexible, stiff, sustainable, and cost effective enough to be built.

In this chapter we will focus on structural analysis models, which are mostly, but not entirely, based on the FEA method. FEA represents structures as multiple stiff elements, then solves how they move when loaded. While FEA is highly successful for computer analysis, as it can handle structures of just about any shape, it is far too involved to use in hand calculations. In these situations, other methods such as standard formulas for beams or method of joints for trusses, are more practical.

Just as the plastic models of aircraft, glued and hung from the ceiling, are simpler than the real thing, yet still containing the desired essence of the original, so must our computer models be simplifications. What aspects remain depend on what questions we are trying to answer, but first you must first ask the right question.

To analyse an engineering problem, you need to simplify it until you can solve it: all simulation is simplification. But **do not oversimplify**: if you oversimplify then you will be insufficiently accurate. This is a difficult rule to follow as there is no definite boundary between 'simple enough' and 'too simple'; you will need to learn this by practicing. For example, Figure 4.1: is not a beam, but it is a useful simplification of a beam from which we can analyse to get useful design data:

Figure 4.1: Simply supported beam with uniformly distributed load

$$\text{Bending moment} = \frac{wl^2}{8} \quad \text{Shear force} = \frac{wl}{2} \quad \text{Deflection} = \frac{5wl^4}{384EI}$$

Where

w = load per unit length
l = span
E = Young's modulus
I = moment of inertia

Beams are not lines, they are three-dimensional solids, but they are (usually) much longer than they are wide or deep. This means we can ignore the beam's dimensions, even its shape and material, when calculating the moments and forces on route to deciding what section we need to carry the load. We can also ignore the support details at this stage. We do need to consider these details though when we get to the deflection and the calculation of any stresses.

But with the connections in mind, how exactly is the span of the beam measured? Is it from the centreline of the support, or the face of the support, or the centreline of the connection bolts, or something else entirely (Figure 4.2)? These are questions you need to address when you model the structure. Incorrect judgement of the appropriate span can result in a 10–20% error in the moments and forces.

Figure 4.2: Analysis conditions compared to actual support conditions

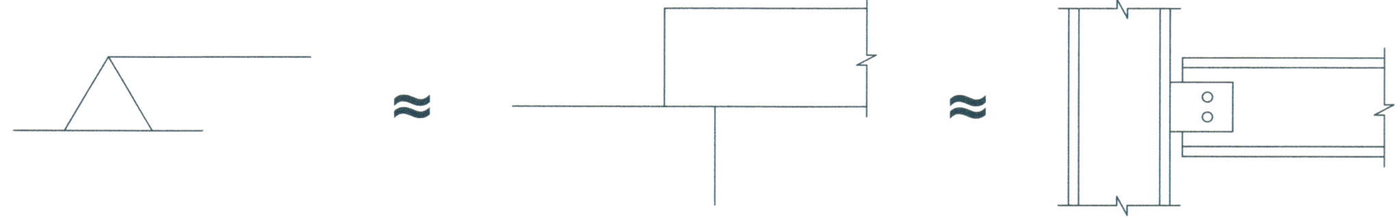

4.2.1 Simplification

"The modulus of elasticity of any substance is a column of the same substance, capable of producing a pressure on its base which is to the weight causing a certain degree of compression as the length of the substance is to the diminution of its length."
Thomas Young's original definition of elastic modulus[52]

The concepts of stress and strain were invented in 1822 by Augustin-Louis Cauchy, allowing Claude-Louis Navier to reframe Thomas Young's 1807 definition of elastic modulus into the more familiar form of $E = \sigma/\varepsilon$ in 1826. This simplification made Young's modulus a lot easier to understand. Likewise, we also need to simplify real structures to a point where we can understand them. To analyse a structure using the finite element methods, we make a model of it — simplifying the complexity of the real structure into a number of elements that represent the parts of the members.

Once we have the element geometry, their properties, and constraints such as foundation locations set in the model, we can apply some actions. These will be forces on the elements or nodes that come from gravity, wind and earthquake. Other actions can include settlements and temperature changes. Temperatures on whole elements will cause them to expand or shrink, and if across the depth of the members will cause them to bow (as experienced by bridge decks under a strong sun).

These actions will be gathered together into load cases, so that you can apply groups of the actions together in the analysis tasks. Note that there is a certain amount of circular thinking here as some loads come from the self-weight, which is calculated from the section size, and the distribution of forces in a structure will be influenced by the different element stiffnesses. Design is an iterative process, but so is analysis. This means that you need to either have a reasonable idea (or guess) what the section sizes are going to be in your structure, or use a script to iterate through the options to find the best choices. The latter option is not as simple as it sounds, unless your structure is statically determinate, so we will return to scripting in the chapter on optimisation. Estimating the correct section size will get easier with experience, but there are also rules-of-thumb, scheme guides, and more experienced colleagues to help.

4.2.2 The stool paradox

There is invariably a mismatch between our analysis models and reality. Our models usually assume perfection, unless imperfections are deliberately added in, such as with notional horizontal loads or adjusting the nodal coordinates. The problem is that imperfections can make differences to the structural behaviour in unexpected ways. As an example, let us look at the 'stool paradox'. Imagine that you have a three-legged stool with a rigid seat. If someone sits centrally on the seat, it does not matter how uneven the floor, each leg will carry one third of their weight. But if you have a four-legged stool then you do not get a quarter of the load in each leg; instead, you get the load carried by just two legs, and the other two carry next to nothing. The reason for this is that if the floor is at all uneven, or if there is any imperfection in the four-legged stool, only two diagonally opposite legs will be in contact with the floor, with one of the other two providing a touch of stability.

Does this happen though? Just as real floors are uneven, real stools have some flexibility, allowing them to naturally redistribute the load[†]. If the stool has enough capacity in each leg and the movement is reasonable, the stool will carry the load equally down each leg. Alternatively, if the stool is stiff and only two legs are carrying the load, we have no idea in advance which of the legs will be working at any time.

The three-legged stool is statically determinate, so we need no knowledge of the floor evenness or the stool stiffness to calculate the stresses within. The four-legged stool is statically indeterminate, meaning that the unevenness of support and construction, and the flexibility of the stool and supports, are needed to calculate the actual stresses. We might make a computer simulation, but without exact knowledge we cannot find an exact answer.

[†] This is especially true for wooden ones with traditional joints. Unlike dovetailed joints, glued joints have no flexibility. This means that while modern furniture is easier to mass-produce cheaply, it is also more likely to be over-stressed and break.

Thankfully, most engineering materials behave plastically to a degree. This means that we do not need to determine all the possible stressed states of the structure but instead determine all the collapse mechanisms. If the structure is strong enough to resist all of these then it will be stable and safe, and the actual stresses are of little consequence.

Similarly, with serviceability states, we do not know exactly how the structure will deflect or vibrate, but if we can demonstrate that the structure is acceptable in standard cases then it can be assumed it will be acceptable in most/all other cases. Occasionally we might find that the standards themselves are inadequate and need updating; the horizontal vibration of The Millennium Bridge in London is an example of this. Vertical human-induced vibration was a known effect and was designed for, while the horizontal version was not. With the global media watching, the opening-day crowds induced lateral vibrations so large that the bridge was closed until the phenomenon was properly understood and dampers installed. Unlike the Tacoma Narrows Bridge, the problem here was not that the Millennium Bridge was at risk of collapse, but it could not be easily walked on. Design codes and guides represent best known practice, but that does not mean they cover all eventualities. They always need applying with careful engineering judgement.

4.3 Software choices

There are so many different programs available to the structural engineer; which one should you use? The answer is to use the right one for the job. Thus, the job should be the main criteria; there are many tasks on a design project and many ways of undertaking them. Each program has its strengths and weaknesses; each might be useful to you at any given moment. There are programs that are dedicated to steel or concrete members, buildings or bridges. There are others that are flexible but then require you to put in more work to create the model. Some programs might only carry out static linear analysis, others might add in nonlinear or dynamic options; some will also design the sections for you. Programs might also have different terminology for similar things, so always read the manuals and tutorials to ensure that you understand what the programs are doing and not doing, to ensure that your models are correct and designs are safe. The CTV Building collapse in Christchurch (which we will look at in more detail in Chapter 7) was caused by the engineer not understanding what the program wasn't doing.

Programs for engineers come in three broad categories: communication, analysis, and section design/detailing. The boundaries between these different tasks can be fuzzy, not least because many programs merge two or even all three aspects. The more dedicated a program is to a particular standard case then the less usable it will be for anything else.

4.3.1 Components, section design and detailing
Individual beams, columns or connections

Section design and detailing entails determining what size section or quantity of reinforcement is required to carry the load. The programs will often be dedicated to particular structural element types and materials, such as steel beams or concrete foundations, or to structural arrangements such as trusses or portal frames. They will typically derive the forces and moments from applied loads, though some will have an application programming interface to allow other programs or scripts to make use of them.

4.3.2 Dedicated whole-structure design
Software for buildings or bridges, but not both

Some programs take the idea of component modelling of the beams and columns and add them together to look at a whole structure, then add in load takedowns and other useful aspects. They are usually efficient at the structural forms they are designed for, such as portal frame sheds, multi-storey buildings, bridges, etc. With a building design program, you might lay out a flat slab or composite steel beam arrangement and then tell the program to repeat the floor the requisite number of times. Conversely, their dedication to singular structural form makes them less than efficient when you stray outside their remit, which is when the general-purpose analysis and design programs come into their own. Sometimes, depending on the program, they can struggle with complications like sloping columns, carpark ramps that join floors, and so on.

4.3.3 General structural analysis and design
Models for any structure — FEA

A common task for engineers is an analysis of the structure to determine the forces, moments, deflections, stresses and strains. There are standard structures that have defined formulas ($M = wl^2/8$ for example) and some programs are dedicated to those, but there are far more structures that are not amenable to such algebra. Also, structures can contain thousands of elements, so even if the individual calculations are simple, automating the maths is a sensible thing to do. FEA has become the industry standard method for such problems.

Many modern FEA programs will also design section sizes and reinforcement, and a few also produce drawings for design communication.

4.3.4 Modelling, documentation and communication
CAD and BIM

The purpose of engineering design is construction, so the engineering design must be communicated with design team partners so that we can check that everyone's designs will fit together, and to the contractor clearly and without ambiguity so that they can build it in the correct way. The communication is usually in the form of dimensioned drawings, though data files of the structure are often used, especially if the geometry is complex. Originally these drawings were produced manually (with pencil and ink on paper, drafting film or linen) but were gradually replaced by CAD drawings. While the initial CAD drawings were only slightly easier to create than doing it by hand, editing the drawings was easier on the computer: we had to erase ink lines off drafting film by carefully scraping them with a scalpel, being careful to not damage the film itself.

3D CAD modelling began to creep into engineering drawing offices in the 1990s. Initially these models were just lines and arcs in three-dimensional space, but structural modelling programs allowed us to draw in the frames and foundations using dedicated 'objects': a beam object knows (from its two end points, section size and orientation properties) how to draw itself and interact with other objects.

Structural engineers led the construction industry in 3D modelling, not least because steel frames lend themselves to the process. Concrete modelling, especially for *in situ* concrete, can be harder to model than steel; partly because the shapes can be non-standard and partly because there is no clear demarcation where one element stops and another starts (for example beams and slabs merge into one another). Reinforcement detailing in three dimensions proved quite difficult and so took a lot longer to develop.

Quite often the structural consultant would build a frame model, extract drawings from it to pass to the steelwork fabricator, who would then build themselves a new model. While t seemed to most people that it was sensible to use a neutral data file to pass the structural details from one program to another, it took a long time to agree on a common format and is taking even longer for the use of electronic data transfer to be accepted in the industry as a standard way of working.

Apart from drawings, many 3D CAD programs can also export geometry and loading data to analysis programs.

4.4 FEA 101

4.4.1 What is the purpose of FEA and how does it differ from CAD?
Many people get confused when working with both analysis and CAD models, trying to make them look the same. This is both a time-consuming and time-wasting exercise, as the two models have different purposes and requirements.

The CAD model first. Its task is to represent the physical detail of the intended structure and communicate that to the other members of the design team, such as the architect or HVAC engineer. It is also important that the CAD model and drawings convey to the contractor how the structure f ts together, so that they can then add their details and requirements, then ultimately build from them. This means that aspects like the 'top of steel level' must be correct to the mm.

The analysis model, however, need give only a passing nod to the dimensions. It will not make a significant difference to the result if the beam is a few mm higher or lower. The analysis model is there, not to represent the physical reality of the building but to model its structural behaviour: How much will it deflect?, Are the sections large enough to carry the load?, What are the loads?, and so on. The CAD model will also contain various pieces of information that we are not at all interested in having in the analysis model. For example, analysis models can ignore tertiary components like handrails, upstands and small openings, or might model just a part of the structure, like a connection or a staircase, and ignore the rest of the building.

4.4.2 Essential aspects of an FEA model

While there are a variety of commercially available analysis programs, they tend to have the same core concepts: that of nodes, elements, loads and restraints.

4.4.2.1 Nodes

Nodes are the connection points for the structural elements and record the coordinates. The nodal coordinates are given in a Cartesian form[†], though some programs will also allow you to define them using polar and spherical coordinates, which are particularly useful for circular structures. The FEA program can then use the coordinates to calculate element dimensions.

The nodes are the points where the program carries out the calculations. Many modern programs are also good at interpolating the results between the nodes to give values such as bending moments and shear forces along the elements.

While you can type the nodal coordinates into your FEA program, or draw them in the graphical view, for more complex structures it is often preferable to create the geometry in a CAD program, possibly with a suitable script, then import the results.

4.4.2.2 Elements

The elements are the beams, columns, and pieces of slab or wall, that make up the structure: these are the eponymous 'finite elements'. In a steel frame the elements might span column to column, but in a slab the situation is more complex and we need to sub-divide the structure into smaller, but not too small, parts. While these elements may look like they are made from steel flanged beams or a thickness of wall, all the program is really interested in is their stiffnesses: how will they bend subject to a moment? how much will they shrink or expand under an axial load? etc.

The elements are connected at their ends, corners, and other locations using the nodes.

4.4.2.3 Members

'Member' is not an officially-recognised FEA term, but I will use it to refer to the original structural beam, column or slab. A steel primary beam is one piece of steel (plus connection plates) but will be broken down into elements at the connections with secondary members.

4.4.2.4 Loads

Loads are the actions on a structure: the self-weight and the load from the occupants, the dynamic excitation from the earthquake, the push from the wind. Some loads e.g. thermal, do not directly exert a force on the structure, though the effects they induce may generate huge stresses in the members.

4.4.2.5 Supports and restraints

Just like real structures, the structural models need supporting or restraining. Our structure would sink into the ground if it had no firm foundations on which to stand; likewise, our structural models need supports to resist the applied loads and thus generate the necessary reactions. These restraints or supports, which are known as 'boundary conditions' in many FEA texts, are points that we set to have either zero movement (translation or rotation), or a restricted movement due to stiffness resistance (known as a 'spring support'). Another option is to force certain nodes to move a fixed amount, to simulate settlement or a known movement on an existing structure.

[†]   After their inventor René Descartes, the 17th century natural philosopher, who is also famous for stating *"I think, therefore I am."*

As a general rule you need at least one restraint for every degree of freedom in the model. To be more exact: the analysis requires that there are no zero stiffnesses in any direction for the mathematics to work: dividing by zero always causes problems. A 2D analysis model allows translation in the two plane axis directions plus rotation around the axis pointing out of the plane — three in total. This means that, for example, a cantilever beam model needs one node to be restrained in the X, YY[†], and Z directions. Alternatively, a simply supported beam needs one end to have X and Z directions, while the other end has only a Z restraint. The two Z restraints prevent rotation and thus provide the YY restraint. Similarly, if you have a 3D model you will need to provide restraint against overall model movement in all six directions (X, Y, Z, XX, YY and ZZ).

4.4.2.6 Constraints
Restraints are a special type of constraint. While a restraint prevents movement, a constraint might just control movement. A typical constraint is a floor diaphragm, which causes all the floor nodes to move together[††] in the horizontal direction, while allowing them to move independently in the vertical direction.

4.4.2.7 Releases
In an FEA model, all elements that share a node are connected and will transfer all forces and moments. Modelling a pinned connection[†††] requires that you add a moment 'release' on the end of the element. Similarly, you can create a movement joint by setting an axial release. Note that these releases will not typically have a movement limit on them like a real structure, which will not be a problem if the moments are small. If you want to model the effect of the connections at their limits then you will need to create a more sophisticated joint, such as linking the end nodes with a 'gap' element or a nonlinear spring that has no stiffness when the movement is small.

4.5 Elements in more detail

Elements are the building blocks of our FEA models. There are a wide variety and it is essential to choose the correct ones if you are going to get accurate results from your model.

In general, we will use 1D elements where one dimension of the structural member dominates: where their length is significantly greater than their depth or width. Note that we mode beams and columns with 1D elements when we are interested in overall behaviour, such as bending moments, shear forces and overall deflections. 2D elements are used for members where their width and breadth are greater than their thickness and where the stress flows are two-dimensional. 3D elements come in to their own with members that have no dominant dimension, or the stresses are in several directions. There is no fixed rule for what element to use where, as all structures are three-dimensional, but as a rule use 1D elements for frames, 2D elements for slabs and walls, and only use 3D elements where you must, such as thrust blocks and steel castings (Figure 4.3). The principle is to keep the model as simple as possible without over-simplifying. For example, 1D elements are good for modelling frames to generate member moments and forces, but you need 2D or 3D elements to analyse what happens within a connection.

The definition of the element properties depends a lot on what kind of element they are. In structural analysis all elements need to provide a stiffness and a mass, which usually come from the combination of their material and dimensions.

Materials properties will comprise the Young's modulus (E) and density, and can also include Poisson's ratio, coefficient of thermal expansion, damping ratio, limiting stresses and strains, etc. For most analyses, the materials might be isotropic but can be orthotropic or anisotropic as required (as allowed by the program). Steel is usually isotropic and timber orthotropic. Reinforced concrete material properties are dependent on a variety of aspects, including reinforcement quantity, location and direction, concrete constituent materials and time since placement, and the stressed state of the member. This means that, for simplicity, they are often just modelled with the properties of the concrete section itself, ignoring any contribution from the reinforcement[††††]. Fabrics are anisotropic, in that they are stiff only in tension in the warp and weft (or fill) directions, have less stiffness on the diagonal (or bias) direction through their shear stiffness, and no stiffness at all when in compression.

[†]   Using a double letter on the axis indicates rotation about that axis, so that a YY restraint stops rotations about the Y axis.
[††]   To be more exact: the distance between each node in the constraint cannot change.
[†††] Also known as 'simple' or 'shear only' connections.
[††††] Later on, when you have analysed the frame and designed the reinforcement in the section, you might update the concrete element stiffnesses to account for the cracked section stiffness.

Figure 4.3: 1D with 2D and 3D elements showing principal stresses

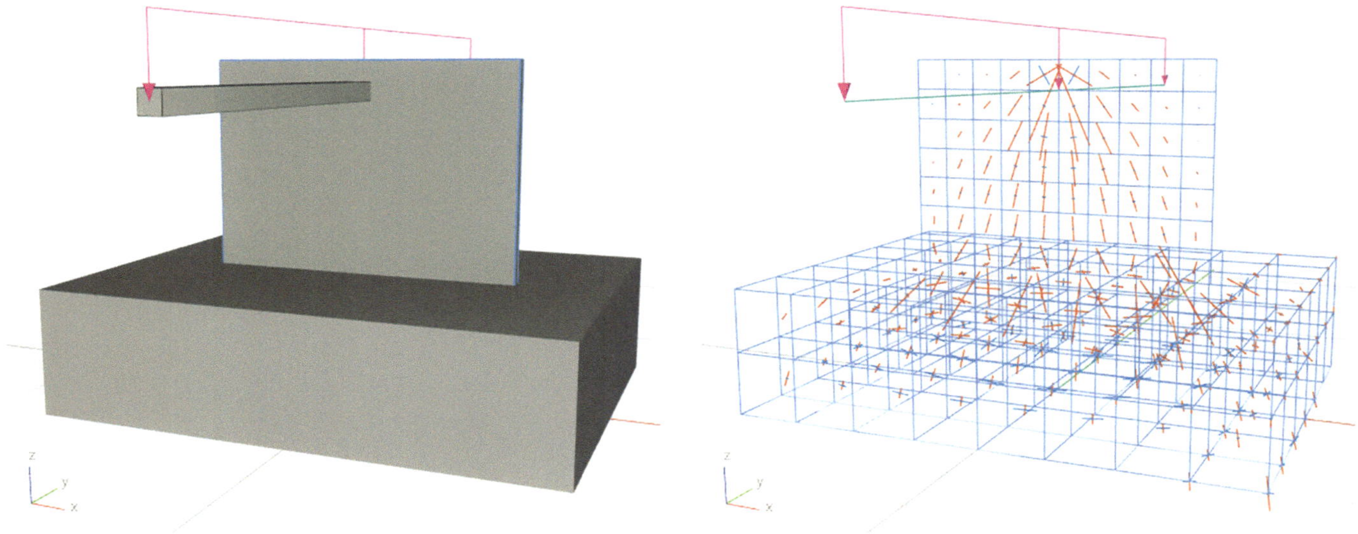

4.5.1 1D elements

1D elements are the most commonly used elements in structural FEA. They are ideal for members that are comparatively long and thin, and thus great for frame and truss models. Their length is defined by their end nodes; the rest of their properties come from assigned attributes. The section shape, e.g. an I beam or rectangle, provides the means to calculate the stiffness values — such as the major and minor axis second moment of area for bending, or cross-sectional area for axial loads. It is common to be able to select standard sections, such as rolled steel profiles, from a catalogue supplied with the program. Their X axis direction derives from the end nodes, the Z axis is as vertical as possible, and the Y axis is then perpendicular to the other two. The element will also have a rotation angle (by default zero degrees) that can change the Y/Z axes directions, for example to allow for column or purlin orientations.

Most programs will show you the 3D representation of the element but do not be fooled: just because it looks like a solid beam, it does not mean that you will get all the beam's behaviours, especially at the connections (Figure 4.4). They are still just lines connecting two points in space but with stiffnesses derived from the assigned cross-sections and materials.

Figure 4.4: 1D frame — members and elements

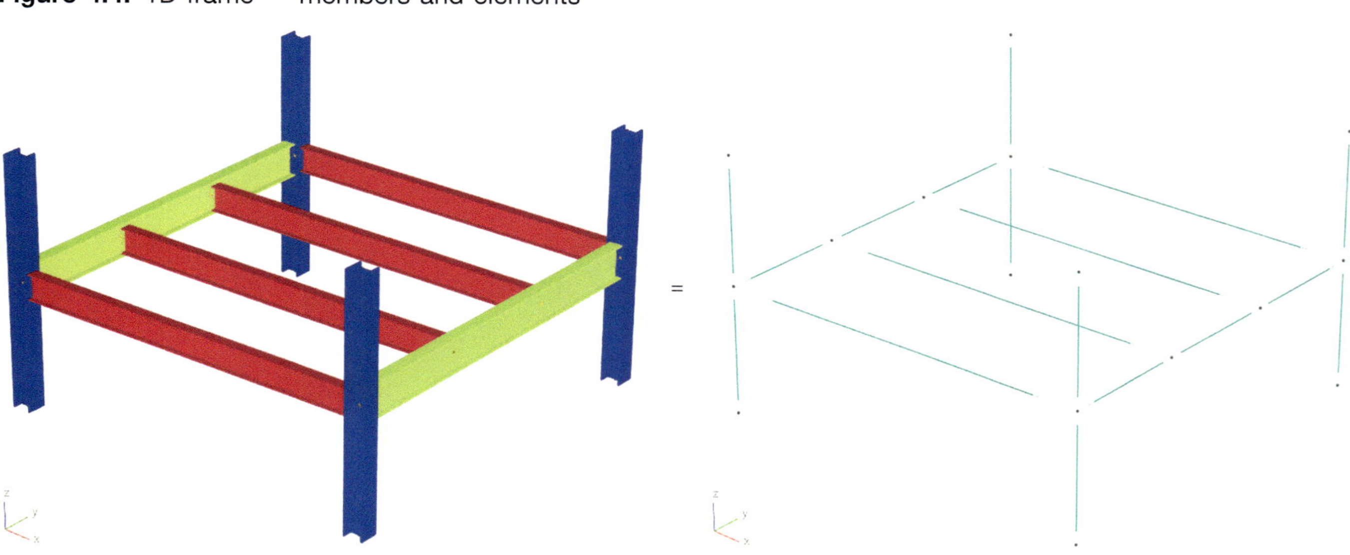

Most programs will supply a variety of 1D element types, but the two main ones are 'bars' and 'beams'. There are a variety of names that are applied to these element types in different programs, but for clarity we will follow NAFEMS terminology, where beams can carry moment and bars cannot[53].

4.5.1.1 Beams
Beam elements[†] are the general purpose 1D element and will be the most common in the majority of frame models. As it has bending degrees of freedom, the beams have full moment and force continuity with their connecting elements, which means that you can safely split steel and concrete members up into several pieces. Where you have pinned connections in the real structure you need to add moment releases on the ends of the beam elements to prevent the moment transfer.

4.5.1.2 Bars
Bars[††] carry only axial load, or in FEA terms: they have one degree of freedom. This means they ignore bending, shear and torsion. Their main use is in 2D models of trusses, and in 3D models for bracing members, space frames and truss diagonals. As they have no bending degrees of freedom you don't need to worry about end releases.

You can also use them in 3D models, but their lack of moment continuity means they are likely to be unstable unless used carefully: every node will need restraint either by other elements or constraints. For example, it is difficult to use bar elements for chords of a truss in a 3D model as they have no out of plane stiffness. In these situations, you should consider using beam elements instead. In some structures, such as suspension bridges, the cables are stabilised by their tension; here you may need to either reduce the model to two-dimensions, or use a p-delta or nonlinear analysis to account for the tension stiffening.

You may have access to special forms of bars, called 'ties' (tension-only) and 'struts' (compression-only), which are nonlinear elements because they will behave differently if the load reverses. This means that if, for example, you model cross-bracing with ties, the load path will change with the lateral load direction (Figure 4.5). Note that some linear analyses may treat ties and struts as bars, so it is worth checking the program's help file to be sure.

Figure 4.5: Load reversal changing load paths in tie elements

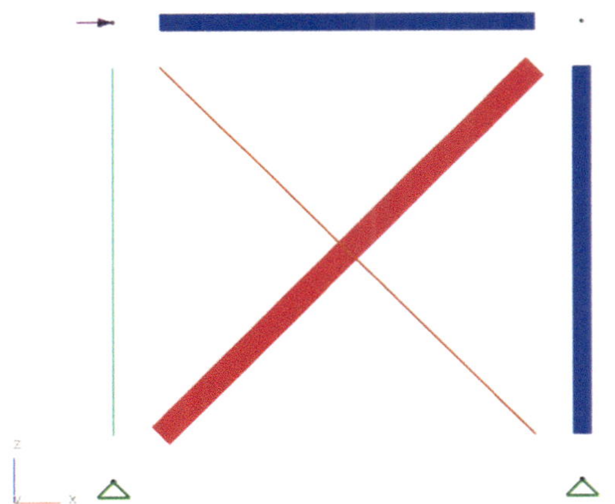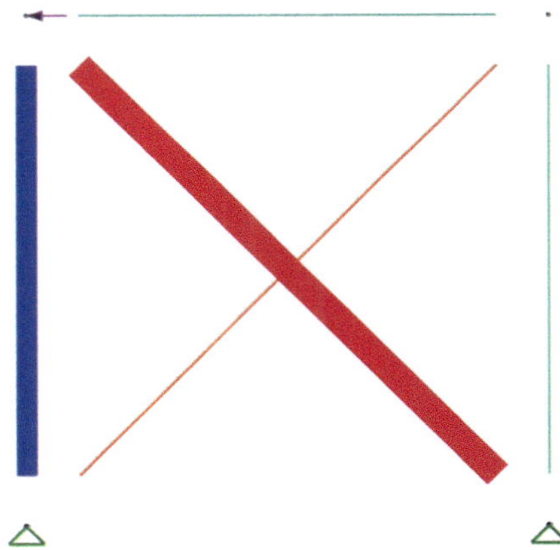

4.5.1.3 Links
'Links' are infinitely stiff elements, or at least appear to be so. If their stiffness was really set to ∞ then the analysis maths would fail through having too wide a range of stiffnesses (Chapter 6). Instead links, like 'rigid constraints', force the displacements of the connected nodes to be the same[†††]. They are particularly useful when you want to model something that is extremely stiff but you're not too concerned with the stresses and strains within in it. For example, we might use link elements to model the offset between a steel beam connection and the column centreline, or where beams are sat on top of perpendicular supporting beams (Figure 4.6).

[†]   Called 'bar elements' in some programs, just to confuse things.
[††]   Sometimes known as 'truss' elements. It goes to show that you need to be familiar with the terminology used by your particular program.
[†††] Strictly speaking, links and rigid constraints ensure that there is no change in distance or rotation angle between the nodes.

Figure 4.6: Connection offsets with links

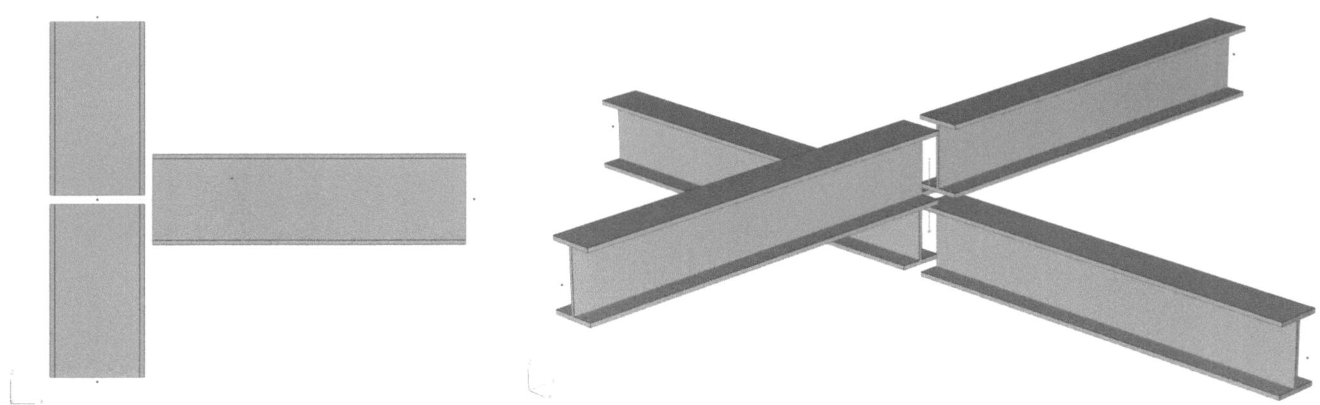

4.5.1.4 Springs
While the stiffness of beams and bars is calculated from their section and material properties, with springs you explicitly set the stiffness properties. This is usually in terms of force/translation or force/rotation.

If you know the section and material, then using a bar or beam might be better, but if you just know the stiffness then use a spring. For example, you might use them for modelling water buoyancy in floatation models, or to model the stiffness of connections between members.

4.5.1.5 1D benefits and limitations
1D elements are for situations where one dimension is much greater than the other two. This means that, in general, the length of a 1D element should be at least twice that of its section's largest dimension. If it is not, then you must question whether the assumptions encapsulated in 1D elements are still appropriate.

While 1D elements are ideal for modelling the overall behaviour of frame structures, providing the moments and forces needed to design the sections, reinforcement and connections, their simplicity means that they cannot capture detailed behaviour, such as that within a connection. If you need to understand what is happening in such areas, you will need a more detailed model made from meshes of 2D or 3D elements.

4.5.2 Meshed elements
While 1D elements can work well when representing an entire beam or column, we need to take a different approach when modelling structural elements like slabs, walls and blocks. The results would be inaccurate if we tried to model a flat slab bay with a single 2D element, or an anchor block with a single 3D element. Also, for nodes to work together they need to be connected by suitable elements. 2D and 3D elements represent just a small area within the real object, so that we can capture the change of stresses and strains within the structure. We need a greater concentration of nodes when the stresses change rapidly and thus need smaller elements so that we can calculate the forces, moments and displacements at these points. Conversely, we need fewer dispersed nodes (larger elements) where the stresses barely change.

Speaking of stresses, the different element formulations are based on how they calculate the stresses within themselves. The original formulations (devised back in the mid-1950s) assumed a constant stress throughout the element; these are now rarely used in modern programs. The most basic elements you are likely to use are 'linear', which have nodes only at the corners and assume a linear distribution of stresses across the element. Elements that have a node on the middle of each edge (and sometimes in the element centre) are known as 'quadratic' elements, allowing them to calculate the stress distribution in more detail (Figure 4.7). Although rare in most commercial structural FEA programs, you can get elements with two nodes along each edge, giving an element cubic stress distribution.

Note: do not fall into the trap of considering only 2D and 3D elements as 'finite elements'. 1D, 2D and 3D elements are all finite elements. On the other hand you can refer to 2D and 3D elements as 'continuum elements'.

Figure 4.7: Stress distribution across element

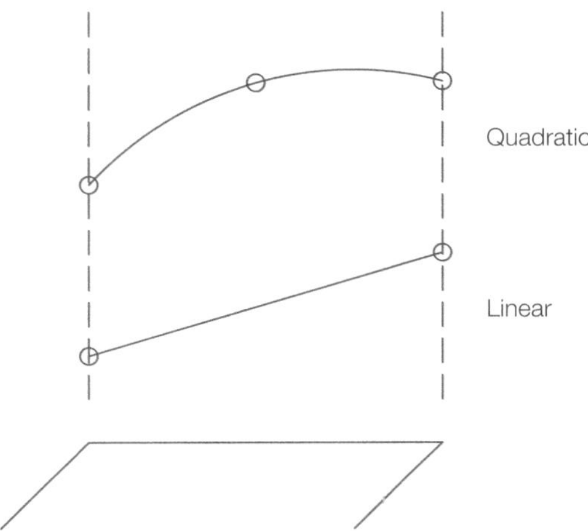

4.5.2.1 2D elements

2D elements (Figure 4.8) are ideal for planar or shell structures where the width is much larger than the thickness (e.g. slabs and walls). With 2D elements you define their dimensions using the nodal coordinates, the thickness and the material.

Figure 4.8: 2D element axes

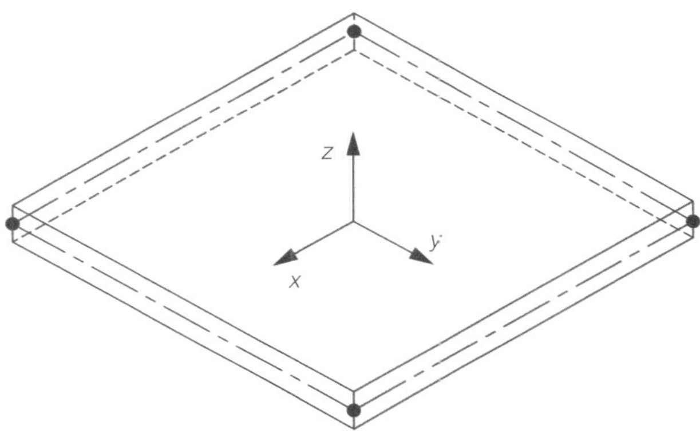

2D element dimensions come from the corner nodes' coordinates plus a thickness property. Their axis defines their in-plane directions e.g. X and Y for orthotropic materials or reinforcement placement; the Z direction will be orthogonal to the element plane.

You have several options when modelling in 2D. The first is the shape of the elements: three or four sided, which are known as 'tri' and 'quad' elements. Next comes the number of nodes, which resolves down to just a node at each corner or an additional one along each edge (Figure 4.9). FEA wil calculate the displacements at these nodes, then use those to calculate the strains (and hence stresses, etc.) within the element. There may also be additional stress calculations at the mid-point or other locations in the elements.

With just a node at each corner the Tri 3 and Quad 4 elements can only calculate a linear distribution of stress across the element; hence they are sometimes called linear elements. The additional nodes on Tri 6 and Quad 8 quadratic elements allow the program to calculate the stresses within the element in more detail. This does not mean that quadratic elements are always the best choice though. Linear shell elements can have the full six degrees of freedom while quadratic shells have only five: they lack the ZZ drilling stiffness. This means that you need to be careful if you are modelling situations like 1D columns connecting to a 2D shell mesh for a flat slab analysis. You can get around the tendency for these beam elements to spin, by connecting them to adjacent mesh nodes with rigid links or constraints. Alternatively you can add rotational restraints directly to some of the column nodes.

Figure 4.9: 2D element options

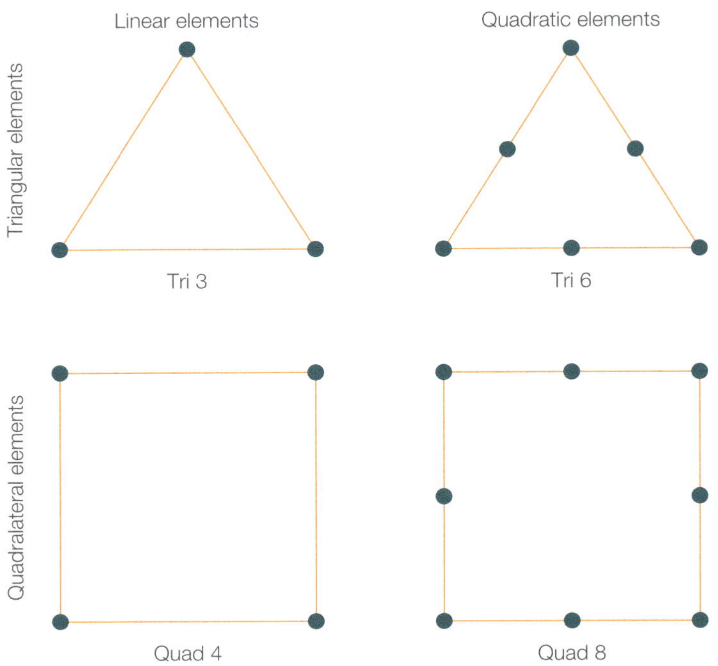

4.5.2.2 2D element formulations

There are several formulations to choose from, with different degrees of freedom, to help you with different problems. Apart from the linear/quadratic variation, 2D elements give us the choice between how many degrees of freedom we want to work with, which you can make by selecting the appropriate formulation for your problem type. Note that sometimes these models also need to be two-dimensional[†].

4.5.2.2.1 Plane stress elements

If we are only interested in in-plane forces then we can use 'plane stress' elements (they only calculate stress in the plane of the element, though they can strain out of plane). Plane stress elements have no bending degrees of freedom which also means they have no bending stiffness and will ignore, or not calculate, the rotations of any connected nodes. This makes them good for modelling structural elements such as deep beams and walls in two-dimensional models, as well as analysing stress flows through floor diaphragms. You do need to ensure that all their nodes are restrained out of plane either by other elements or global restraints.

4.5.2.2.2 Fabric elements

Fabric elements are, as the name suggests, for modelling fabric structures. These are essentially plane stress elements that can only carry tension.

4.5.2.2.3 Plane strain elements

The plane strain elements are used in plane strain analyses and are not available in any other model type. Plane strain elements have strain only in their plane, though they can experience stress out-of-plane. This makes them particularly good for analysing cross-sections through long structures such as tunnels and dams, so they are also popular with geotechnical engineers. You can only use them in two-dimensional models.

Plane stress and plane strain elements have just two degrees of freedom: X and Y translation.

4.5.2.2.4 Plate elements

Plate elements only have bending degrees of freedom and thus have no in-plane stiffness. They have no practical use on their own, but if you combine plate and plane stress formulations you can create shell elements.

[†] As with 1D elements, I will ignore the names given in some programs and follow NAFEMS industry standard terminology.

4.5.2.2.5 Shell elements
Like the beam formulation for 1D elements, the shell is the workhorse of the 2D element family. They have five or six degrees of freedom to allow you to calculate both bending moments and in-plane forces. They combine the in-plane stiffness of the plane stress element and the bending stiffness of the plate. They are ideal for modelling flat slabs (or 'flat plates' in some parts of the world), walls and cores. I have also used them for modelling plate girders to examine web buckling and the effects of stiffeners.

Shell elements are sometimes divided into 'thick' and 'thin'. Thick shells take shear deformation into account while thin shells do not.

4.5.2.3 3D element types
In some respects, the 3D element is the simplest of the finite elements: unlike 1D and 2D elements it doesn't need extra properties to define its geometric stiffness. The dimensional properties of 3D elements (and hence their stiffnesses) come directly from the corner nodal coordinates[†] and materials. If you are using non-isotropic materials, you will need to make certain that the element axes are set to ensure that the material properties are correctly aligned.

How many nodes we need to use will depend on the element type (Figure 4.10).

Figure 4.10: Tetra 4, Pyramid 5, Wedge 6 and Brick 8 elements

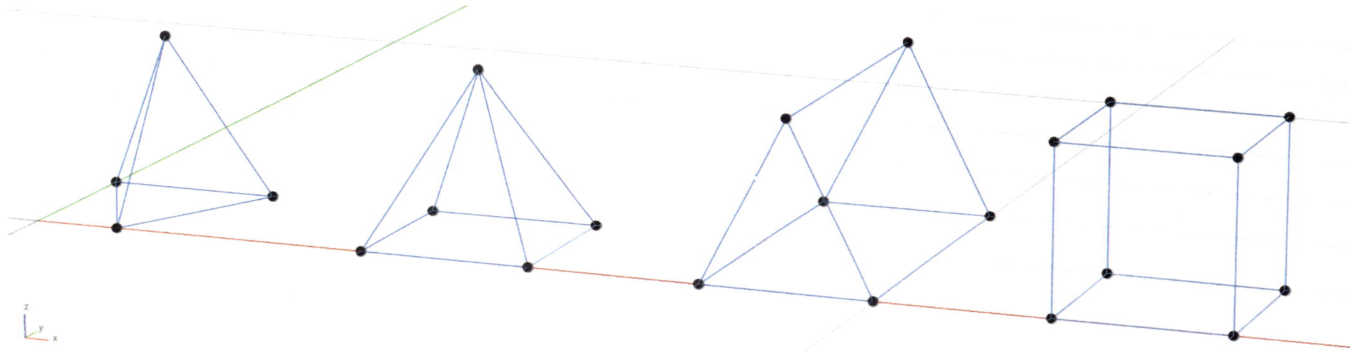

- Tetrahedral elements have four triangular sides, four corners and four edges, so need either four (one at each corner) or eight (corners plus mid-edge) nodes
- Pyramids have four triangular sides and one square side, so need either five or thirteen nodes
- Wedge elements have two triangular faces and three rectangles, so need six or fifteen nodes
- Brick or hexahedral elements have six rectangular sides so have eight (one at each corner), sixteen (corners plus mid-edge nodes on top and bottom layers), or twenty (one on each corner and edge) nodes

You may also come across a form of brick called an 'infinite' element. As the name suggests, the infinite side effectively extends forever and thus acts as a restraint to the model without you needing to restrain the nodes. This makes them ideal for modelling the edge elements in a soil model.

4.5.2.4 3D benefits and limitations
If all models are simplifications and abstractions of the real structure, 3D elements are the least abstracted. They represent an actual region in the solid part of the structure where we want to calculate the deformations and stresses. Apart from nodal coordinates, their main properties are their material and axis (for reporting results and defining orthotropic or anisotropic material directions).

There are limitations on the usefulness of 3D element models. The first is that you need to know the exact shape and dimension of the structure. This is tricky early in the design when you are trying to determine the section sizes. The second is that you only get deflections and stresses from the model without a lot of post processing of the results. For mechanical engineers who want to check the stresses within a component this is fine, but if you want the bending moment on a beam they are not so useful.

[†] You can sometimes get 3D elements with mid-side nodes.

All structures are three-dimensional but creating and maintaining a detailed model made with 3D elements for most structures would be more trouble that it was worth. Each beam would be made up of hundreds of elements and require a complex remeshing when the section sizes or any other significant aspect of the structure changes. Considering that one of the main functions of the analysis is to determine or prove the section sizes, this would be an exceedingly difficult way to work. Luckily, we can usually simplify our structures into a series of sticks (1D elements) and surfaces (2D elements), to which we apply properties such as section sizes. This makes the creation, changing and analysis of the model a practical option.

Bending moments are, of course, an abstract concept, where you have tension and compression stresses on opposite faces of a section. The total moment is the integration of all the stresses. Similarly, 'shear stress' is tension and compression stresses at the same location but at right angles to each other; 'shear force' is shear stress that is equal on both sides of a member; while 'torsion' is shear stress that is equal but opposite on opposing sides. If you want to extract abstract results, then you may need a more abstract element.

3D elements are good for modelling homogeneous materials like soil or steel but less so for composite materials like reinforced concrete, unless you want to ignore or smear the influence of the reinforcement on the structural element. For complete accuracy you would need to include the rebar themselves, including the bond strength between the bars and concrete if you want to analyse the section to failure. A simpler approach might be to define a concrete material with 'smeared' reinforcement, so it takes on the average stiffness of the reinforcement layer.

4.6 Model types

While reality is three-dimensional, it can make your life easier to analyse just a single frame or even a single beam. One of the arts of engineering is to take a complex three-dimensional structure and reduce it down to its essential aspects, whether in three dimensions, two or even one.

> Try not to get confused between 1D, 2D and 3D elements and models: a three-dimensional model can contain 1D, 2D and 3D elements, a two-dimensional model can have 1D and 2D elements, and a one-dimensional model only 1D elements. To improve clarity, I will keep '1D', '2D' and '3D' descriptions for elements and 'one-dimensional', etc. for the models.

4.6.1 One-dimensional models

One-dimensional models are mostly the province of the dedicated programs looking at columns or multi-span beams, but do have their place in FEA. Dedicated beam design programs, such as for steel or concrete beams, tend to be dedicated to standard problems. FEA is fully flexible and allows you to address the cases that the dedicated programs cannot.

Note that FEA models tend to be three-dimensional, so you need additional restraints on the model to ensure they will analyse. A common mistake is to not recognise that a pin-supported beam has no resistance to spinning; while there might not be any torsional loads on the beam, the maths of the analysis will not accept any zero stiffnesses. You can resolve this either by adding torsional restraints to some of the nodes or include global restraints on the model.

4.6.2 Two-dimensional models

Two-dimensional models are the point at which FEA begins to be useful, though there are specialist two-dimensional programs, such as for designing portal frames or concrete slabs. Note that we are not referring to 2D element types here (though some two-dimensional models include 2D elements) but that the model is defined in just two dimensions.

Like the one-dimensional models, you need to include restraints on the model to prevent out-of-plane movement. There are important considerations to bear in mind though. The first is that over-constraining the model will hide problems, such as minor axis buckling, or over-stiffen to incorrectly predict the natural frequencies. The next is to forget that while the frame model is two-dimensional, the real frame needs restraint out-of-plane and it will not stand for long without it.

While I did use a high-end three-dimensional FEA program at university (on the mainframe, which was far slower than any modern phone) I only had access to two-dimensional analysis when I started with my first consultant. While most of the problems I was involved with did squeeze comfortably into two-dimensions, there were some that required a bit more imagination. I was designing a lot of portal frame buildings for retail parks, and one option involved leaving out half the internal columns on alternate frames (Figure 4.11). This meant that I had a multi-span portal frame (itself statically indeterminate) being supported occasionally on beams spanning at right angles (a hit-and-miss frame; even more statically indeterminate). I couldn't model the support beams in the right place, and there were no spring support options available to me. I realised instead, that I could model the perpendicular beams in the same plane as the portal frame, with a stiff bar element to share the vertical load (Figure 4.12). Sometimes you need to be creative in your analysis to get the answers you're looking for.

Two-dimensional models are useful for problems where you can extract the key attributes down to just a single plane. Portal frames, trusses and floors, as mentioned previously, are ideally suited to this treatment.

Figure 4.11: Hit-and-miss portal in three dimensions

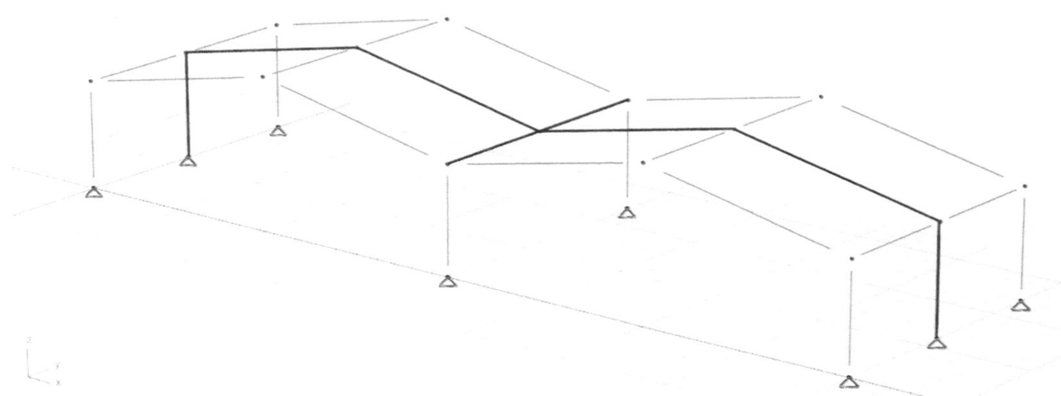

Figure 4.12: Hit-and-miss portal in two dimensions

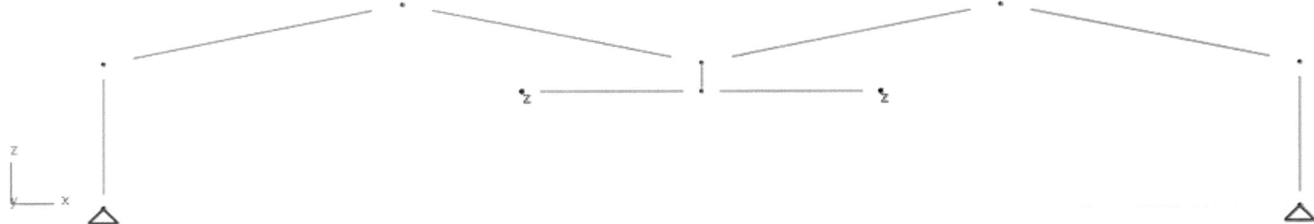

4.6.2.1 Plane stress
In this situation, there is only stress in the plane of the model, though strain can occur out-of-plane. Typical applications include shear walls and deep beams (Figure 4.13). Plane stress models use plane stress elements.

4.6.2.2 Plane strain
These are similar to plane stress models — except here, strain is only in the plane of the model, while stresses can be generated out-of-plane. These models only use plane strain elements. They are often used for long structures such as earth dams, tunnel cross-sections and retaining walls, so are frequently used by geotechnical engineers (Fig. 4.13).

4.6.3 Three-dimensional models
Finally, we come to the most common model form today: the three-dimensional model. Here you are free to model anything you want. A word of warning though. Just because you can build the model in three-dimensions, it does not mean that you should. The principle of simplicity still applies. Consider whether you can model the required behaviours in two-dimensional models — and only build a three-cimensional model if you can't.

Figure 4.13: Plane stress and plane strain models

4.7 Design the analysis — analyse the design

Like all engineering practice, analysis models require thought, planning and design to ensure the end result matches the requirements (Figure 4.14).

Figure 4.14: Modelling process

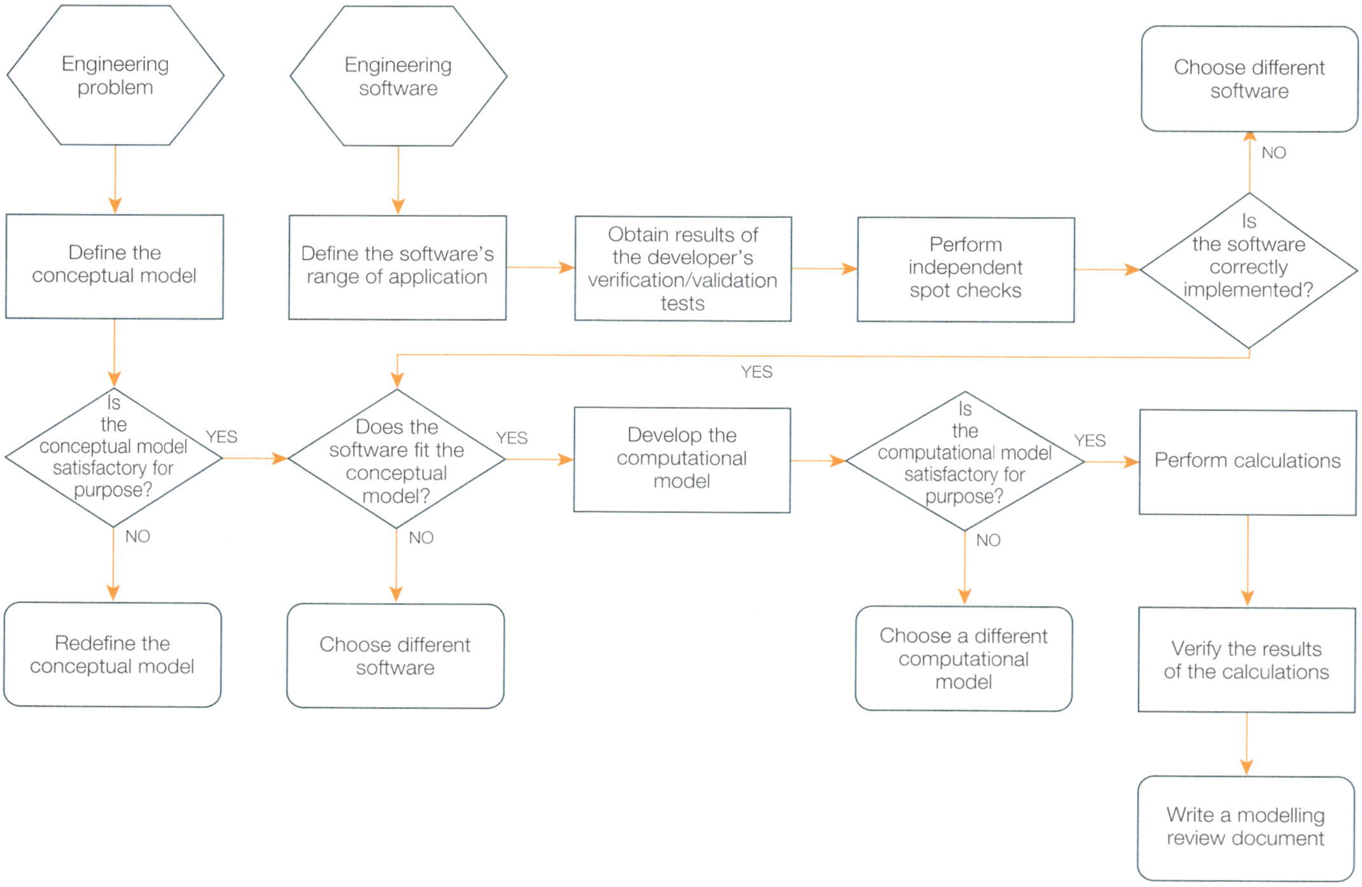

4.7.1 Plan

The first step is to review the design itself. What behaviour needs to be predicted? What kind of actions must it resist (gravity, wind, water, seismic...)? What criteria must it meet (deflection, vibration, constructability...)? What type of analyses and models will you need to build to analyse these criteria? Who in the team is best placed to build the models, to check them, to supervise the work? Do you need to seek advice from specialists?

You need to choose some software to address these requirements. Is the software capable of solving these problems — does it have the features you need? Is it trustworthy and reliable? Do staff know how to use the software correctly?

4.7.2 Do

When you build the models, are they suitable for solving the problems? Is the model efficient — could they have been built better?

4.7.3 Check

Having run the analyses, do the results look right? Do they broadly align with predictions? What do the results tell you about both the model and the original design?

4.7.4 Act

Does the design need updating? If so, then so do the models. Do the models need updating or do they need to be entirely recreated?

Now that we have seen what goes into a model, we need to look at what makes a **good** model.

5 Modelling structures

> *"All models are wrong, but some are more useful than others."*
> McCullagh and Nelder[21]

5.1 What is a good FEA model?

The answer to this question seems straightforward: a good finite element analysis model is **accurate**, **realistic**, **simple** and **useful**. There are many problems with achieving this though. **Accuracy** is difficult to determine without knowing the answers in advance, yet we are using FEA to get the answers. **Realism** requires a lot of engineering knowledge of unknown (even unknowable) aspects — yet the modelling work is often given to relatively inexperienced graduates. And isn't **usefulness** just a value judgement?

So, what makes a model realistic, accurate, simple, or even useful?

5.1.1 Accurate (as necessary)

> *"It is better to be vaguely right than exactly wrong."*
> Carveth Read[54]

Which is the more accurate: a clock that gains one minute a day or a clock that has stopped? The answer is the clock that has stopped, as it is accurate twice a day, while the other is correct only once every four years. But a stopped clock is no use for telling the time, while the fast clock gives you a good indication of the time, especially if you adjust it regularly. The fast clock is thus the more useful even if it is not as accurate as we would like.

> *"Everything is vague to a degree you do not realise till you have tried to make it precise..."*
> Bertrand Russell

When we are modelling a steel frame, we can be confident of the properties of the structure, as the components are both formed and fabricated in a factory. But how confident? If we look at a steel universal beam, depending on the size, the European standards allow it to vary in thickness by 2% and mass by 4%. This means that the self-weight and axial stiffness (ignoring bucking) can vary by $\pm4\%$ and the bending stiffness by up to $\pm10\%$. Concrete structures are even more variable due to their site casting and changes in material properties over time. What this means in practical terms is that if your deflection is >10mm there is no point in looking beyond the decimal point.

Similarly, construction tolerances will mean that your building is not built quite straight or exactly to the dimensions that you specify, but should be close enough. This does not mean that you can be inaccurate in your design knowing that they will be inaccurate in the construction, but there comes a point when accuracy to another decimal place will make no difference to the result.

But there are ways to get some measure of the accuracy of the model. One is known as 'verification', which is checking that the model is correct. But how do we do this?

First, we need to ensure that the dimensions, sections, materials and loads are all correct. If you get these wrong, you have no hope of generating the correct answers.

Next, check the 'total loads' and 'reactions' output. These should sum to close to zero[†], or at least be insignificant compared with the original numbers. If they don't, there is a problem in your model serious enough to lose loads but not enough to stop the analysis. You should also carry out an estimation of the total loads by hand and check that the answer is similar to the FEA output (Chapter 7).

[†] The loads and reactions on a model in equilibrium should exactly cancel out, but there are often rounding errors and other approximations in the analysis.

Look at the deflection of the structure: is it the right shape? Then look at the magnitude, making note of the display units, and check if it is moving as much as you expected. Too much or too little should be a warning.

Look at the meshing. 2D and 3D elements work best when they are closest to square or equilateral. Conversely their results suffer if they are stretched or the corners have extreme angles. Look at how much the stresses vary between the elements across the boundaries[†]. Be wary if the nodal results are always averaged: while it may be good practice to do this for the result, it can hide problems in your mesh.

Check for extremely short (and therefore high stiffness) elements, as well as exceptionally long (low stiffness) ones. An excessively wide range of stiffnesses will reduce the accuracy of the FEA results. This means you should consider using links and rigid constraints if you have parts of the model that are considerably stiffer than the rest, or reduce the stiffness to nothing (or remove all together) if it is exceptionally low.

Then, if your model will not run because of element or nodal instabilities, run a stability analysis to find the areas of lowest or zero stiffness[55]. If you do not have access to that, try a modal analysis but be warned that it may not run with zero stiffnesses in the model.

Einstein's Theory of Relativity has proved itself and predicted the existence of black holes, yet there comes a point when the theory is no longer valid. Relativity breaks down when you reach the event horizon of those same black holes. Similarly, representing building structures as 1D lines in an FEA model is good for analysing the flow of forces around a building frame but is hopeless when considering the three-dimensional nature of the connections. You need a different model for that, comprising 2D or 3D elements. While you could model an entire building using 3D elements, and it might be more accurate, it also might be so detailed that you miss important errors. It will definitely be difficult to make, extremely difficult to edit, and painful to extract integrated results like bending moments and shear forces.

> *"We demand rigidly defined areas of doubt and uncertainty."*
> Douglas Adams[56]

Accurate models correctly predict the behaviour of the structure when loaded, within appropriate bounds — as no model can predict exactly what will happen to a structure or how it will be constructed. All structures and components are built to tolerances, whether straightness, verticality or size. Two significant figures is about the practical limit of accuracy for a structural model[††]. Ultimately, ensure you validate and verify your model results before using them for the detailed design.

5.1.2 Realistic (as appropriate)
Without realism our structural models are useless, yet there is also a limit to how realistic our models can be and, indeed, need to be. Our models must use the correct stiffnesses of the structural members and supports if we are to calculate the correct moments and stresses in the structure, yet all structures are built approximately in accordance with our instructions (hopefully within tolerances). There is a limit to the certainty we can have in some things.

> *"Models are simplified reproductions of portions of reality that, if validated,*
> *are still able to capture a few of its essential properties."*
> Guido Fioretti[57]

This means we need to be aware of the unrealistic nature of our models but validate them to ensure they are sufficiently realistic. Validation is a determination of how realistic the model is. Realism in models comes not from making the model look like the actual physical structure but ensuring that it behaves like it.

Good loading is managed risk. If our structure must stand for 50 years, then what is the chance that a 1-in-50-year event will occur in that time? 1-in-50-year events have a 2% chance of happening in any year, meaning there is a 64% chance that the structure will have to withstand it at some point in its service. A 1-in-500-year event, on the other hand, has a 10% chance of happening in the structure's 50-year life. Loading can be code-prescribed, but always requires engineering judgement.

[†] You may have a 'stress error' output option, which calculates the differences in the stresses at each node.
[††] The partial safety factors are partly to address this problem.

Realistic models are those that reflect the reality of the structure and its supports. Its materials are sufficiently close to the real materials. The geometry includes the critical elements. The loads are not likely to be exceeded in the structure's lifetime.

5.1.3 Simple (as possible)

"Perfection is achieved not when there is nothing more to add,
but when there is nothing left to take away."
Antoine de Saint-Exupéry[58]

'Simple' is a tricky thing to achieve and often comes with experience. The inexperienced engineer who doesn't really understand the structure needs to include more detail in the model to ensure that the important aspects and behaviours are covered. The more experienced engineer knows what is unimportant and therefore what can be left out.

For example, when looking at the stability systems in a steel framed building, the slabs and beams might only transfer the load to the bracing or cores but not otherwise stiffen the stability elements. In these cases, consider replacing the floor diaphragms with rigid constraints and leave out hundreds of elements.

"Make things as simple as possible, but no simpler."
Albert Einstein[59]

There is, of course, the danger of oversimplifying the model, but simple models can be useful for checking complex ones. For example, you can check trusses by replacing them with equivalent beams and simplify 3D models to 2D or even 1D. You can also model many structures as single beams: a truss is a simply-supported beam, a tall building is a cantilever, and so on. You can then use standard formulas (Appendix F) to determine the bending moments and shear forces, then back-calculate what the element forces are likely to be.

You need to consider the level of detail appropriate to the model. A forensic analysis to determine why a structure failed will require a particularly detailed model running a high-end nonlinear analysis engine. But in forensics the unknown is the failure mechanism while the structure itself is known. More usually, engineers are trying to determine the structural details such as section sizes, and thus a simpler, more flexible model is often the right approach. For example, when looking at the behaviour of the whole structure, use a frame of easily modified 1D elements with 2D elements for any slabs and shear walls. A study on a single plate girder to check the web-buckling requires only that member in the model, so a mesh of 2D shell elements is the obvious choice. Zooming closer to look at the stress paths in a connection or casting requires a very restricted but highly detailed model made with 3D elements.

Simple models include only what they need to give accurate results; they contain no superfluous elements or loads. For example, industry guidance such as the *Steel Designers' Manual*[60] suggests that when you have more than eight equal point-loads uniformly spaced on a beam you can consider them as a uniform load (Figures 5.1 and 5.2). The result is not the same, but the error is <10%.

Figure 5.1: Point loads = UDL?

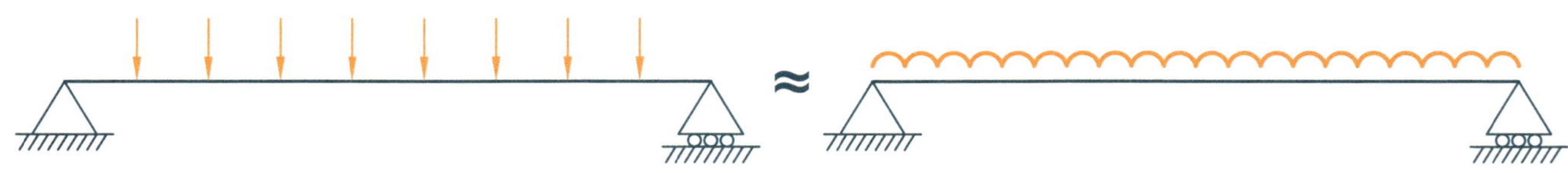

Simple models may also enable you to reduce the number of dimensions. You can check the stability of a three-dimensional steel frame by analysing just the braced bays, or analyse a symmetrical structure by taking a half or quarter and adding suitable restraints on the join lines. Overly simple models are an excellent way to verify complex ones; the results will not be exact but hopefully close enough to give you confidence. Simplicity is the hardest discipline to master with structural analysis.

Figure 5.2: Point loads compared to UDL

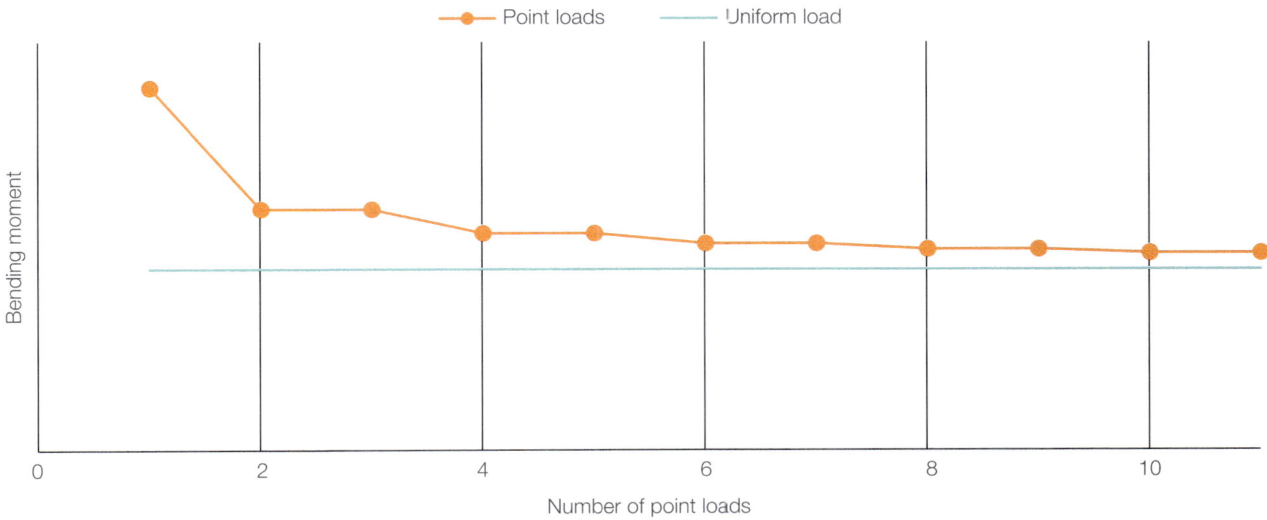

5.1.4 Useful (and relevant)

Simulations cannot totally represent reality and it is generally far too expensive to build a structure just to see if it will work. Instead, simulations are used to inform the design decisions, to test them to see if they are likely to work in a wide variety of expected, and perhaps more particularly, extreme situations. For this reason, simulation models need to adopt a 'Confidence, Risk and Sensitivity' methodology, where the input values are pushed to sensible extremes and the models are then tested for sensitivity to these values by varying them over multiple runs to see if the models give consistent results.

> *"The purpose of computing is insight not numbers."*
> Richard Hamming[2]

While the clock example (Section 5.1.1) was a little extreme, it does highlight the need for usefulness. Computer modelling may be fun[†] but we do it to answer specific questions about the behaviour of structures. This means that before you start to model a structure, there is a question you must ask yourself: What is the problem that I am trying to solve? You may be trying to determine the exact bending moments and forces, or find the wind deflections, or work out the likely extent of damage to a building from an earthquake. It might also be to find the most economic steel sections or reinforcement that carry the load. Might we improve the structure by adjusting its geometry? In general, you should have a reasonable idea of what the answer is (or how will you know if it's correct?) but you need to find the detail: the model is there to prove your design. Alternatively, you may have a structure that you just do not understand and need insight into its behaviour. These are trickier to check, so be prepared to ask for a second opinion.

What your question is will determine the type of model that you build. If you want to check a structure in its services condition then a linear analysis model may be appropriate, but an extreme case such as loss of a column means that you will need a nonlinear one. Static and dynamic analysis likewise often require different models with different properties for the same structure. Design your model to give you the answers you need.

Useful models will focus on relevant behaviours which will depend on the use of that structure. Strength and stability are always essential, but how important is the deflection? Do you need to consider vibration?

[†]   Well I think so.

5.1.5 Checking

With all these questions in mind you must validate and verify your model:

- Have you included what you need to include?
- Have you removed what you can remove?
- Is what is modelled sensible and appropriate?

5.1.6 Saint-Venant's principle

When modelling structures, it is incredibly useful to remember Saint-Venant's Principle, which in full is:

> *"If the forces acting on a small portion of the surface of an elastic body are replaced by another statically equivalent system of loads acting on the same portion of the surface, this redistribution of the loading produces substantial changes in the stresses locally, but has negligible effect on the stresses at distances which are large in comparison with the linear dimensions of the surface on which the forces are changed."* [61]

Or in summary:

> *"... the difference between the effects of two different but statically equivalent loads becomes very small at sufficiently large distances from load."* [62]

In other words, loads or reactions might be a mathematical point or spread over an area as a pressure; it will make a difference locally to the structure, but once you move far enough away, they become indistinguishable. This means we can simplify things locally if it does not influence the overall structure. This is not a hard and fast rule, so must be applied with judgment, but is very useful. One example is that of loads applied to a beam. They may fall on an area of the top flange, but we can model the load as a point load on the beam's neutral axis when considering the beam's bending moment and shear forces. On the other hand, we will need to model the beam in full with the actual loaded area if we want to explore the bending of the flanges themselves and the stress flow though the web.

Likewise, if a small area of a structure is loaded with a set of forces that sum to zero, you can ignore them when analysing the larger structure. This means that when we are modelling a whole building or bridge, we don't need to worry about the detail of connections or exactly how loads are applied to the structure if the behaviour of the connection and the equilibrium of the loads is the same. Also, modelling the supports as a single point can give unrealistic stresses local to that support but will not affect the rest of the frame.

Saint-Venant's principle does not apply to the design though, as a local failure can cause a global collapse if there is a lack of redundancy and robustness. For example, both the Hyatt Regency Hotel[63] and Florida University Bridge[64] collapses were caused by under-strength connections plus a lack of alternative load paths.

Where self-weight plays only a small part of the total load on the beam, such as in steel or timber framed buildings, a reasonable allowance is usually sufficient for the analysis. The problem becomes more complex when the self-weight of the beam forms a significant part of the load, such as with concrete beams or long-span bridges. This requires either a reasonable guess of the section size (and thus the self-weight) in advance of the analysis, or an iterative analysis and design cycle that adjusts the beam size until one is found that works.

5.2 Applying good modelling

5.2.1 Structural types

Like their models, structures themselves can be thought of in three principal forms:

- Stick
- Shell
- Mass

Stick structures will include steel, concrete, and timber frames and trusses; you can usually model these with 1D elements.

Shell structures include domes and others that are predominantly flat, such as slabs or walls. These have the stresses or strains flowing in only two dimensions and thus you can model these with 2D elements or grillages of 1D elements.

Mass structures have significant stresses flowing in all directions and thus require 3D elements to capture their behaviour. Examples include anchor blocks, wind turbine bases, and flat slab-to-column connections.

Let's look at how you should model some typical structures.

5.2.2 Meshing
As the FEA model does all its calculations at the nodes, you must ensure that your model has nodes at all the critical points:

- Where structural members join
- Where there is a significant change in the structure or to describe the geometry (e.g. if your beam is cranked or curved)
- Where any change in stress needs to be captured
- On the boundaries of the problem
- Where concentrated loading is applied[†]

And in addition:

- Ensure that the array of nodes in a continuum can be joined with elements that are reasonably square or equilateral

In a 1D element frame model this means that you need nodes at the connections, the free ends of any elements, and where it connects to the foundations (Figure 5.3).

Figure 5.3: 1D element frame and 2D element mesh

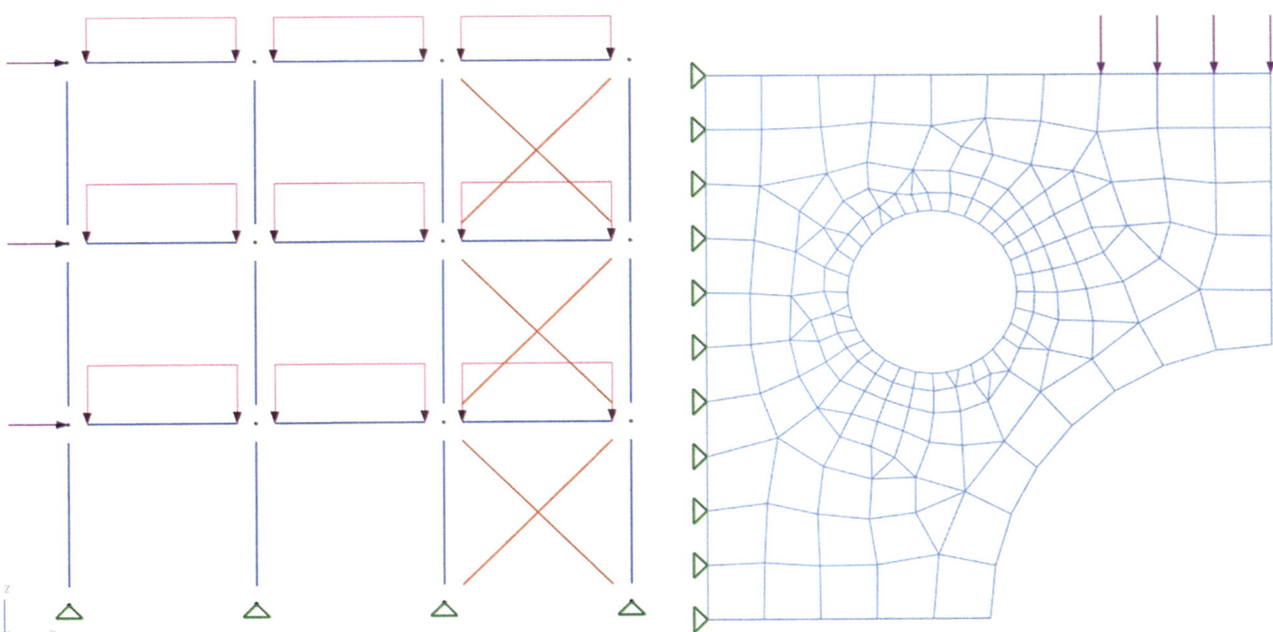

[†]   Loads on elements are usually converted by the program into equivalent nodal loads during the analysis.

Many programs have meshing tools to assist you in creating the model, or you can use third party programs to create the mesh and import the result. A word of warning though: meshing is done for many purposes, not just FEA. A program that creates meshes for rendering is unlikely to create a good mesh for analysis.

I am often asked what size the mesh elements should be. This is a tricky question as in many cases there is no definitive answer. There are, however, some rules of thumb that you can follow to achieve a good mesh.

The first consideration, and the easiest one, is that of minimum size. A good rule is that the element's width should be at least twice its thickness. This also means that features that are smaller than this dimension can usually be ignored in the analysis model as they are often insignificant. Unless you know them to be significant, of course. Small pipe penetrations or localised thickness changes might be ignored in a slab in bending, but openings where you have high tension need careful consideration, especially where you have stress concentrations from sharp-edged openings (Figure 5.4).

Figure 5.4: Stress concentration at square deck opening — causing S.S. Schenechtady to crack in half

As a general rule you should use as large an element as you can get away with. Large elements are more suitable where the stresses do not change much: e.g. in the middle of the bays of a flat slab subject to bending. Locations where the stresses are changing rapidly, such as at column locations on flat slabs, need a finer mesh to capture the behaviour correctly.

Meshing for a vibration analysis is the reverse of static analysis: not much vibration occurs next to the columns — it is the middle of the bays where the action is. A general rule here is that major spans (however that phrase relates to your project) should be broken up into 8–10 elements. Of course, slabs have many different spans so there is no one span that you can use. Meshing engines tend to ask for a target element size, so base that on the largest span. Interestingly, if the slab span to depth ratio is 20, both the 'span/10' and 'thickness × 2' rules give the same element size, so an element size of twice the thickness is a suitable place to start.

While FEA wants all the loads applied directly on the nodes, many programs allow you to load the elements instead. It then calculates the equivalent node loads for you, based on the elements' dimensions.

5.2.2.1 2D mesh vs 1D grillage

2D element shells are the standard option for flat slabs, but there is an alternative worth considering: the grillage. In the early days of FEA the focus was on the analysis and design of jet aircraft wings. Until the end of WWII, aircraft wings were cantilevered perpendicular from the fuselage, meaning that the maths was straightforward.

The advent of the jet engine pushed the speeds such that swept-back wings were required, which was a much more difficult calculation and informed a lot of the FEA funding and research in the 1950s. The aeronautical engineers were also reducing the weight of the aircraft by using the metal skin as part of the structure, further complicating the analysis.

Initially the wings were modelled using 1D beam elements in a regular grid (a grillage) to represent the wing structure (Figure 5.5), where each element section gave the stiffness of the wing in that location.

Figure 5.5: Grillage analysis of aircraft wings

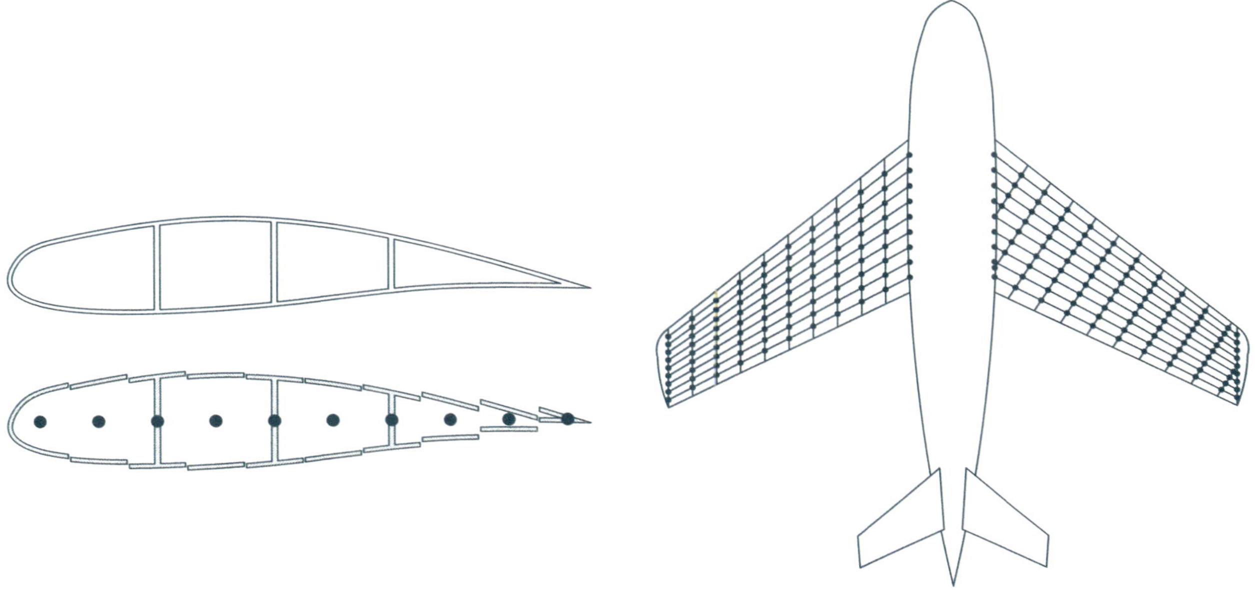

Grillages worked well if the real structure had parallel edges. Triangular areas or tapering structures deviated from the assumptions and thus reduced the accuracy of the analysis.

You can make grillages detailed if you maintain the rule about the element length being at least twice the thickness. Alternatively, you can directly model the 'beam strip' approach to flat slabs and use a chain of 1D beams for each beam strip (Figure 5.6).

Figure 5.6: Flat slab grillage model

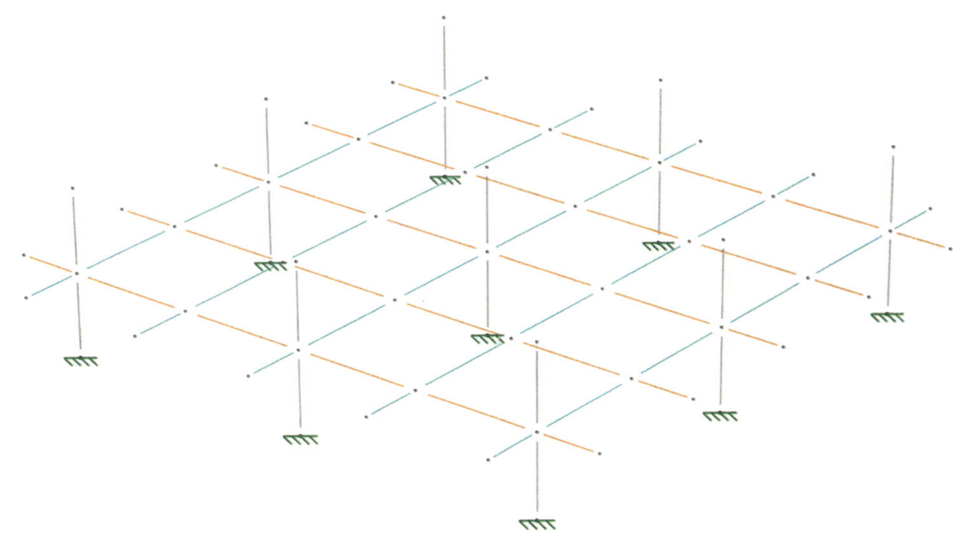

Load a grillage using an area load (if you have access to one) or uniform beam loads. Beware of self-weight calculations as these will overestimate the load if you are not careful. Because you have modelled 100% of the slab one way and 100% in the other, you now have twice the volume of material in your model compared to the real structure. Likewise, be careful with beam loads that represent the area load. To avoid over-designing the structure you must either apply the full load to only half the elements (say in one direction), or half the load to all the elements.

Grillages are advantageous because they are simple; they automatically redistribute the moments across strips because the beams directly represent the strip, it is easy to create orthotropic behaviours, and you can show bending moment diagrams across the floor. Their disadvantages include the difficulty in modelling around openings, edges, non-rectangular areas, and areas that do not work with a regular spacing of elements. The question of torsion is also tricky: flat slabs do generate torsional moments, and slabs handle torsion differently to beams. Should you then work with the reported moments, or set the element torsional stiffness to (virtually) zero and make all the torsional effects change to major/minor axis bending? The choice is yours.

2D shells are the most common way of modelling flat slabs and walls. With suitable meshing you can model almost any floor plan with any number of openings. It is worth remembering though, that you don't need to model everything. The guidance that the element's width should be at least twice the thickness still stands. This means that features smaller than this size, such as holes or protrusions, can usually be ignored. This highlights one of the problems in linking the analysis model directly to the CAD/BIM model: the drawings need to include everything so that the contractor will build it correctly, but too much detail will slow the FEA model down while having a negligible effect on the results.

Many programs will allow you to extract beam strip results from the 2D mesh by collating and integrating the stresses into overall shear forces and bending moments. This assumes that your columns are all in a neat grid layout and thus you can find beam strips; one of the advantages of flat slabs is that the column layout can be arranged to suit the building requirements, which may not be in neat rows.

5.2.3 Trusses

Trusses, with a few exceptions, work by triangulating the structure so that no bending in the elements is required for stability. Bar elements are excellent for truss models as their lack of bending degrees of freedom mean the structure will work under pure axial load.

Position the nodes at the intersection of the centroid of the truss members as this avoids local effects. In reality there will be some lateral stiffness at every support, but if the truss is carrying only vertical loads, you only need to provide one node with a lateral restraint. It is often extremely difficult to calculate the lateral stiffness, plus it will have a significant effect on the load distribution within the truss, so it is safest to assume that it is not there. If you use this assumption, then the resulting design will be strong enough without the lateral restraints. If you do have lateral loads on the truss then you will need to consider the supports in more detail, but with care: consider using lateral spring supports.

We won't know the exact sizes of these truss members before we run the analysis, but we can make informed decisions. In general, and especially at the initial stages of the project, you should use bar elements in the model as these carry only axial load. This will keep the model statically determinate, and the element forces won't be influenced by the individual element stiffnesses.

Be careful using bar elements in 3D models as their lack of moment capacity can make them unstable. They work very well modelling trusses if the model is restrained to just two dimensions but modelling in 3D means that every node will need restraining out-of-plane. When modelling trusses in 3D, use beam elements with carefully thought-out releases: the booms or chords often need bending continuity out-of-plane to keep them stable.

As a rule, you should use beam elements in truss models where you have continuous members, as that reflects the reality of the final structure. This will make the truss statically indeterminate so the relative stiffness of the continuous members to the truss will affect the force distribution. The actual set-out of the connections can also induce forces and moments on the members, but you need to know the section sizes to model these correctly. This means you are best modelling the truss initially in 2D with bar elements so that you can determine the best element sizes.

Model the continuity and connections in more detail at a later stage of the project, when all the section sizes are known, and you want to make detailed checks on the structure.

If the truss is a sensible one (and I have seen some lacking in sense in my time) the beam stiffness will have a negligible effect on the overall truss behaviour if the members are sufficiently slender. Do note that at ultimate limit state, plastic rotation of the connections can dissipate moments but it requires joint rotation and ductility. Overall buckling of members will need checking, as will local and lateral torsional buckling.

Modelling trusses entirely with beam elements and moment-resisting connections will result in only a small reduction in axial forces and deflections but will necessitate more expensive connections and bigger section sizes to deal with the moments. Question whether they are worth it.

5.2.3.1 Connections and eccentricities

While trusses might be analysed with all pin-ended beam or bar elements aligned to common nodes, connection detailing can mean that it's not always possible to keep the member centroids meeting at a single point. The resulting eccentricities of the connection work-points can induce moments in the members. In these cases you will need to model the chord members with beams, then connect the bracing work-points to the chord work-point using links[†] to generate the additional moments (Figure 5.7).

Figure 5.7: Offset bracing work-points

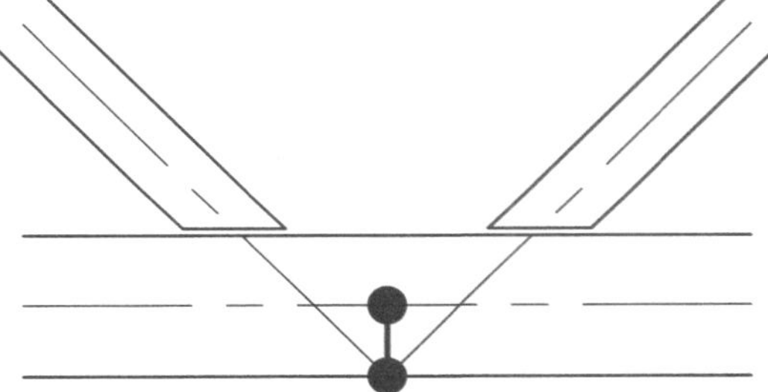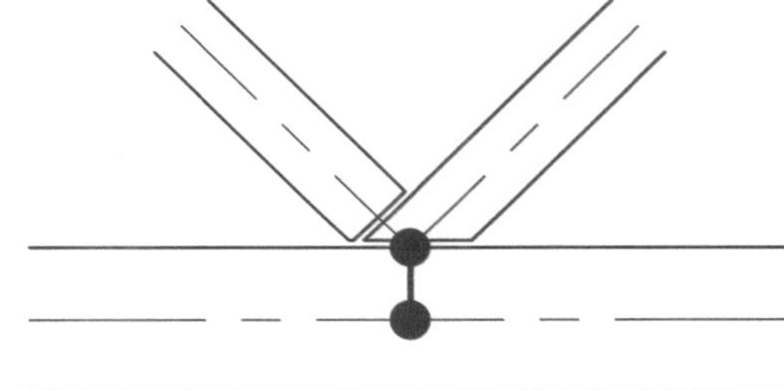

5.2.3.2 Validating a truss

To validate the truss analysis, we might assume that the truss is a simply-supported beam carrying the loads as points or distributing it uniformly along its length. Once we have used beam theory to calculate the bending moment, calculate the top and bottom chord axial loads by dividing the moment by the truss depth. Likewise, the truss diagonal members will carry the shear force so convert it to an axial force using trigonometry.

Construction tolerances, erection sequencing, welding, connection slip, uncertainties in the loading, etc. will all mean that, even with the most sophisticated analysis models, we are not 100% sure of the forces in the truss elements. This means the analysis must include a sensitivity analysis to bracket the range of results, and the design must be robust — ideally with allowance for plastic behaviour.

5.2.4 Steel frames

Steel frames are typically modelled using 1D beam elements to model the beams and columns, and bar or tie elements to model any bracing (Figure 5.8).

[†] Links have infinite stiffness but do not destabilise the analysis matrix (Chapter 5).

Figure 5.8: Steel FEA model

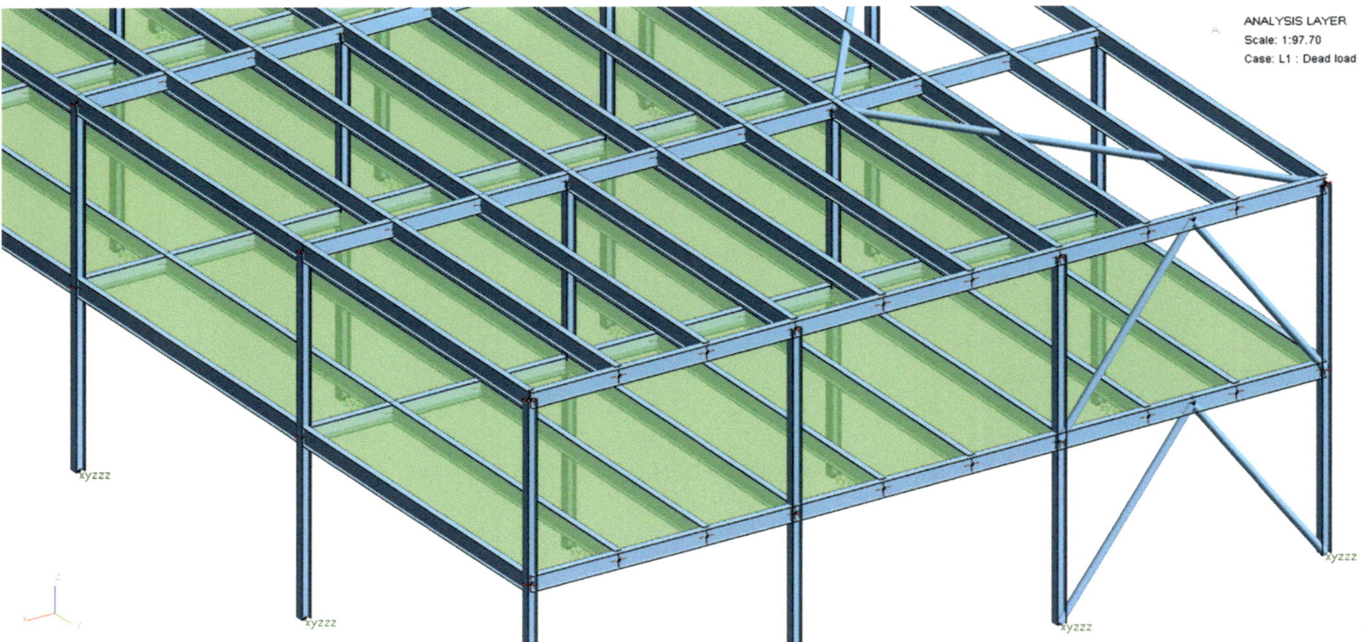

5.2.4.1 Stability

In the UK it is typical to stabilise multi-storey buildings using either bracing or concrete cores. You have a range of options for modelling cores (Section 5.25). You should model bracing explicitly so you can derive the axial loads. Note that compression-capable bracing (e.g. when using bar elements) will pick up some gravity loads from the structure. If you use X-bracing its stiffness will probably be much less than the columns and so not attract too much load, but still needs checking. If you have slender members such as 'flats', you can model these as tie elements and avoid gravity effects. Note also that the beams and columns in a braced bay will have axial forces induced by the lateral loads, so make sure you account for that in your design calculations: a braced bay is just a truss in the vertical direction.

While on the subject of gravity on slender braces: be careful of adding self-weight to such elements as they tend to have minimal bending capacity. In reality they might be tension-stabilised or allowed to buckle when not loaded. The self-weight of the bracing is likely to be tiny in the grand scheme of the building but check it as it might matter more than you think. Plan bracing will usually have to span, so consider using beam elements with end releases rather than bars, to account for any self-weight induced effects.

Except for some roofs and single-story buildings, stability forces tend to be transferred horizontally though concrete floor slabs. In general, you can avoid modelling the slab explicitly, as it is much simpler to use UDL or area actions to load the beams, and a rigid constraint to generate the floor diaphragm action.

There are a couple of things to watch out for with rigid constraints. The first is that they must only constrain in the horizontal direction: beam deflections may be prevented and column/core loads will get very strange if you accidentally include the vertical direction. The second is that rigid constraints will prevent any axial load in horizontal elements, which may be ok but note that you will get no load at all in trusses that form part of the floor or beams in braced bays; the analysis and design of the truss will need a separate model if you are to get the right member sizes.

If you want to model the actual horizontal stiffness of the slab, you can use plane stress 2D elements. If you are only interested in loads into the stability systems, try to keep the 2D elements large. You might do this by connecting them to the column locations and the corners of significant openings. You will need a finer mesh if you want to analyse the stresses within the slab though. As plane stress elements have no out-of-plane stiffness you'll need to ensure that either every node is attached to the frame (say one element per bay where possible), restrain all the slab nodes vertically, or model the floor in a separate 2D model.

Your model will become complicated when you want to share the horizontal loads between a concrete deck and composite beams. Do the beam connections allow some axial movement, such as fin plates, in which case the deck will carry the load first, or have minimal movement capacity, e.g. end plates, when the beams will take up the load more rapidly? Are there significant tension forces in the system? How are the beams and slab connected to the core? Will it make any difference? Hybrid systems are much harder to analyse as the relative stiffnesses will change the load paths.

To keep things simple, you can break the building down into separate floor and stability models. You can ignore diaphragms in a 2D floor model and instead restrain the beams at the column locations. With stability models you can include just the bracing or cores, connected by a rigid constraint, and forget the other beams and columns. As before, ensure that the constraint only acts in the horizontal direction — as you do not want it to artificially stiffen the braced bays or cores out-of-plane.

5.2.4.2 Connections
5.2.4.2.1 Releases
When modelling a steel frame with 1D elements you need not model the end connections in any detail, but you do need to include their effect by adding end releases. The requirement for releases in our models is a consequence of the finite element method: elements are fully connected to their end nodes to the limit of their degrees of freedom unless they are explicitly released. Modelling pin-ended beams does require additional thought and work, but it is an essential process as many buildings in the UK and other parts of the world are designed this way.

Remember to not release the ends of all the beam elements. Members are often modelled from many elements, and releasing them all will result in a catenary that will be rejected by the linear solver. Only release the ends of the members where there are simple connections in the real structure.

BS EN 1993[65] refers to three principal types of joint:

- Simple — there is no transfer of moments in the model
- Continuous — moments are transferred but the joint behaviour has no effect on the model
- Semi-continuous — you need to take the behaviour of the joint into account in the model

Simple connections in a steel frame need to be able to rotate only about 2–3° to act as a pin under all service loads. They will have a small moment capacity, but this can be ignored for most analyses as the service rotations are less than the limit. This means you should model beams with simple connections with major and minor axis moment releases.

Footfall vibration analysis is one exception, as the rotational strains are not sufficient to overcome the friction within the connection, making them behave as continuous. Analysing accidental loads and buckling are similar, as connections under extreme conditions can reach the limits of their travel and lock-up, giving them some effective moment capacity[†].

As most steelwork connections and open sections have little torsional capacity, you do have the option to avoid all torsional moments by adding a torsional release at one end of every beam. Be careful though, as the chances of accidentally releasing both ends of a beam increases with the number of beams in the model; personally, I would avoid this approach. Most connections have some torsional stiffness and so this should be included by not releasing the beam in the torsional direction. This means that you'll need to check to see if any torsional moments are induced in the beams and then design the members and connections for it, adjust the layout to avoid torsion, or deliberately disregard the torsion if it's nominal — as slabs and secondary members can deal with small amounts. Also, if your model shows torsion that should not be present, there may be some unrealistically reduced moments elsewhere in the model[††]. If there is significant torsion in the model you need to address it, as ignoring torsion is risky.

[†] How they cope with that moment is another matter…
[††] I was checking one user's model that had unexpected torsion in a simply connected primary beam. Careful investigation revealed that the beam was not actually straight.

Continuous connections are the simplest to model as you don't need to do anything; end offsets can be helpful though (Figures 5.7, 5.9 and 5.10).

Semi-continuous connections are the hardest to address, as they have some stiffness but not enough to treat as continuous. Exactly how you model these will depend on your software. If it doesn't directly address connection stiffness, then you'll need to have separate nodes for the primary and secondary elements that have full connection in some directions (e.g. shear) and a spring stiffness in others (e.g. major-axis rotation). Parametric tools will come in useful here, for both connecting the nodes and defining the stiffnesses. You should also do a sensitivity analysis and investigate what difference the stiffness of the connections makes to the results — you may learn what needs including and what you can leave out.

5.2.4.2.2 Modelling connections in detail
How detailed your model gets depends on how much detail you need; not a helpful statement but worth remembering. Note that if you include any level of connection detail, the element count will grow rapidly. The modelling can also get complex if there are aspects like stiffener plates to include. It is good practice to build separate connection models away from the main building model, otherwise you will get bogged down in complex modelling and slow analyses. Using parametrics to build the model can also be sensible as they can automate the joint modelling.

1D elements work well for the beams and columns, but for the connection you need to understand how the stresses are distributed between the web and flanges. At this level it becomes useful to mesh the steel members and plates using 2D shell elements. Node sharing usually suffices for welds as that gives you the overall forces for the weld design.

While 'simple' connections can be straightforward to analyse by hand, they are actually exceedingly difficult to model explicitly with finite elements due to their nonlinear behaviour. Many of the connections between parts are in bearing and so cannot take tension. Modelling bearing behaviour, such as end plates to column faces or bolts in tolerance holes, is not something that standard matrix methods can model without a lot of monitoring and iteration, and a nonlinear solver is recommended. An alternative is to use lots of compression-only 1D elements to model the contact, but the results are rarely satisfying.

It may be preferable to model bolted steel connections on paper (or the digital equivalent) unless the design does not fit with the standard assumptions, at which point you need a specialist analysis.

Welded connections are much easier to model as all parts can take both tension and compression, meaning that you can use a linear analysis solver as long as the stresses are in the elastic range. If you need to factor in local stresses and eccentricities around single-sided welds or bolt bearing, you will need to model the connection with 3D elements. If not you can use 2D shell elements to model the flanges, webs and plates. Other methods, such as explicit solvers or hand calculations, are more suitable for this sort of problem.

3D elements will allow you to explore finer detail in the connection, but the number of elements will grow exponentially. Don't try modelling detailed connections in the complete frame, as the complexity can get overwhelming; keep the models separate and transfer loads and stiffnesses in each direction as appropriate.

5.2.4.2.3 Inclusion of end offsets
While the end connections of the steel beams might prevent moments being transferred from the beams into the columns, the connections will still induce moments in the columns. This is because the connections will transfer the shear forces eccentric from the column centreline. If you have an I section column, the eccentricity in the minor direction may be small, but so is the moment capacity (and *vice versa* in the major direction). The design codes tend to make recommendations for this offset: if in doubt try half the column thickness plus 150mm (Fig. 5.9).

It is not recommended to model this small offset using beam elements, as exceptionally short elements will have an extremely high stiffness relative to the rest of the model and thus will reduce the analysis accuracy. It is better to use short link elements or end offsets if they are available in your software.

Figure 5.9: Offsets for pinned connections

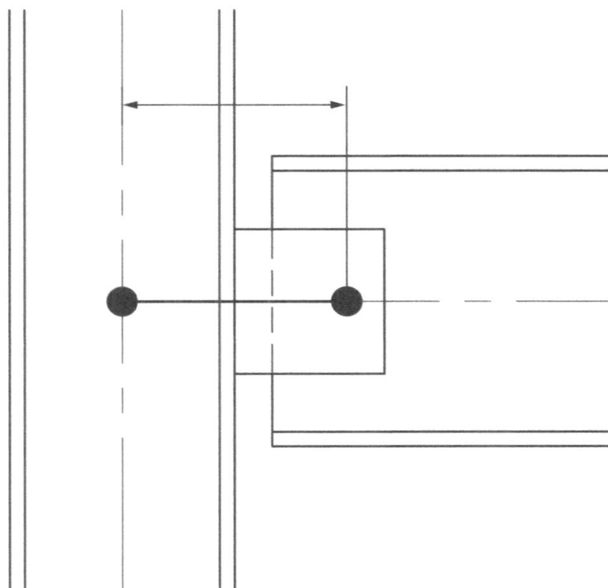

The connection eccentricities will induce moments only if the loads are asymmetric. This means that you need to use patch or pattern loads (load only one side) to calculate the design moments on the columns. Building design programs might do this automatically for you, while FEA programs will require you to specify the loads.

Connection offsets are also recommended for the analysis of moment frames. They will not create any additional moments but do allow you to easily extract the moments at the connections.

5.2.4.2.4 Bracing work-points
Just as beam connections can induce moments in the supporting members, so can bracing members. Ideally, the work-points of the bracing (the intersection of the centroids of the bracing, beams and/or column) should all meet at a single coordinate (Fig. 5.10). Practical considerations often make th s impossible, meaning that additional moments will be induced in the connection. Model these connections with links in the same way as the truss connections.

Figure 5.10: Bracing work-points

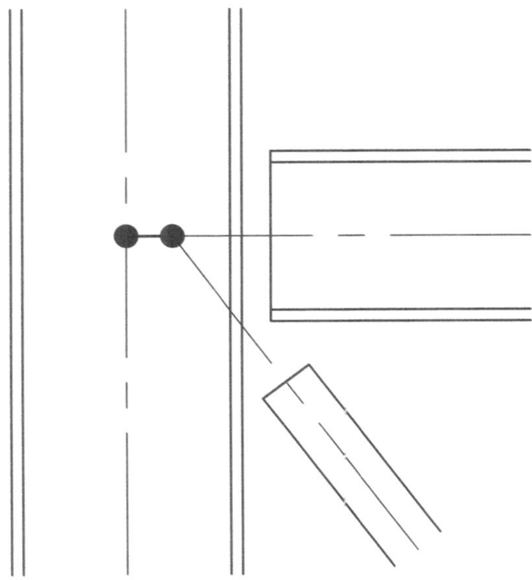

5.2.4.3 Tapered elements

While the finite element method struggles to model 1D tapering elements, several programs do offer them. It is likely that the tapered element is broken down into half a dozen prismatic elements in the analysis run, then the results are joined back up again for the output. If your program doesn't include tapers, you can reproduce them with a few short prismatic elements. As per the connections, avoid excessively short elements as their high stiffness can upset the analysis accuracy.

The easiest way to create the properties of a tapered member is to average the dimensions of its ends, but while this might be reasonable for axial loads where the stiffness is in proportion to the area, for elements in bending the stiffness is in proportion to the cube of the depth. The most accurate method is to provide sections that have an average of the stiffnesses, but this can be difficult. Thankfully, with enough elements, the difference in results between elements with average stiffness and dimensions becomes small[†].

5.2.4.4 Composite floors

Concrete is good in compression and is commonly used to make suspended floors. The top flange of a beam is usually in compression, so using the concrete floor as the top flange of the floor beams makes for a very efficient structure. Accurate modelling of such composite structures does pose a few problems in FEA but nothing that cannot be overcome with a little thought.

The first problem to consider is the physical relationship between the steel beam and concrete slab. The normal detail is for the concrete slab to sit on top of the steel beam, meaning that the slab centroid is offset from the beam centroid. If the slab is cast *in situ* on profiled metal decking, there is also a zone above the beam where the slab is only partly present and must be ignored (Figure 5.11).

Figure 5.11: Modelling composite beams and slab (physical vs. FEA)

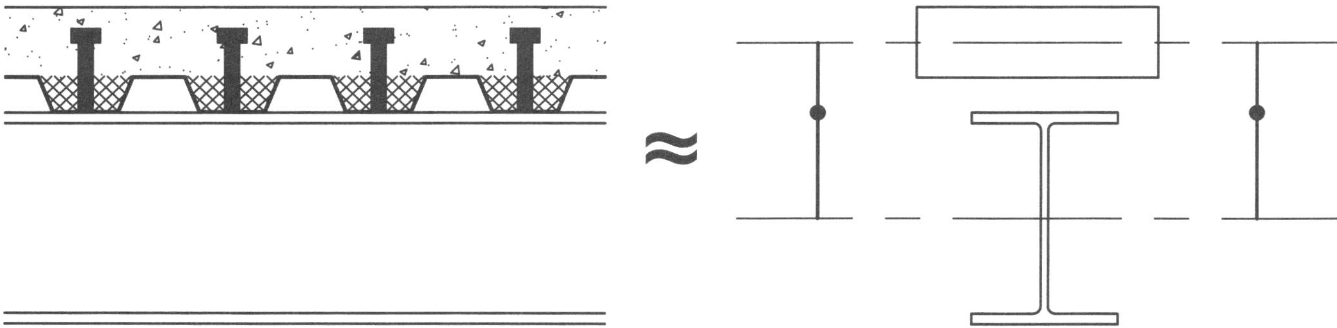

The connection between the real slab and beam is achieved with shear studs that are welded to the top flange. If you are modelling for floor vibrations, the strains will be small and slippage negligible, so link elements are an acceptable way to represent the studs. A suitable alternative is to have the slab shells and beams share nodes, and offset the sections. For general service and ultimate limit states you do need to consider the shear interaction and this slippage between the beam and slab. If you use links here, the forces and moments induced will be unrealistically high; modelling the studs with beam elements set to the studs' dimensions will not take the stiffening effect of the studs' embedment in the concrete into account. In this situation it is best to represent the studs with spring elements at a suitable spacing.

Another option is to include the slab directly in the steel section profile, making use of the modular ratio to give the effective width. Defining all these sections with all the various effective widths will be a lot of work (unless you automate the section definitions) but might save you work elsewhere in the model.

Finally, if the composite beams are all statically determinate, you might just export the forces and moments from the FEA model to a dedicated composite beam package for design and deflection calculation.

[†]   Do not use too many elements in the taper or too few. 4–5 is probably about right.

5.2.5 Concrete
5.2.5.1 Beams and slabs

How you model concrete beams and slabs will depend on how you want them to work together. You might have them work independently, so that the slabs span on to the beams and the beams span onto the columns, but without them interacting in any way. You can make the beams work compositely with the slab, so that rectangular beam sections are effectively T-shaped by using some of the slab as a top flange. The beam might locally reinforce the slab. Or you can have the beam as a thicker part of the slab.

If I need to model the slab, I would use 2D elements (though a grillage is also an option). You might leave the slabs out of the model if they have no moment continuity with adjacent slabs; instead, apply their loads direct to the beams.

While precast slabs will rarely have moment continuity, *in situ* slabs are usually continuous over internal beams. In fact, with the exception of the outer edges of the structure, it is exceedingly difficult to avoid continuity with an *in situ* slab, and usually undesirable. Continuous slabs are more efficient and so can be thinner, plus they minimise the cracking likely to occur over the supporting beams.

How should we model the structure if we assume that the beams support the slabs but are not composite with them? The easiest way is to model the beams with 1D elements and (usually) rectangular sections and mesh the slab with 2D elements (Figure 5.12). Although the physical beam is below the slab, keeping all the elements on their centroids will mean the beam is doing the majority of the work spanning between the columns and primary beams due to its higher stiffness.

Figure 5.12: Beams arranged on slab centroid

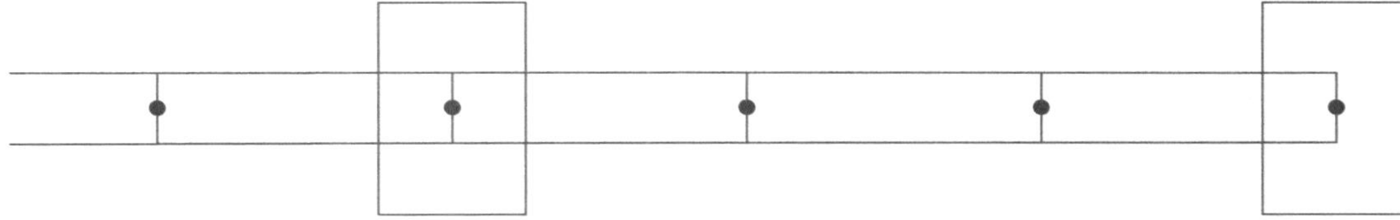

There will still be a small amount of moment sharing between the beams and slab. This will mostly show itself in the slab: the deflection of the beams will redistribute some of the slab hogging moment to the middle of the bays.

There is also a slight overestimation of the self-weight of the structure due to the slab/beam overlap. You can counter this by offsetting the slab edge to the beam face. By offsetting the beam down, forming a T-beam, you will generate composite action between the beam and slab (Figure 5.13). The result is that the beam will be more efficient and there will also be axial force in the slab adjacent to the beam. The reinforcement in the slab must reflect this.

Figure 5.13: Beam offset to slab to generate composite action

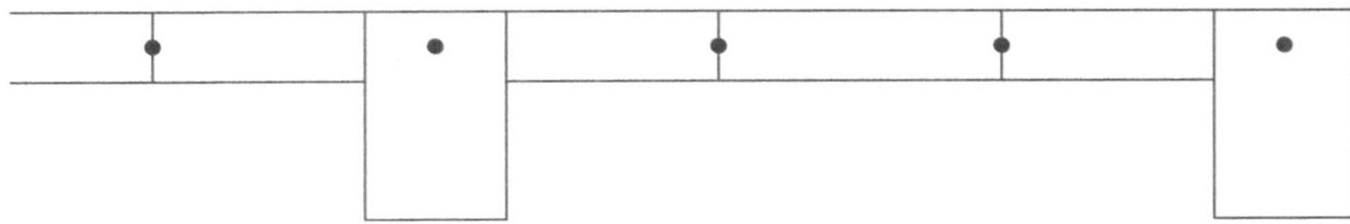

The alternative is to model the beam explicitly with a T-section but be careful not to include too much self-weight (Figure 5.14).

Figure 5.14: T-beam central on slab

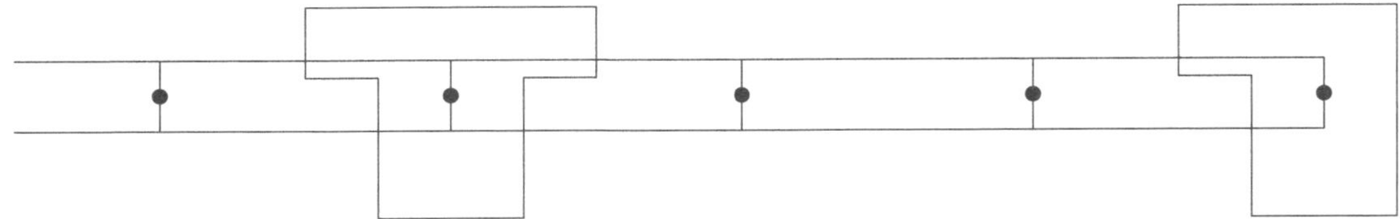

Or you could make the slab elements do most of the work and have a beam section that matches the downstand (Figure 5.15). I have never seen this option used though, so perhaps it should remain theoretical.

Figure 5.15: Downstand beam on slab

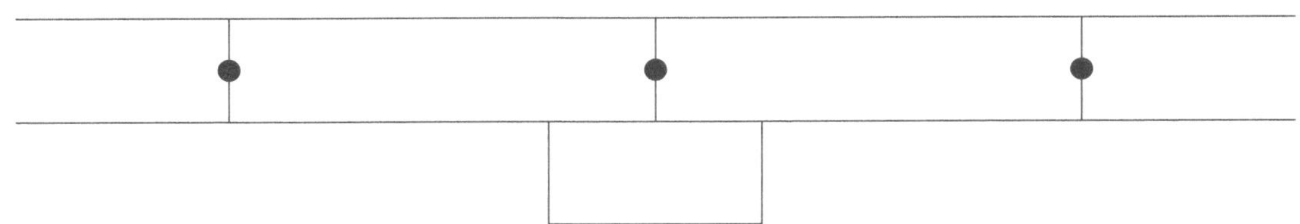

If you have ribbon beams, (also known as 'band beams') modelling entirely in 2D elements is the best approach (Figure 5.16). I wouldn't bother offsetting the slabs to align the top surface, but feel free to conduct a sensitivity analysis to check what difference it actually makes to the results.

Figure 5.16: Ribbon beam

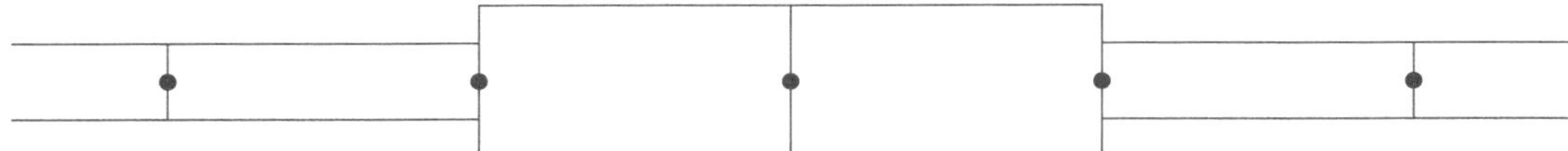

5.2.5.2 Flat slabs

With flat slabs there are no beams, or just some beams around the outer edge: the slab does all the work. There may be some thickenings in the slab at the column locations to deal with the punching shear. You can usually ignore any upstands on the slabs, especially if they are jointed to ensure they avoid attracting load. Follow the rules in Section 5.2.2 for meshing the slab, remembering the minimum element size is twice the slab thickness and that you can ignore any details (such as openings) smaller than that.

If there are steps in the slabs, you should include them in your model. Model minor changes of thickness with a change in element property but do not worry about offsetting the elements — while there will be a disruption in the stresses local to the connection, there will be no global effect[†]. You should model significant steps in the slab as these will generate a line of higher stiffness — modelling the elements at the appropriate levels and connect the nodes with links if the step is moderate. If the step is significant (the soffit of one side is above the top surface of the other) you need to explicitly model it with a vertical 2D mesh.

Flat slabs are popular with contractors as they are simple to build. The shuttering for the slab base is all at one level and the reinforcement is defined in bands; there may be some additional shear reinforcement around the column connection. Flat slabs are less efficient regarding use of concrete as the constant thickness means that much of it will be underutilised. The cost of any construction is a combination of both material and labour, which means that the cheapest forms will change over time. I would expect that robotic construction and the quest for zero carbon construction will put future emphasis back onto the material.

[†]   Saint Venant's Principle.

5.2.5.3 Connections

2D shells work well for most of a flat slab, where the assumption of the width being much greater than the thickness is valid (Figure 5.17). It is only close to the column supports that the assumptions start to break down. Here, 'through-thickness shear' starts to dominate the design, which is a 3-dimensional phenomenon. 2D shells can only give an approximation of this, but there is the further problem in the column itself. Using a 1D element for the column is good for modelling the way the column stiffness adjusts the bending moments in the slabs, but this creates a point support. While you may have set the column dimensions when defining the column section, the program only uses that to generate the stiffness of the element: 1D elements are 1-dimensional — they have no width or thickness! In technical terms what we have here is a singularity in the matrix. We have modelled a 3-dimensional structure using 2-dimensional elements (i.e. no thickness) supported on 1-dimensional supports (no dimensions other than length). If the structure was truly 1- and 2-dimensional this should produce an infinite stress at the connection; the more you reduce the mesh size at the connection the higher the stresses will become. If you could achieve an infinitely small mesh then you would get infinitely high stresses, which would, of course, be nonsense. The 1- and 2-dimensional elements represent 3-dimensional reality: they are an analogy and, like all analogies, break down if you take them too far.

Figure 5.17: Flat slab meshing

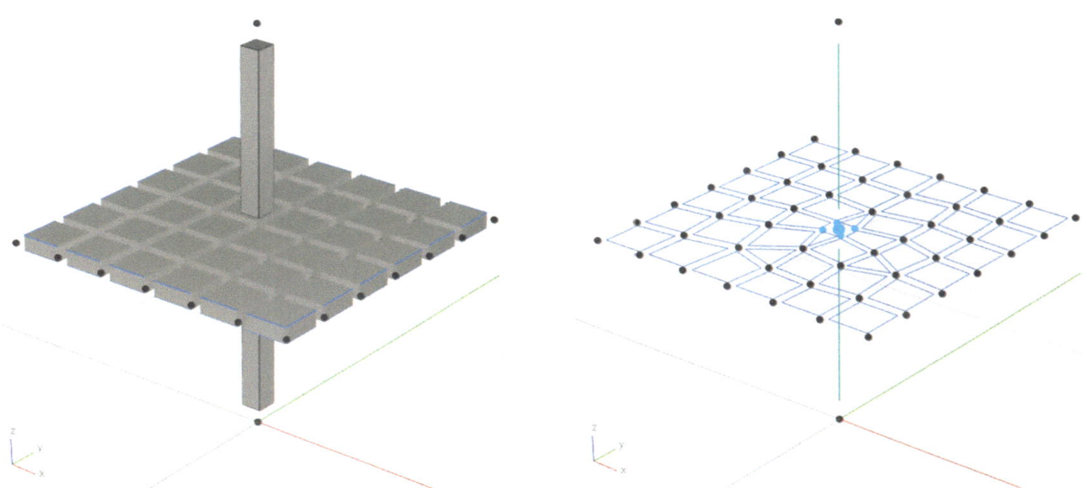

While refining the mesh normally gives more accurate results, this is not one of those times. Instead you have several modelling choices:

1. Ignore the slab moments within the column perimeter. They stop increasing within the connection so do not worry about them being in the model. The problem with this is that it underestimates the stiffness in the connection leading to a slight error in the bending moments.
2. Thicken up the slab in the column zone. The structure here is stiffer than the slab in general, so boost the 2D elements to match. You will still get the extreme moments within the column but the deflections will be more realistic. You should do this if your slab has drop heads or column heads.
3. Stop the slab at the column perimeter and connect to the column node via a link or constraint (Figure 5.18). This removes the unrealistic moments within the column but causes some strange ones on its perimeter.
4. Model the connection zone in 3D elements. You will then get the full picture of the stress flows in the area, but with all the problems that come from interpreting the results from an isotropic mesh into an anisotropic material like reinforced concrete[†].

But which option to use? Number 1 is the easiest; 4 if you want to understand what is happening and thus check the design rules; a combination of 2 and 3 is possibly best.

[†] Strictly speaking reinforced concrete is not a *material*. For that matter, neither is concrete: it is a *composite material*, which we can shorten to *material*. The problem with computer programming is that it demands pedantry, and this can spill over into everyday life.

Figure 5.18: Flat slab connection moments

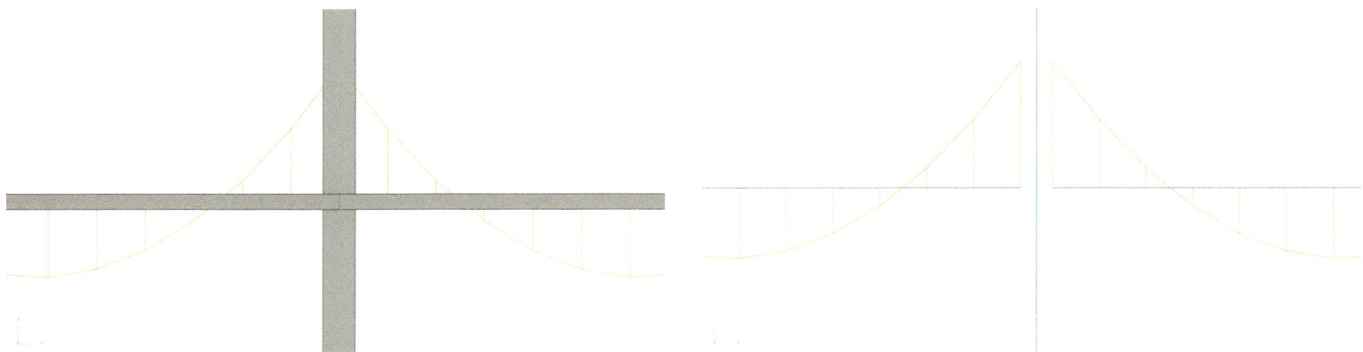

To check the punching shear, *How to design reinforced concrete flat slabs using Finite Element Analysis*[66] recommends ignoring the punching shear force output from the 2D FEA model as there can be inaccuracies in the reported values due to both moment-shear effects not being included correctly and the 1D/2D connection not modelling the 3D behaviour. Small openings close to the column might have been excluded from the analysis but must be included in the design checks. Instead take the reactions from the slab model (or the difference in axial load in the column above and below the slab) and use that in the hand/digital calculations. Make sure that you account for any openings in the shear perimeter!

Connections between beams and columns are usually simpler, as there are fewer 3D aspects to worry about. You might offset the ends of the concrete beams to the column faces, but this will have a negligible effect on the result other than removing a slight over-calculation of the self-weight, shear force and end moment.

The reinforcement detailing of the beam-column connection is a different matter. The first professional 3D model that I made was of a complex reinforced concrete connection, where an eaves, floor and rafter member all met at the same point at the top of the column. The reinforcement was so congested it was not possible to fit it all in, never mind the rainwater hopper and pipe the architect wanted straight down the middle. 30 years later, 3D modelling of complex reinforcement has become commonplace.

5.2.5.4 Stiffness and deflections
One of the main purposes of FEA (and indeed its main function) is to calculate deflections of structures; everything else comes from that. The distribution of bending moments in a structure is dependent not just on the local member stiffness but also the member's stiffness relative to those nearby: stiff members attract more load. This happens when disparate parts of the structure are forced by the geometry to have a similar deflection; deflection is proportional to load over stiffness. But with reinforced concrete we have a problem, in that we do not know how stiff the sections are.

Section stiffness, of course, comes from the geometry of the section and its material stiffness. The stiffness of the reinforcement bars is known as they are forged in factories, but its exact location within the section is down to the skill and whim of the rebar bender and fixer. I have seen reinforcement mats bent under foot and reinforcement cages being put in upside down; I have even seen reinforcement being left out altogether. You can never be 100% certain of what has been built once it's been cast.

The concrete members themselves are often cast in wooden moulds made by carpenters on site, where it is difficult to achieve the same accuracy as might be possible in a factory. The formwork must then be held in position to stop the sides bulging by means of 'falsework' — note that the depth of the beam or slab depends on how much concrete is poured into the moulds.

The location of the work is going to have an effect as well: precast concrete members are made in factory conditions with regular staff and so are likely to be more consistent and of a higher quality than those made on site.

The concrete's particular properties — a conglomeration of stone, sand, cement, water (and sometimes other chemicals) that strengthen and stiffen over time — will also affect stiffness. This happens not just over the first few hours, when the jelly-like paste turns to pseudo rock, but over the days, weeks and months after casting. This is influenced by the exact properties and mix proportions of the natural materials used, plus the weather on-site (heat, cold and rain will all affect the concrete both during and after construction).

The concrete member's bending stiffness is also dependent on the bending moment and axial load vs. its capacity. When the tensile capacity of the concrete is overcome, causing it to crack, the stiffness reduces. The reinforcement then takes up the load until it fails or when the concrete crushes (you need to avoid this failure mode as it will fail fast[†]).

With all these variables it is not surprising that most concrete deflection is based on simple rules of thumb to determine whether it will be ok. But increasingly, more buildings are clad with modular facades, which have a low tolerance for deflection and require more exact calculations.

Where does this leave us when the FEA method works by predicting the structural deflection? The default for most FEA programs is to use the gross concrete section, which we calculate by multiplying the material's Young's modulus (E) by the second moment of area (I), giving us the EI or stiffness of the section. This means the section stiffness ignores any contribution from the reinforcement, which at first sight might appear conservative, but the picture is more complex (Figure 5.19). If we take an example beam section with minimum reinforcement, the reinforced section stiffness is only slightly higher than the gross concrete section. If we add the maximum amount of rebar, we are initially underestimating the stiffness, but as the moment increases the section cracks and the gross concrete section overestimates the stiffness.

Figure 5.19: Short-term cracked reinforced concrete section stiffness

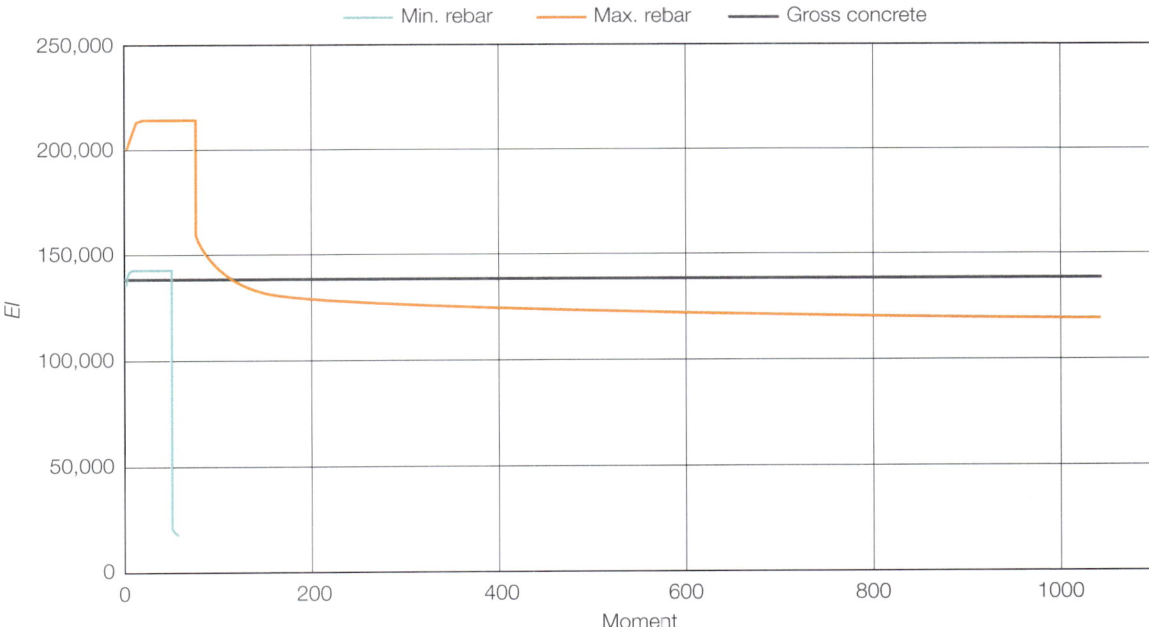

The simple approach to long-term behaviour is to reduce the concrete Young's modulus; this does not affect the stiffness of the reinforcement but does change the reinforced EI. Updating the stiffness diagram to use the long-term Young's modulus shows that the section stiffness is boosted by the reinforcement and the gross section EI is now much lower than the real section (Figure 5.20).

[†] Some business gurus say that it is good to fail fast and fail often, but neither are acceptable in a structure.

Figure 5.20: Long-term cracked reinforced concrete section stiffness

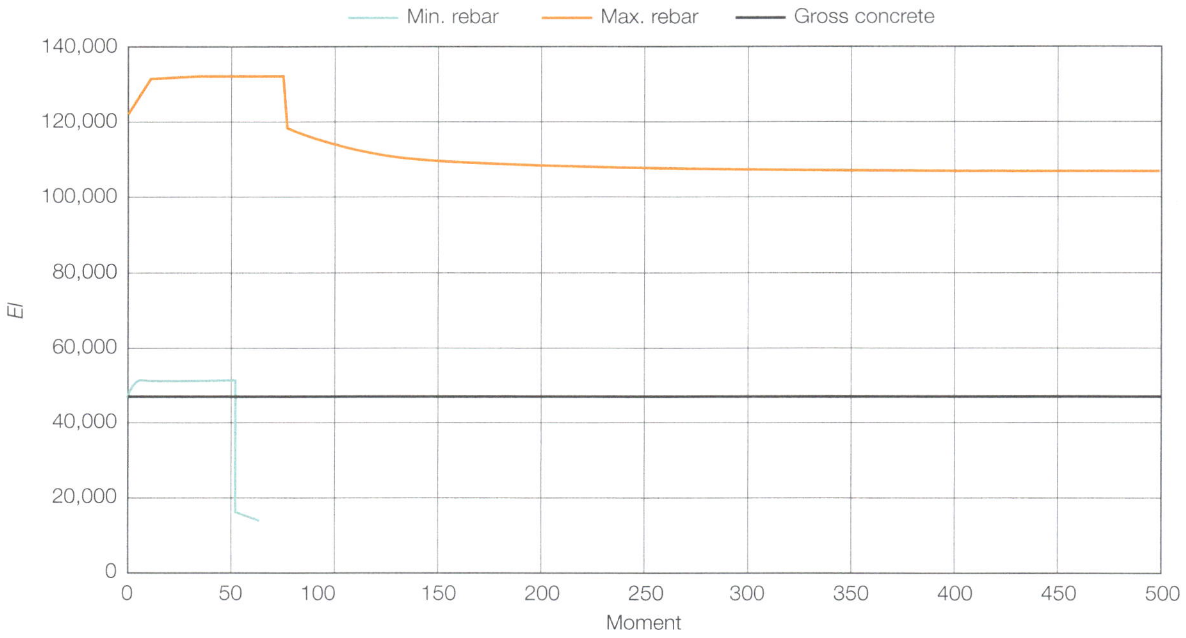

Finally, noting that the stiffness changes with the moment, remember that ULS sway stiffness will generate larger moments than the SLS deflections.

5.2.5.5 Stability

Concrete buildings have a variety of options when it comes to stability systems. The most efficient usually depends on the size of the structure, though stiffness requirements can influence the selection. As a rule, you can stabilise the structure with:

- Moment continuity
- Shear walls/cores
- Bracing

As with steel frames, you need to include a rigid diaphragm in your model if you don't want to include the slab explicitly. It is true that the moment continuity of the concrete beams will mean that the model will be stable without one, but the minor axis stiffness of the beams will not be a good representation of the slab stiffness. You'll also find there are lots of fake minor axis moments to either design for or ignore/explain away.

One question you need to answer is whether you restrain the nodes at the foundations as pinned or fixed. This is because some foundations can act as fixed under short-term loads but will rotate, and as such are pinned, when the loads are applied for a long time.

You may want to model the rotational stiffness of the foundations, especially if the loading is overly sensitive to the support stiffness. I was once working on the design of a large multi-storey car park over a shopping centre over a bus station. The complexity was such that we could not fit in movement joints at the recommended 50m spacing and I had to carry out a nonlinear thermal analysis to prove that the structure could withstand the resulting forces. One of the crucial aspects of the analysis was the rotational stiffness of the stability core's piled foundations as they were nonlinear and direction-dependant. This required several iterations with the geotechnical engineers until we got to the final design.

Hindsight is a wonderful thing, so it is always worth reviewing your most difficult calculations and considering how you might do it differently next time. At the time I was not familiar with nonlinear analyses having only used linear ones, but the nonlinear base stiffness meant there was a lot of uncertainty in the results, as well as hassle for my colleagues calculating multiple foundation stiffnesses. Today, I would just ask for the stiffnesses over a range of loads, calculate the nonlinear spring curves from that, and do a single static nonlinear analysis. Also, with the intervening decades of development, the analysis software itself is much better at dealing with nonlinear problems. Experience tells us how to make things simpler next time: sometimes that means a simpler model, sometimes it means a more complex analysis (but only once).

5.2.5.5.1 Moment frames
Modelling moment frames in FEA is an easy option as elements automatically assume moment continuity unless told otherwise. As a rule, these will be beam and column arrangements, though moment frames for flat slab structures are known.

5.2.5.5.2 Walls and cores
5.2.5.5.2.1 1D cores
The simplest way to model a stability core is as a single 1D element (or chain of elements if it changes with height, intersects with floors, or you need a dynamic analysis). It is appropriate if you are interested in the overall behaviour of the core, such as natural frequency for wind excitation analysis. The downside is that openings are exceedingly difficult to include, so while you might extract overall bending moments, it is difficult to investigate local effects.

For example, if you want to predict the seismic behaviour of the building and stability core and quickly explore design options, use 1D elements and add masses at each floor level to represent the local structure: if the floor and columns do not contribute to the overall lateral stiffness then leave them out of the model. Changes to the core are thus quite simple to enact and so you can quickly explore different options with minimum effort.

If you have a wall, rectangular or C-shaped core you can define it as an appropriately shaped section in the profile table of your FEA program. For other shapes without a suitable parametric wizard you might draw it in a CAD program and import it. The program will define the section about its centroid, so be careful where you position asymmetric cores.

You can also use 1D elements when you have multiple cores or shear walls stabilising a building. As before, do not model the floor structure but represent the diaphragm using a rigid constraint. Ensure it is set to work in the horizontal directions and not in the vertical for the most realistic behaviour; as the floor diaphragm will make the lateral movement consistent, but its bending stiffness will be too negligible to contribute. Include rotation about the vertical axis in the rigid constraint if possible, but not at the expense of including rotation about the horizontal axes as they will also add an unrealistic stiffness.

Add a node in the centre of each floor and include it in the rigid constraint. You can then apply the wind load on each storey as a nodal point load on this node or add a mass to replicate the weight of the floor. The position might be the floor centroid for seismic analysis or the centre of wind action.

To find the centre of wind action consider all the wind loads in a direction. Resolve all these down to a single vector. Repeat in the orthogonal direction and mark where the two vectors cross. Some floor plan shapes may necessitate a variety of centres for winds in different directions.

5.2.5.5.2.2 Cores made of 1D elements
You can also use 1D elements to model the individual walls of a stability core[67]. Make the size of the sections either match the overall wall dimensions (with reductions in stiffness) or area (where there are openings). Alternatively, model the individual sub-walls either side of the doors and windows (Figure 5.21).

Cores are more than the sum of the shear walls but instead work as a whole; this means that you need to connect the parts together with stiff elements to generate the shear transfer between walls. You can do this with very stiff beam elements (ensuring that you include their shear stiffness) or link elements. If using beam elements, ensure they follow the wall lines and thus connect where the real structure does. The beam properties will be that of the wall panel, so its depth will be the floor spacing if there are no openings, or less where there are openings. Ensure they have no mass so that you do not double-count the self-weight or modal mass. Links do not need to follow the wall lines as, like vectors, they are equal to the sum of the parts, but the substructure will be easier to understand and thus check if you do keep them within the walls. Links don't have a self-weight and so you don't need to worry about over-counting the mass.

Figure 5.21: 1D element core

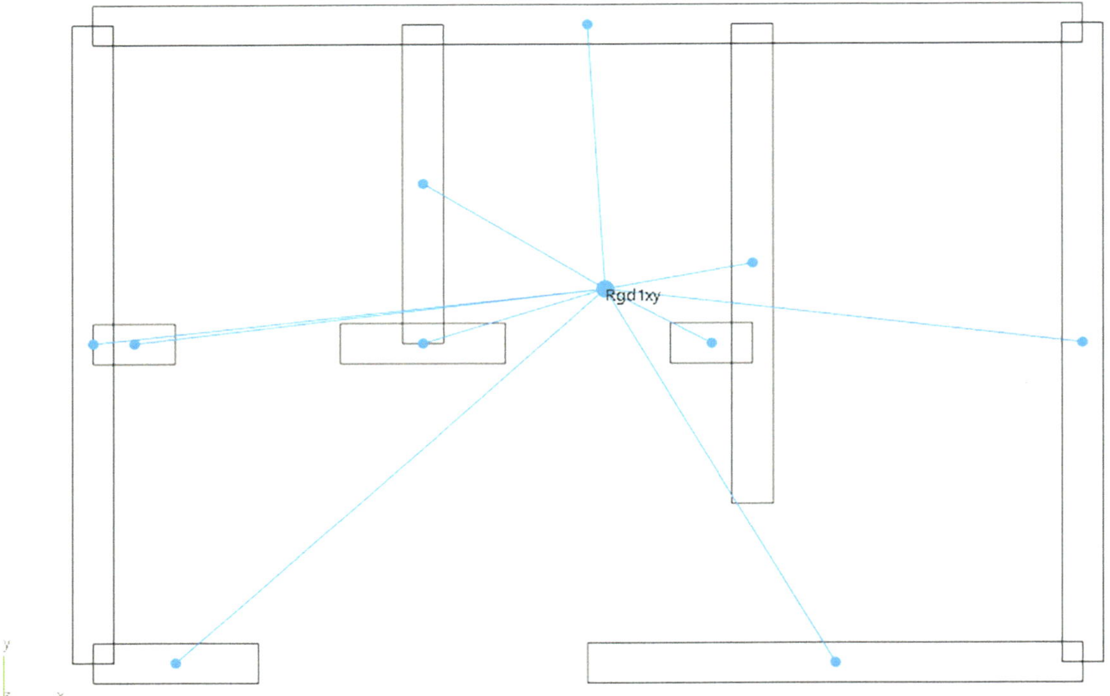

The problem with the horizontal elements is that you can quickly get lost if you're not careful, especially when working in 3D. Switching on the element solid display normally helps visualise a structure but that is not the case here. As with grillages, you have the structure represented twice: once horizontally and once vertically.

An alternative to horizontal elements is to use a rigid constraint that works in all six directions. This assumes the shear transfer elements are stiffer than the walls. This has the advantage of being easier to model as there are no horizontal elements. Note that if you also have a rigid constraint for the floor diaphragm you need to be careful how you connect them. Only include the primary node of the core rigid constraint in the floor diaphragm — otherwise you will get an error, as a secondary node cannot be connected to two primary nodes.

Like grillages, the 1D element approach for cores struggles with non-regular openings. It is an abstraction of the real core and some structures are easier to abstract than others. Also, the 1D elements are great at giving you the overall force or moment in a wall but are not so clear about how the forces flow between the sections.

5.2.5.5.2.3 2D element cores
While you may need 10–100× more elements, using shell elements to model cores is the best way to understand how the stresses flow through the structure. As discussed previously, they are too detailed to use with a whole-building model at early design stages, but excellent later when you need to focus on stress concentrations and reinforcement around openings, or cast-in plates for beam connections.

As with flat slabs, the rule of 10 elements across while maintaining the minimum size of twice-thickness is a reasonable one. The corners of openings will generate stress concentrations that will grow as the element sizes reduce.

Remember to use spring supports as the base if your core is bearing directly onto the ground, as full translational restraints will generate unrealistic local stresses. While the footing might provide out-of-plane moment restraint, you should usually leave out any in-plane moment restraints as this will be provided by the vertical reactions along the wall.

Watch out for significant moments on the core that put the soil supports into tension, which it cannot take unless they're piled. Remember that overturning forces from wind are resisted by the structural gravity load, so check load combinations not individual load cases. The quick hand calculation approach is to divide the moment by the vertical load to find the eccentricity of the line of thrust and ensure that it is within the middle third of the base. It is an easy calculation when the base is rectangular and the loads are orthogonal but gets tricky for other situations such as loading on the diagonal. The core might still be stable if the eccentricity is a little outside the middle third but the contact stresses with the soil will grow rapidly with the eccentricity, so be careful. In these situations, replace linear spring supports with compression-only ones. This will make the analysis non-linear, so also check that the core is stable (is there much point in having a stability core if it is not?) and thus that the non-linear analysis will converge.

Following the analysis, display the principal stresses and check that the forces are flowing through the walls as you expect. In general, the reinforcement will be vertical and horizontal, but there may be locations, such as around openings, where diagonal bars are also required.

Always remember to compare the results of your 1D element stick model with the 2D shell element version to ensure they are giving comparable results. If not, you know there has been a problem in modelling or loading one of them (possibly both).

5.2.5.5.3 Bracing
Bracing of concrete frames is normally reserved for very tall buildings, as sloping RC members are not easy to cast. When you get to the mega-frame scale the effort becomes worthwhile due to the savings on floor space (smaller cores).

Due to the size of the bracing on very tall structures, it might be expressed (i.e. made visible) in the building cladding. You should also try to optimise the bracing layout as it is such an important part of the structure and difficult to construct. Simple X-bracing is now giving way to evolutionary topology optimised (ETO) methods, resulting in asymmetric but more efficient layouts (Chapters 8 and 9).

5.2.6 Substructure
If the frame is statically indeterminate the support stiffness can affect the bending moments and other results. This means you should use spring supports, unless the stiffnesses are significantly greater than those of the structure, when the support stiffness will not make a significant difference to the results. If in doubt, undertake a sensitivity analysis to get the range of results: supports pinned, supports fixed, model with an estimate of the support stiffnesses.

You can make a quick guess at the soil stiffness by assuming that the foundations will settle 10mm under full, unfactored load. If the spring stiffness of the support is measured in KN/m (the force to generate 1m of movement) it will be 100x the load. You will not see any effect on the moment distribution for the fully loaded condition as the settlements will all be the same, but you may see a redistribution under partial loads.

Differential settlement will have a major effect on the moment distribution, so you should model it as accurately as you can. Bear in mind that soil is even more variable than concrete so a sensitivity analysis covering the range of possibilities is sensible. You might also look at soil-structure interaction, as the soil will tend to displace more under the centre of the structure than at the sides (Figure 5.22).

The 'dishing' of the soil is due to the overlapping stress bulbs under each footing. This means that a constant stiffness under a building, and especially under a raft foundation, is inaccurate. Instead, the foundations have a lower effective stiffness under the middle of the structure.

When modelling the bending moments and shear forces in a raft foundation, do not automatically include the self-weight of the slab itself as it will be poured wet. This will add stress to the ground so do consider it if that is of interest to the design. Conversely, you do need to consider self-weight if the slab is likely to lift and lose contact with the ground at certain points e.g. asymmetrical loading or ground heave.

Figure 5.22: Soil-structure interaction

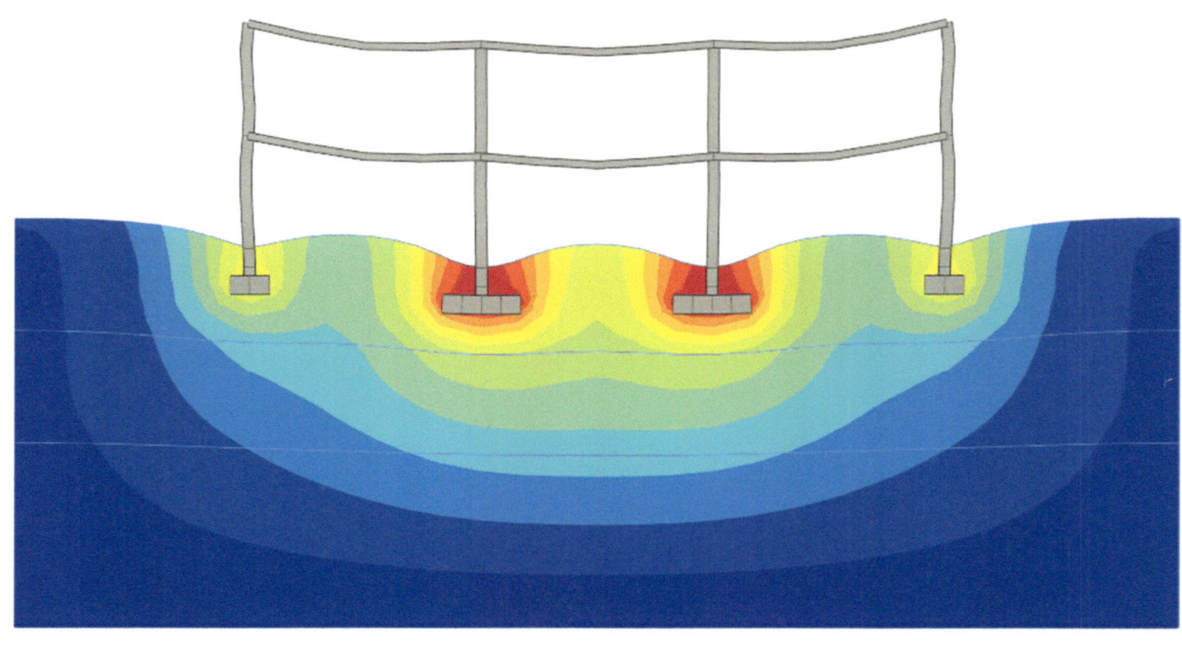

5.2.7 Timber
When timber structures are frames you can follow a similar modelling approach as that for steel frames, but if you are using cross-laminated timber (CLT) slabs then concrete modelling should be your guide.

Timber has orthotropic properties and you should use orthotropic materials for 2D and 3D elements, but you only need isotropic materials for 1D elements.

Timber member sizes are often governed by the connections, so their design must be included early on as part of your workflow.

Remember that these are generic suggestions; you need to consider very carefully what is the best way to model *your* structure.

5.2.8 Masonry
Masonry is quite different to steel and concrete. It is a composite material comprising stones, bricks, or concrete blocks held together (or apart) by mortar made from sand plus cement or lime.

Masonry is stiff, can be strong in compression (especially stone masonry), but has virtually no tensile strength due to the weakness of the mortar joints. For this reason, the primary requirement for masonry design is to keep it in compression, which we achieve using two principal methods:

1. Shape — arches for spanning gaps, buttresses for turning lateral loads into vertical ones, etc.
2. Prestress — while this might come from tension rods, the standard method is to use the masonry's self-weight plus any load coming from elsewhere in the structure.

It is also possible to give masonry tensile capacity by adding reinforcement, but the scope for this is limited. Any significant tension will indicate cracking and thus a change of load-path, so remember to include permanent and variable actions with wind loads to get reasonable results.

Linear finite element analysis works on the assumption that materials are not limited in either tension or compression, which makes it less than ideal for masonry — all models are wrong and masonry models are more wrong than most. That said, there are some applications where you can use it, under consideration. One option is to use it for load takedown or principal stress analysis on walls and other gravity-loaded structures (Figure 5.23). Another is to get moment and shear distributions on wind-loaded panels.

Figure 5.23: Principal stress analysis of masonry cathedral

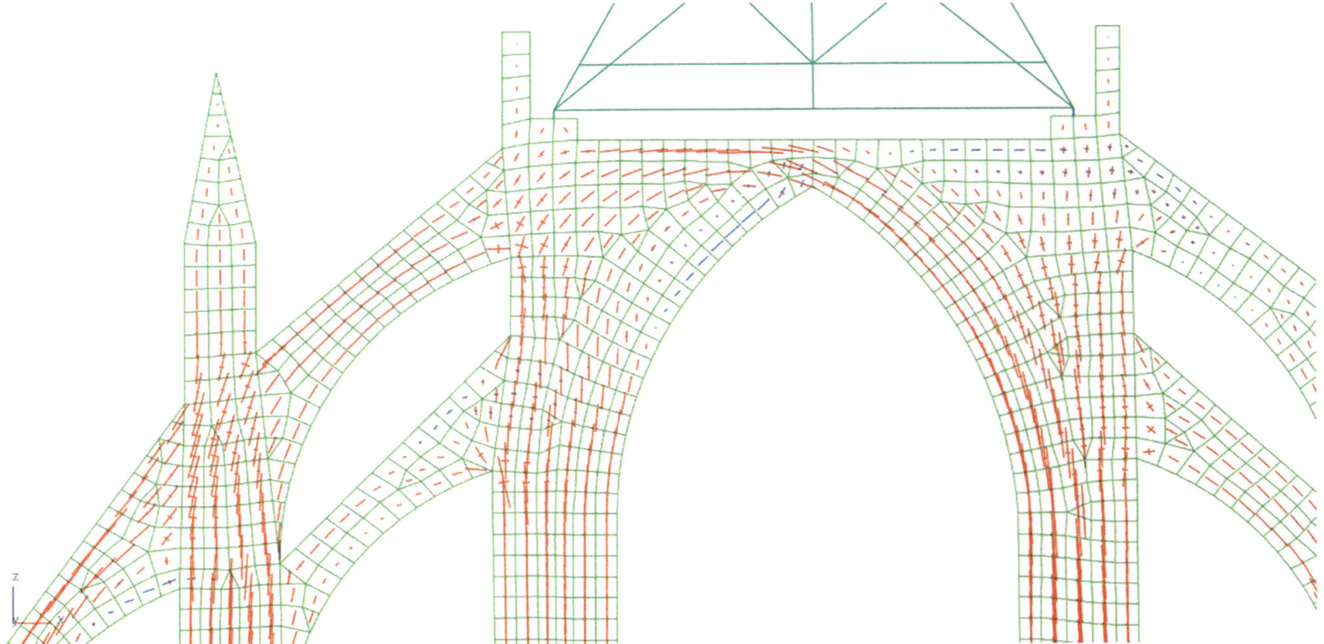

The essential point for modelling masonry in FEA is to remember that it is nonlinear, so you need to either model it in a very high-end analysis package, use linear analysis and treat the results with a lot of scepticism, or use a dedicated masonry package that adopts a completely different analysis method to FEA. I would also recommend reading specialist masonry design guides[68].

5.2.9 Modelling for vibration

Although the actual structure does not change when subjected to dynamic rather than static loads, you often need a different analysis model to capture the relevant behaviour. Exactly how these differences manifest themselves will depend on whether you are modelling for extreme events such as earthquakes or low intensity vibrations such as those from machinery or people walking.

Strictly speaking, all loads are dynamic, at least when they are first applied; most loads are there long enough for the induced vibration to stop. I was told once that even if a weight is placed onto a beam and released, the initial deflection will be twice the final, static deflection. If the structure is linear then twice the deflection means twice the strain and hence twice the stress[†]. The loading on a structure tends to be applied gradually: one piece of furniture at a time. Loads that are applied instantly, such as explosions, tend to be short duration ones, so the structure does not always have time to react and thus develop the stresses or strains sufficient to break it. Suddenly, applied loads that remain for a period of time can be difficult for the structure to withstand.

When conducting a vibration analysis for steel frames with simple connections, the convention is to model the connections as continuous. The reason for this is that under serviceability vibrations the strains are usually too small to overcome the friction in the joint, and thus act as if they have a full moment connection. Likewise, it is appropriate to model the floor as composite with the beams, even if it is not designed for this in the ultimate limit state.

[†]   Structural safety factors do a lot of work.

For ULS we might also model the connections as fixed because they have only a limited rotation before locking up. The real connections during an earthquake might be constantly slipping and locking, which is difficult to model, but might release some energy; ignoring this aspect in the design makes it a little more conservative. Generally, seismic zone steel connections will be designed as continuous.

Concrete structures are usually moment frames, but the time-dependant aspects of the material mean that you should use the dynamic modulus, which is stiffer than the regular Young's modulus value.

What we include in a vibration model may differ from a static model. Under ultimate loads we would ignore fixtures such as handrails and glazing, but for serviceability vibration their stiffness can have a significant impact on the response. While balustrading will not strengthen an edge beam it will help to reduce local vibration excitation.

How much of the structure should we model? The answer is, as always, as little as possible while still obtaining sufficiently realistic behaviour. For example, if we are modelling the overall behaviour of a tall building that is stabilised by just its concrete core, we can dispense with the floors and just add the relevant masses at each level. Likewise, if there are multiple cores or shear walls we can again dispense with the floors and instead link them with a horizontal diaphragm plus the missing mass.

Appropriate modal mass is a key feature of a good vibration analysis because, along with stiffness, it resists the induced vibration; the worst case can be when the loads are at a minimum. This means you need to include the permanent and variable actions that will actually be there and not the extreme required for the ultimate limit design. You will need a sensitivity analysis (testing the range of loads) unless the loads are given in the relevant design guide or code.

5.2.9.1 Floor vibrations

When modelling the floor vibrations[69], you can usually model just the floor itself plus the columns to give some additional stiffness. The question is then: How much of the column should you model? Opinion varies on the best approach here: some say "model the whole column and give it a fixed end"; others say "model half the height and pin the end". Making the columns full height with pinned ends also gives reliable results. The least accurate option is to leave the columns out altogether. The basic concept is that a floor will vibrate independently of adjacent floors, so the columns will deflect with the excited floor but not the unexcited one.

If we're looking at human or machinery induced vibrations, the excitation occurs at a particular location but can also resonate in the adjacent bays. This means it is very conservative to model just a single beam, as it will not include the mass of the bays that resist the excitation. Instead you must model the beams and the slab on the floor or bridge deck. You will also need to include the columns to the next floor. Unlike static analysis, model the walls and cladding as restraints (fixed or spring depending on the design guide[70–72]) rather than loads.

If you have a very regular frame, the modal analysis may show that the whole floor has the same natural frequency (Figure 5.24). This may result in an over-estimation of the modal mass as damping will prevent the vibrations from travelling too far, and thus be unconservative regarding the predicted floor accelerations. You will get a more accurate vibrating machinery or footfall analysis if you only model the adjacent bays, as then you will avoid counting mass that will not have a chance to resist the vibrations. The downside to this is that you will have to create a model for each bay. Luckily, most buildings are not regular and thus the natural frequency modes are already quite restricted in extent. This means you can often get away with modelling the entire floor in one model and not worry about over-counting the mass.

Watch out for bays that have only local vibrations (Figure 5.25) as these will have a low modal mass and thus can experience a high response. The simple hand calculations in the footfall design guides cannot cope with these bays and FEA is the only sensible way to analyse them.

Footfall vibrations are linear, so a modal dynamic analysis is sufficient to find the natural frequencies. You need to then follow it with a dedicated footfall analysis (Appendix E) or set up a number of harmonic (steady state vibration) analyses to see how each location responds to footsteps. With bridges you can have a reasonable idea of the paths people will take across the structure (much harder on a floorplate) so you can set up time-history analyses to get the effect of people walking and running over the bridge.

Figure 5.24: Floor vibration modes across bays

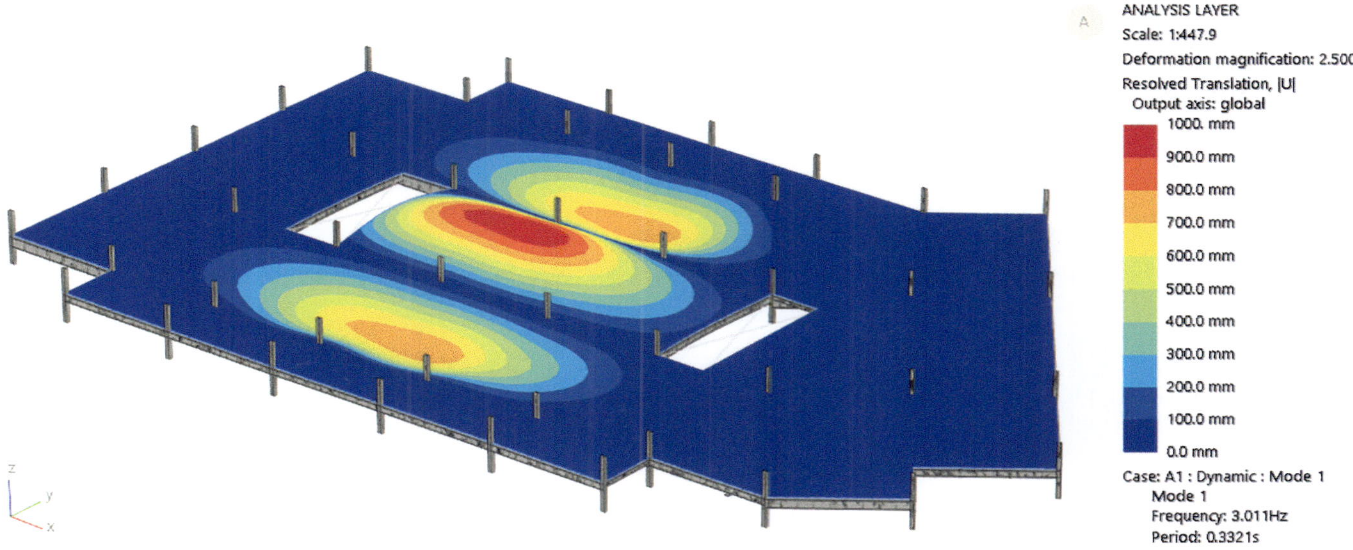

Figure 5.25: Floor vibrations in single bay

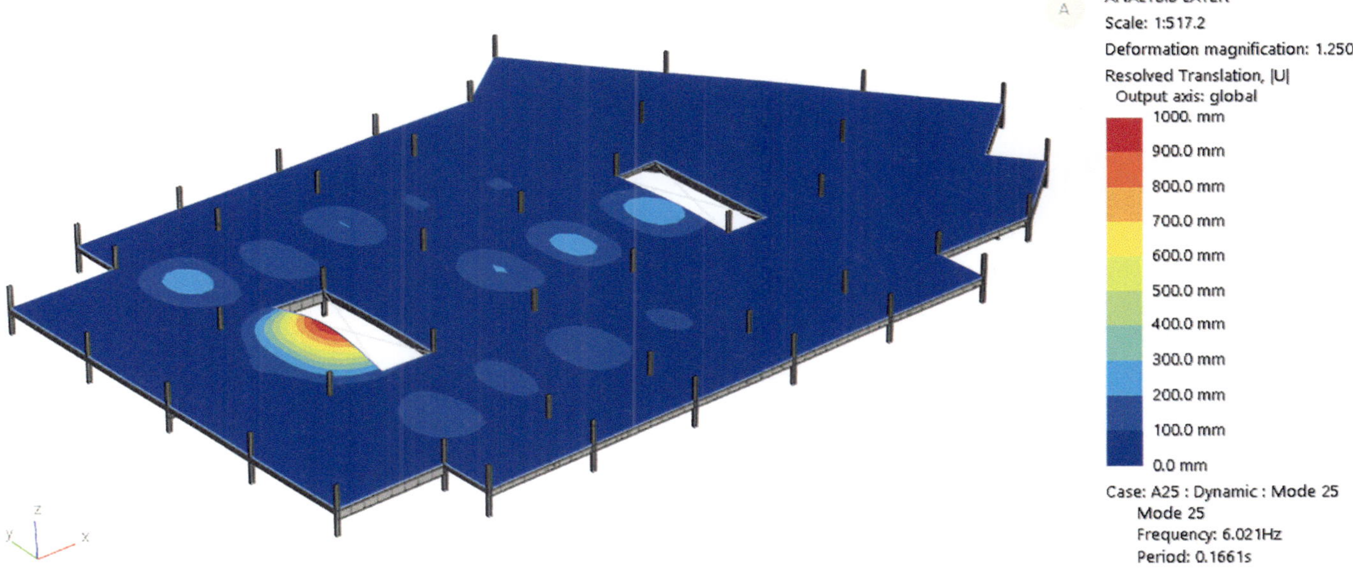

5.2.9.2 Seismic

With seismic design we know acceleration and the mass, and we are trying to find the resulting forces. So, while we resist floor vibrations by adding mass, a good engineering response to seismic loads is often to reduce the mass as much as possible. Curiously, another way to reduce the earthquake (and explosion) loads can be to reduce the stiffness. This is because earthquakes induce deflection in the structure. It takes little force to deflect a low stiffness structure, so an induced deflection in something that has little stiffness induces little stress.

Behaviour will change if the structure is damaged by the event. The formation of plastic hinges under seismic load can be an important part of the design but will alter the dynamic behaviour of a steel structure. Likewise hinges, spalling and cracking will change the frequencies of a concrete frame.

You have a number of options for analysing seismic behaviour. The simplest is to use a pushover analysis, which assesses the frame for a number of equivalent horizontal static loads. You can run a linear time-history analysis following the modal analysis to see what the peak forces are likely to be. The most advanced option, used for tall buildings and other particular cases, is to use a nonlinear explicit solver with time-histories of real and simulated earthquakes, to see how the structure will actually perform, included as hinges form and parts break. The last option is time-consuming, not only because the analyses are slow, but you also need to run many different cases to ensure you have captured the most significant structural behaviours.

5.2.9.3 Explosions

"The green reed which bends in the wind is stronger than the mighty oak which breaks in a storm."
Confucius

Explosions exert huge forces but only for few milliseconds. An explosion initially causes high pressures on structures, pushing away from the blast, then there is a negative load, a suction, as air rushes back in. This causes a lower pressure on the structure but over a comparatively longer time; it can be this part of the explosion that breaks the structure.

force $\rightarrow$ acceleration $\rightarrow$ deflection $\rightarrow$ strain $\rightarrow$ stress

If you compare structures of high and low stiffness for a given deflection, the stiffer structure will experience greater stresses and strains, and so is more likely to fail. The best thing is to have a heavy structure with low stiffness.

A linear time-history with a modal dynamic analysis can provide some insight into explosion behaviour, but I would expect to use nonlinear time-histories with explicit solvers to see what breaks and where it lands.

6 Analysis methods

> *"Unfortunately, no one can be told what the Matrix is.*
> *You have to see it for yourself."*
> Morpheus, *The Matrix*[73]

6.1 Introduction

Now we come to the heart of the book, but it is arguable whether this is the most important chapter or the least. As engineers we must always focus on how to bring about the best result for our project. Most structural engineers' jobs are to create the best structure possible, though what we mean by 'best' will vary from project to project and client to client. Others, like my colleagues, have the job of producing the best structural analysis and design software. My job here is to ensure that you know what you need to know to use the software to its best, to create the best structure.

You do not need to understand the engine to drive a car. If, on the other hand, you want to drive that car optimally and get the maximum performance from it, you do need to understand it in more detail. Even more so if you want to tinker with the engine and make modifications. If you want to build your own engine from scratch, you need to know it best of all. This chapter will go into some detail, enough perhaps to give you the understanding needed to build good models. If you want to tinker and create your own analysis engine, or want to learn more, this should be a good primer, but you will need to follow up with more specialised tomes.

In my time, I have been sent numerous models with the complaint that "it is not working, can you fix it?" The errors contained in those models often reveal a lack of understanding of the analysis engines. While you do not need to know how to invert a matrix[†] to answer the question: Why do I need restraints in directions where there is no load?, knowing why the lack of a restraint stops the analysis might help you to remember to add it. Likewise, why are trusses made of bar elements ok in some models but not others?

This chapter is the most mathematical in the book, which will be a horror to some and a relief to others. My colleagues, whose job it is to write these programs, will no doubt find it over-simplified, which is fine as its purpose is not to teach you how to write an FEA program but only to have an insight into how they work. I will stick, in the main, with bar elements, as they illustrate the principles. Elements that bend are far more useful and far more complex; but while the detail changes, the principles stay the same. If the challenge of writing a program for yourself excites you, I hope this text will be a springboard to more detailed works. For other readers, it may provide insight into why your model did not give you the results you were hoping for, and how to avoid this in the future.

The key question that all the FEA methods are trying to solve is: How do the nodes move relative to each other? Though the detail and implementation of the methods are complicated, the principles are usually quite simple.

We will start by looking at the most common method: static linear analysis using stiffness matrices.

Next, we will add in the geometric stiffness matrix, giving us P-delta analysis. This matrix enables loads to stiffen or soften the structure.

If you apply too much load to a P-delta analysis, the stiffness and geometric stiffness matrices can cancel out: the stiffness falls to zero, and the structure buckles. Modal buckling analysis uses Eigenvectors and Eigenvalues to track down what those critical loads are.

[†] I will let you into a secret: many FEA programs do not actually invert the matrix as it is exceptionally difficult and time-consuming for large models, but use other methods to produce the same answer. The principles remain the same though.

Adding a mass matrix to the stiffness matrix and running a modal analysis gives us modal dynamic analysis, which tells us how the structure wants to vibrate. Dynamic response analysis then uses the modal dynamic results to predict how it will actually move — from minor wobbles to seismic events.

Next, we will look at nonlinear methods. The Newton-Raphson method extends P-delta analysis by iterating more, as well as considering material nonlinearity and other effects. Explicit solvers then dispense with matrices and simulate how structures move and even fail under load.

Finally, we will take a brief look at meshless methods, which dispense with elements and allow for some very advanced analyses.

6.2 Static linear analysis and the stiffness matrix

The finite element method (FEM) is the most common way to analyse the physical behaviour of a structure when actions are imposed upon it; these may include gravity loads, wind forces, thermal expansion and earthquakes. We give the model boundary conditions (typically restraints at the foundation or other locations where the model connects to the rest of the world) and material and section properties. We can then analyse the structural model to find its static or dynamic behaviour. Other options for FEM include thermal flow, electromagnetic flux, and computational fluid dynamics (CFD), which are occasionally used by structural engineers.

The basic concepts[53] are:

- The whole structure is in equilibrium — the sum of all external forces and moments is zero
- Every node is in equilibrium — the sum of all forces and moments on a node is zero
- Every element is in equilibrium — the sum of all forces and moments on an element is zero
- The force (f) in an element is equal to the deflection (u) multiplied by the stiffness (k): $f = ku$. Similarly, a force on a node will result in a deflection of that node in inverse proportion to the stiffness of the connected elements
- We know the external forces and can calculate the element stiffnesses, so we just need to find the nodal displacements: $f = ku \Rightarrow u = f/k$
- If we write the equations for every node and group them into a matrix[†], we get $\{f\} = [K]\{u\}$
- To solve this, we multiply both sides by the inverse of the stiffness matrix $[K]$: $[K]^{-1}\{f\} = \{u\}$

The FEM follows this basic process:

Pre-processing (model creation) — user

1. Define model as a series of nodes connected by elements.
2. Apply properties to elements.
3. Apply enough suitable constraints on model.
4. Apply actions on model, typically forces and moments.

Processing — solver

5. Calculate individual element stiffness matrices from model properties and assemble into global stiffness matrix.
6. Apply boundary conditions by striking out lines and columns relating to zero-translation or rotation nodes to give reduced stiffness matrix.
7. Invert reduced stiffness matrix and multiply it with reduced load matrix to obtain reduced displacement matrix.
8. Multiply full displacement matrix by full stiffness matrix to find reactions.
9. Use displacement matrix to find element strains and hence stresses, moments, etc.

[†] There are several writing styles for matrix maths. Here we are using braces $\{f\}$ to indicate a vector or column matrix, square brackets $[K]$ for a square matrix, and if there are no brackets then it is a scalar.

Post-processing (results) — user

For example:

10. If analysis is linear, multiply and combine load case results as needed for serviceability and ultimate limit state cases.
11. Run design checks on elements/members and change sections or reinforcement as needed

    a) Rerun from #5 until no further changes are required.

12. Output results for calculation record.

6.2.1 Stiffness matrices

The concept of the stiffness matrix method for analysing structures has been around for a long time, but it was not a practical option until the invention of the programmable electronic computer. The intention here is not to cover all the mathematics of the stiffness method as there are plenty of textbooks that go into detail. Instead I will show you how the finite element or stiffness method works in principle, with some simple examples.

Consider this structure (Figure 6.1), consisting of several bars supporting a single force F.

Figure 6.1: Statically indeterminate structure

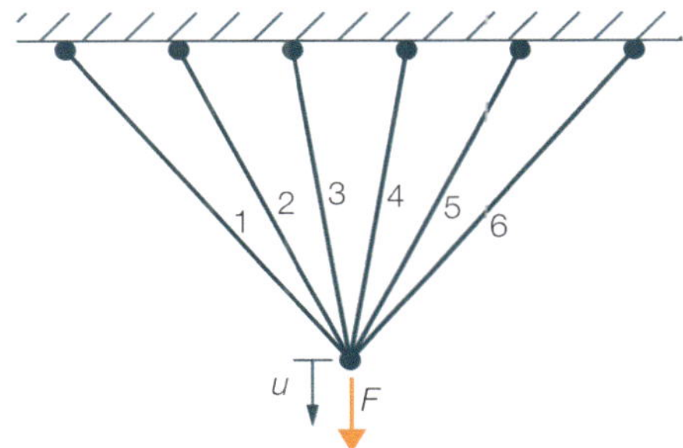

While we do not know how the bars share the force, we do know that the node loaded by the force will move by a certain amount. We do not know what that deflection will be at first, but we do know that it will be governed by the ratio of the force against the total stiffness of all the elements. As the node is connected to all the elements, the node's displacement will cause a strain in the each of the elements and that strain will induce a stress. When we know the stress we can calculate the force in each element. If the structure is in equilibrium, the total force in the elements must equal the applied force.

6.2.1.1 Bars in one-dimensional models

To start with let us look at the single bar in Figure 6.2. This element has one degree of freedom allowing us to focus on the principles.

Figure 6.2: Bar loaded axially at one end

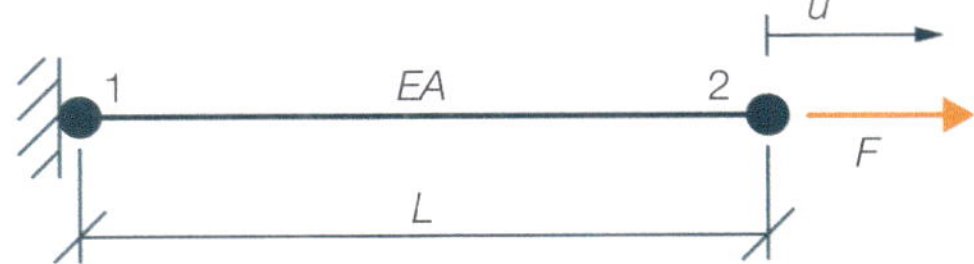

The stress on the bar is equal to the Young's modulus multiplied by the strain:

$$\sigma = E\varepsilon$$

The stress is equal to force over the bar's cross-sectional area:

$$\sigma = \frac{F}{A}$$

The strain is the extension over the length:

$$\varepsilon = \frac{u}{L}$$

Substituting these into the first formula we get:

$$\frac{F}{A} = E\frac{u}{L}$$

Thus:

$$F = \frac{AE}{L}u$$

Which tells us that the force is equal to the bar stiffness multiplied by the end displacement. The conventional variable to represent the stiffness AE/L is k, giving us the fundamental equation of the stiffness method:

$$F = ku$$

We know the applied forces and the element stiffnesses, so to find the nodal displacements we need to divide both sides by the stiffness — or in matrix terms, multiply both sides by the inverse of the stiffness:

$$k^{-1}F = u$$

Once we have the nodal displacements, we can calculate the relative displacement between the element ends, which gives us the element strains. From the strains we calculate the stresses and from the stresses we get axial forces. If we were using beam elements (Figure 6.3), we would also have nodal rotations and lateral translations, enabling us to calculate the bending moments and shear forces.

Figure 6.3: Bar loaded axially at both ends

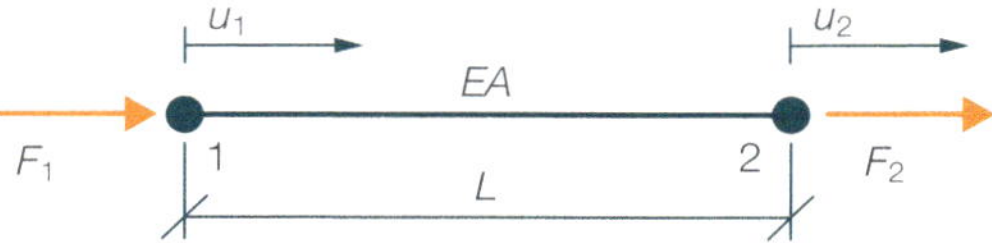

In general terms the end forces are:

$$F_1 = \frac{AE}{L}(u_1 - u_2)$$

$$F_2 = \frac{AE}{L}(u_2 - u_1)$$

Which we can write in matrix form:

$$\begin{Bmatrix} F_1 \\ F_2 \end{Bmatrix} = \frac{AE}{L}\begin{bmatrix} 1 & -1 \\ -1 & 1 \end{bmatrix}\begin{Bmatrix} u_1 \\ u_2 \end{Bmatrix}$$

Rewrite this equation in matrix shorthand:

$$\{f\} = [K_e]\{u\}$$

Where $[K_e]$ is the element stiffness matrix:

$$K_e = \frac{AE}{L}\begin{bmatrix} 1 & -1 \\ -1 & 1 \end{bmatrix}$$

Note that the element stiffness matrix is symmetrical.

The individual element stiffness matrices K_e are combined to form the global stiffness matrix K. To show this let us consider a slightly more complex example (Figure 6.4), though still keeping things in one dimension and thus one degree of freedom.

Figure 6.4: Two-bar element

We will also set the stiffness of each element to be the same: EA/L.

The stiffness at each node comes from the combined stiffnesses cf the connecting elements. This means the stiffness at node 1 is the stiffness of element from 1 to 2, at node 3 the element from 3 to 2, and at node 2 it is the combination of the stiffness of both elements. We can combine the element stiffnesses (defined previously) with these nodal stiffnesses to make the model stiffness matrix:

$$\begin{bmatrix} k_{e1} & k_{e1} & 0 \\ k_{e1} & k_{e1}+k_{e2} & k_{e2} \\ 0 & k_{e2} & k_{e2} \end{bmatrix} \begin{Bmatrix} u_1 \\ u_2 \\ u_3 \end{Bmatrix} = \begin{Bmatrix} f_1 \\ f_2 \\ f_3 \end{Bmatrix}$$

Substituting in the stiffness k and force f values, and extracting the common denominator[†] for clarity, gives us:

$$\frac{AE}{L}\begin{bmatrix} 1 & -1 & 0 \\ -1 & 2 & -1 \\ 0 & -1 & 1 \end{bmatrix} \begin{Bmatrix} u_1 \\ u_2 \\ u_3 \end{Bmatrix} = \begin{Bmatrix} 0 \\ 0 \\ P \end{Bmatrix}$$

We cannot solve this yet because we have not taken the boundary conditions (the restraints) into account. In mathematical terms we can see that this will not solve because the stiffness matrix is singular (a matrix that we cannot invert), which we know because its determinate is zero[††]. One-dimensional problems like this need a minimum of one restraint, two-dimensional problems need at least three restraints (two to prevent translation and one to prevent rotation of the body), and three-dimensional problems need six or more (three translation directions and three rotation).

In this example, the boundary condition is that node 1 is fixed in position and thus $u_1 = 0$. As we know that multiplying the first column by zero will give zero, we can remove it from the equation. As a result, we also need to remove the first row of the matrices. This gives us the equation with the reduced stiffness matrix, which now has a non-zero determinate:

$$\frac{AE}{L}\begin{bmatrix} 2 & -1 \\ -1 & 1 \end{bmatrix} \begin{Bmatrix} u_2 \\ u_3 \end{Bmatrix} = \begin{Bmatrix} 0 \\ P \end{Bmatrix}$$

$$K_r u_r = f_r$$

While finding the determinate and inverse of larger matrices is complicated (Appendix C), for a 2×2 matrix

$$\begin{bmatrix} a & b \\ c & d \end{bmatrix}$$

[†] In a real structure the stiffness for each element will be unique, so the FEA programs will not extract the common denominator but instead maintain the numerical values in the matrix.

[††] The determinate is a function of the matrix (Appendix C).

the inverse is:

$$\frac{1}{ad - bc} \begin{bmatrix} d & -b \\ -c & a \end{bmatrix}$$

The inverse of our reduced stiffness matrix is:

$$K_r^{-1} = \frac{L}{AE} \begin{bmatrix} 1 & 1 \\ 1 & 2 \end{bmatrix}$$

Multiplying both sides of our model equation gives us:

$$K_r^{-1} K_r u_r = K_r^{-1} f_r$$

$$u_r = K_r^{-1} f_r$$

$$\begin{Bmatrix} u_2 \\ u_3 \end{Bmatrix} = \frac{L}{AE} \begin{bmatrix} 1 & 1 \\ 1 & 2 \end{bmatrix} \begin{Bmatrix} 0 \\ P \end{Bmatrix}$$

Multiplying this out tells us that the displacements are:

$$u_2 = \frac{PL}{AE}$$

$$u_3 = \frac{2PL}{AE}$$

That the displacement of node 3 is twice that of the middle node should come as no surprise but is a welcome sign that we are on the right track.

Now that we know the three nodal displacements, we can substitute them back into the original equation to find the external forces, which are the reactions and the applied forces. Recalculating the applied forces will be a good check to ensure that we have our sums correct.

$$\frac{AE}{L} \begin{bmatrix} 1 & -1 & 0 \\ -1 & 2 & -1 \\ 0 & -1 & 1 \end{bmatrix} \begin{Bmatrix} 0 \\ PL/AE \\ 2PL/AE \end{Bmatrix} = \begin{Bmatrix} F_1 \\ F_2 \\ F_3 \end{Bmatrix}$$

We can cancel out AE/L

$$\begin{bmatrix} 1 & -1 & 0 \\ -1 & 2 & -1 \\ 0 & -1 & 1 \end{bmatrix} \begin{Bmatrix} 0 \\ P \\ 2P \end{Bmatrix} = \begin{Bmatrix} F_1 \\ F_2 \\ F_3 \end{Bmatrix}$$

Calculate the forces at node 1, which is our reaction:

$$F_1 = (1 \times 0) + (-1 \times P) = -P$$

Node 2:

$$F_2 = (0 \times 0) + (2 \times P) + (-1 \times 2P) = 0$$

Node 3:

$$F_3 = (0 \times 0) + (-1 \times P) + (1 \times 2P) = P$$

This is our original value and thus confirms that we have not made a mistake. In larger matrices with elements at angles, the values can be subject to rounding errors so the applied loads and calculated loads are unlikely to be exactly the same, though they should be very similar.

We can now calculate the forces in the elements from the relative nodal displacements. The strain in the first element is:

$$\varepsilon_1 = \frac{u_2 - u_1}{L}$$

$$\varepsilon_1 = \frac{PL/AE - 0}{L} = \frac{P}{AE}$$

The force in an element is:

$$F = A\sigma = AE\varepsilon$$

So, the force in the first element is:

$$F_1 = AE\frac{P}{AE} = P$$

Which is what we would expect. Repeating the calculation on the second element will produce the same value.

6.2.1.2 Bars in two-dimensional models

While bar elements have only one degree of freedom, they can exist in a two- or three-dimensional model. This means that for models to be useful we need to consider the bars' stiffness in two and three directions (Figure 6.5).

Figure 6.5: Bar at angle α

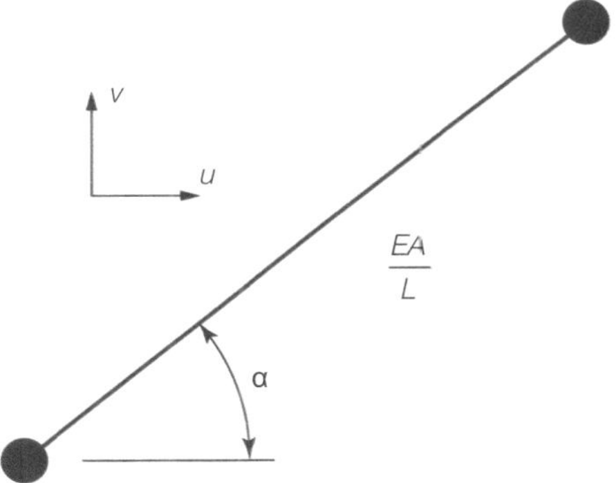

To do this we need to transform that stiffness into the two (or three) global directions. If the element's angle to the plane is α then the two-dimensional stiffness matrix for the bar element is:

$$[K_e]\{u_e\} = \frac{AE}{L}\begin{bmatrix} \cos^2\alpha & \cos\alpha\sin\alpha & -\cos^2\alpha & -\cos\alpha\sin\alpha \\ \cos\alpha\sin\alpha & \sin^2\alpha & -\cos\alpha\sin\alpha & -s n^2\alpha \\ -\cos^2\alpha & -\cos\alpha\sin\alpha & \cos^2\alpha & \cos\alpha\sin\alpha \\ -\cos\alpha\sin\alpha & -\sin^2\alpha & \cos\alpha\sin\alpha & \sin^2\alpha \end{bmatrix}\begin{Bmatrix} u_{1x} \\ u_{1y} \\ u_{2x} \\ u_{2y} \end{Bmatrix}$$

If we abbreviate $\cos\alpha$ as c and $\sin\alpha$ as s we can write it in a shortened form:

$$[K_e] = \frac{AE}{L}\begin{bmatrix} c^2 & cs & -c^2 & -cs \\ cs & s^2 & -cs & -s^2 \\ -c^2 & -cs & c^2 & cs \\ -cs & -s^2 & cs & s^2 \end{bmatrix}$$

While I am not going to give you the derivation of this matrix[†] you will find that if you set the angle α to 0° or 90°, you will get the one-dimensional bar stiffness matrix here, though with some zero rows and columns. This is to be expected as the bar elements have no stiffness in the orthogonal direction.

The principles are the same for other element types, though their stiffness matrices are far more complex, not just taking bending and shear into account, but also accommodating the additional nodes on some element types.

To demonstrate the method in two-dimensions, let's carry out a similar exercise using Figure 6.6.

Figure 6.6: Two-bar truss

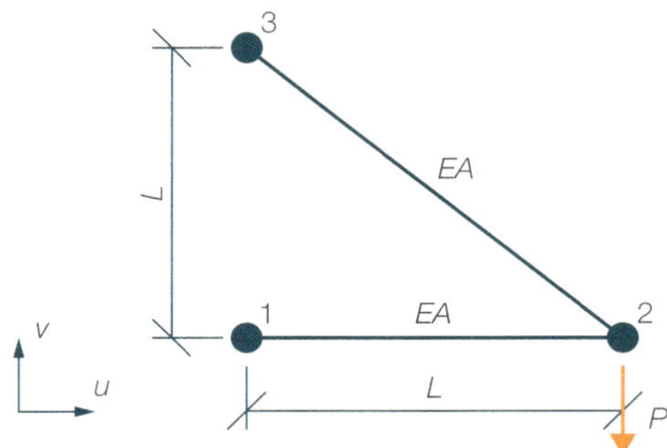

First determine the stiffness matrices:

$$K_{12}u = \frac{AE}{L}\begin{bmatrix} 1 & 0 & -1 & 0 \\ 0 & 0 & 0 & 0 \\ -1 & 0 & 1 & 0 \\ 0 & 0 & 0 & 0 \end{bmatrix}\begin{Bmatrix} u_{1x} \\ u_{1y} \\ u_{2x} \\ u_{2y} \end{Bmatrix}$$

$$K_{23}u = \frac{AE}{L}\begin{bmatrix} 0.35 & -0.35 & -0.35 & 0.35 \\ -0.35 & 0.35 & 0.35 & -0.35 \\ -0.35 & 0.35 & 0.35 & -0.35 \\ 0.35 & -0.35 & -0.35 & 0.35 \end{bmatrix}\begin{Bmatrix} u_{2x} \\ u_{2y} \\ u_{3x} \\ u_{3y} \end{Bmatrix}$$

The second matrix values take both the element's angle and extra length into account.

Assemble these element stiffness matrices into the global stiffness matrix $[K]\{u\} = \{f\}$:

$$\frac{AE}{L}\begin{bmatrix} 1.0 & 0 & -1.0 & 0 & 0 & 0 \\ 0 & 0 & 0 & 0 & 0 & 0 \\ -1.0 & 0 & 1.35 & -0.35 & -0.35 & 0.35 \\ 0 & 0 & -0.35 & 0.35 & 0.35 & -0.35 \\ 0 & 0 & -0.35 & 0.35 & 0.35 & -0.35 \\ 0 & 0 & 0.35 & -0.35 & -0.35 & 0.35 \end{bmatrix}\begin{Bmatrix} u_{1x} \\ u_{1y} \\ u_{2x} \\ u_{2y} \\ u_{3x} \\ u_{3y} \end{Bmatrix} = \begin{Bmatrix} f_{1x} \\ f_{1y} \\ f_{2x} \\ f_{2y} \\ f_{3x} \\ f_{3y} \end{Bmatrix}$$

[†]   Feel free if you are up for the challenge.

Reduce the matrix by applying the boundary conditions at nodes 1 and 3 and adding the external forces. This reduces the equation to:

$$\frac{AE}{L}\begin{bmatrix} 1.35 & -0.35 \\ -0.35 & 0.35 \end{bmatrix}\begin{Bmatrix} u_{2x} \\ u_{2y} \end{Bmatrix} = \begin{Bmatrix} 0 \\ -P \end{Bmatrix}$$

Note this now only has two degrees of freedom: x and y translation for node 2.

Inverting the residual stiffness matrix gives us:

$$\begin{Bmatrix} u_{2x} \\ u_{2y} \end{Bmatrix} = \frac{L}{AE}\begin{bmatrix} 1.0 & 1.0 \\ 1.0 & 3.83 \end{bmatrix}\begin{Bmatrix} 0 \\ -P \end{Bmatrix}$$

$$u_{2x} = \frac{-PL}{AE}$$

$$u_{2y} = \frac{-3.83PL}{AE}$$

Substitute the displacements back into the primary equation and cancel out AE/L to find the external forces:

$$\begin{bmatrix} 1.0 & 0 & -1.0 & 0 & 0 & 0 \\ 0 & 0 & 0 & 0 & 0 & 0 \\ -1.0 & 0 & 1.35 & -0.35 & -0.35 & 0.35 \\ 0 & 0 & -0.35 & 0.35 & 0.35 & -0.35 \\ 0 & 0 & -0.35 & 0.35 & 0.35 & -0.35 \\ 0 & 0 & 0.35 & -0.35 & -0.35 & 0.35 \end{bmatrix}\begin{Bmatrix} 0 \\ 0 \\ -P \\ -3.83P \\ 0 \\ 0 \end{Bmatrix} = \begin{Bmatrix} f_{1x} \\ f_{1y} \\ f_{2x} \\ f_{2y} \\ f_{3x} \\ f_{3y} \end{Bmatrix}$$

$$\begin{Bmatrix} f_{1x} \\ f_{1y} \\ f_{2x} \\ f_{2y} \\ f_{3x} \\ f_{3y} \end{Bmatrix} = \begin{Bmatrix} P \\ 0 \\ 0 \\ -P \\ -P \\ P \end{Bmatrix}$$

The forces on node 2 match the known forces, plus the forces all sum to zero showing that we have static equilibrium. I will leave it to you to check the element forces.

So far, we have only dealt with axial load in bar elements. If we were to change the bars into beams we would need terms for lateral movement and rotation at each end to give shear and moment, so while the bar matrix is 2×2 for one-dimension, 4×4 for two-dimensions and 6×6 for three-dimensional models, the beam will need the rotation displacement terms, so needs a 12×12 in three-dimensional models.

While the detail and solution of finite element analysis is complicated, it is in essence quite simple:

$$Ku = f$$

6.3 P-delta analysis and the geometric stiffness matrix

We have seen that the stiffness matrix approach can deal very well with statically indeterminate linear structures by finding the nodal displacements that generate the element forces to keep every part of the structure in equilibrium, but this assumes that the equilibrium of the structure is based on the undeformed geometry. This approach is sufficiently correct if the resulting displacements are small. If, on the other hand, the loads produce large eccentricities, the results will be inaccurate and we need to use a more complex analysis.

Let us consider a cantilever with a parallel load P and perpendicular load F (Figure 6.7) If the displacements are small the axial load and vertical reaction will be P, the shear and lateral reaction will be F, and the bending moment $F \times L$.

Figure 6.7: P-delta effects

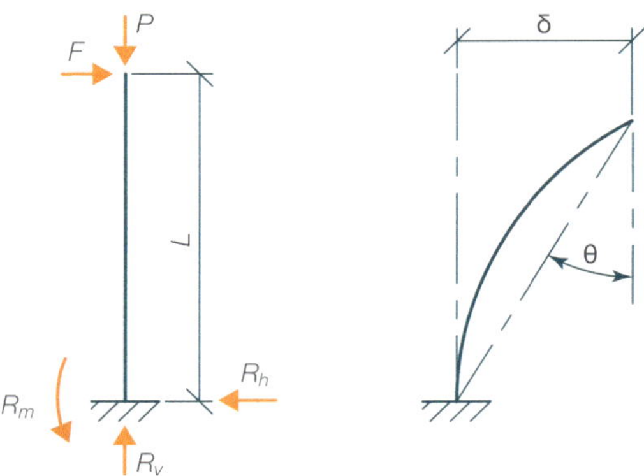

$R_v = P$

$R_h = F$

$R_m = FL$

But if the deflection δ is significant then the eccentricity of the load P will cause an addition moment on the base, equal to $P\delta$, hence the name of the method. If the angle is θ then the axial and shear loads will also change:

$R_v = P\cos\theta + F\sin\theta$

$R_h = F\cos\theta + P\sin\theta$

$R_m = FL + P\delta$

There is a useful convention where we represent all the stabilising stiffness in the elements by replacing the beams with bars and adding a spring (stiffness k) at the free nodes which matches their lateral stiffness (Figure 6.8).

Figure 6.8: Cantilever stiffness

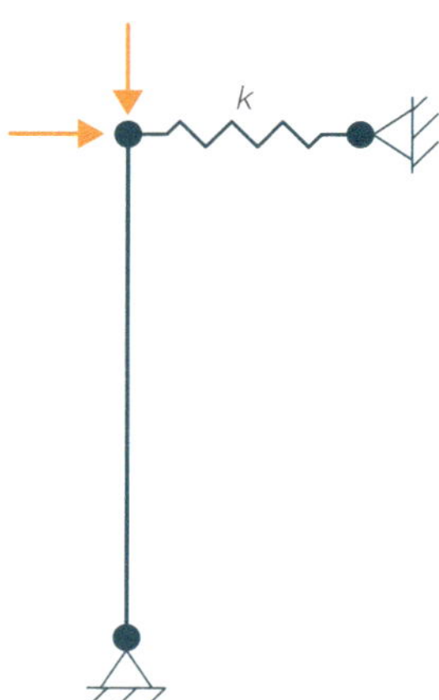

Now let us make use of this and consider another example, this time with two bar elements. If there is no spring then the system is unstable, but with a spring stiffness it can resist the lateral load, again giving a deflection δ (Figure 6.9).

Figure 6.9: Beam stiffness

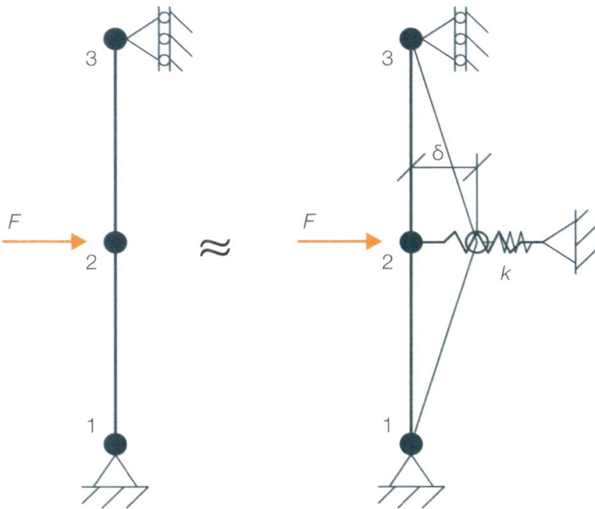

$$\delta = F/k \text{ or } F = k\delta$$

If we add a vertical load P to node 3 it will resist the deflection at node 2 by inducing tension in the elements, but the resistance will be in proportion to the deflection at node 2 (Figure 6.10).

Figure 6.10: Tension stiffening

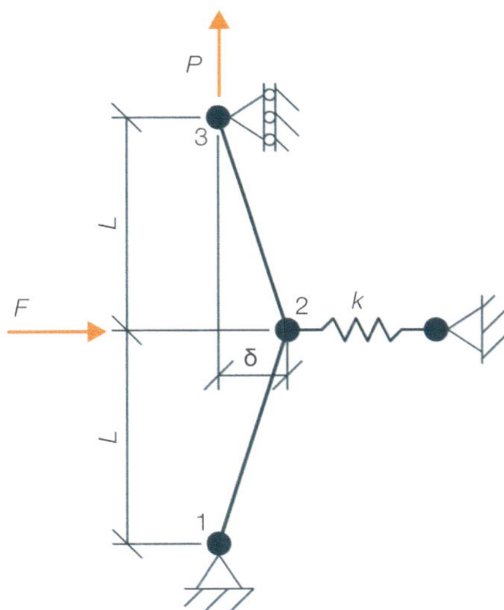

The horizontal force resisting F from both bar elements induced by the load P is $2P\delta/L$ so combining them with the system stiffness k gives:

$$F - \frac{2P}{L}\delta = k\delta$$

Rearrange to give:

$$F = \left(k + \frac{2P}{L}\right)\delta$$

The load P has added stiffness to the structure, which is a factor of the force and geometry. This is known as the 'geometric stiffness', or K_g for short.

From these equations, the stiffness matrix of a horizontal bar element is:

$$[K]\{u\} = \frac{AE}{L} \begin{bmatrix} 1 & 0 & -1 & 0 \\ 0 & 0 & 0 & 0 \\ -1 & 0 & 1 & 0 \\ 0 & 0 & 0 & 0 \end{bmatrix} \begin{Bmatrix} u_{x1} \\ u_{y1} \\ u_{x2} \\ u_{y2} \end{Bmatrix}$$

The geometric stiffness matrix of the same bar element is:

$$[K_g]\{u\} = \frac{P}{L} \begin{bmatrix} 0 & 0 & 0 & 0 \\ 0 & 1 & 0 & -1 \\ 0 & 0 & 0 & 0 \\ 0 & -1 & 0 & 1 \end{bmatrix} \begin{Bmatrix} u_{x1} \\ u_{y1} \\ u_{x2} \\ u_{y2} \end{Bmatrix}$$

As required, the geometric stiffness adds stiffness perpendicular to the element in proportion to the element tension and length. A compression load will reduce the stiffness of the element.

Taking the geometric stiffness into account, the equation we now need to solve is:

$$\big[[K] + [K_g]\big]\{u\} = \{f\}$$

The first step is to find the element forces by running a standard linear finite element analysis $[K]\{u\} = \{f\}$ and use the results to create the geometric stiffness matrix $[K_g]$. The stiffness and geometric stiffness matrices are added together and used to find the new displacements and forces.

One of the points to note here is that while we can multiply and add results from a linear analysis (e.g. to find ultimate limit state forces and moments), it is incorrect to do this for P-delta analysis as the results are nonlinear.

6.4 Buckling analysis and the geometric stiffness matrix

We have all, I am sure, picked up a long thin rod or bamboo pole and noted that a little compression causes it to bow outwards (or 'buckle' to use the technical term). We will also have noted that we do not see the same effect if the rod is short or thick. This is an effect that can occur at any scale on any part of a structure, whether a beam flange, a column, a truss or even an entire tall building. In 1787 Leonhard Euler found that the relationship between the length, material and cross-section of the column, and the load needed to buckle it was:

$$P_{crit} = \frac{\pi^2 EI}{(nL)^2}$$

Where:

P_{crit} = critical axial load
E = Young's modulus
I = second moment of area
L = length of column
n = effective length factor based on the end conditions
(1 for pin-ended, 0.5 for fixed ends, etc.)

To avoid confusion with other uses of n in the equations, we can refer to nL as the effective length L_{eff}:

$$P_{crit} = \frac{\pi^2 EI}{L_{eff}^2}$$

This approach is suitable for hand calculations on a single strut but is less so when we have a large finite element model. What we can do instead is extend the P-delta analysis and find the critical loads that cause the geometric stiffness to cancel out the element stiffness.

Let us revisit the column that we looked at earlier, but this time reverse the load, giving us an axial compression load of $-P$ (Figure 6.11).

Figure 6.11: Compression buckling

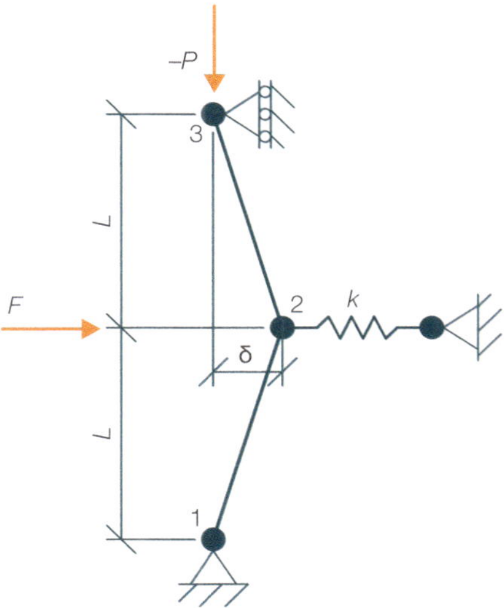

In the previous example, the axial tension load increased the effective stiffness, now it is reducing it:

$$k - \frac{2P}{L}$$

There will be a critical axial load P where:

$$k - \frac{2P}{L} = 0$$

Which is where the axial load reduces the system stiffness to zero:

$$k + \frac{2P_{crit}}{L} = 0$$

$$P_{crit} = \frac{-kL}{2}$$

At this point we can say that $P = P_{crit}$ and the system is unstable. You may also notice that the P_{crit} is independent of the lateral load F[†] and, as the stiffness is zero, the deflection will be, at least in theory, infinite.

If the lateral load was zero, would the structure still buckle? While a pair of perfectly straight bar elements would be in equilibrium, it would be unstable as the slightest eccentricity or imperfection would cause the middle node to buckle out. Even if our models are perfect[††], real structures always include imperfections, whether from the factory manufacture or site erection tolerances.

[†]   Or is it?
[††] Some programs can include imperfections automatically, whether based on a height, a formula, or randomly within tolerances. You can also sometimes update the geometry in line with the linear buckling analysis prior to running a nonlinear analysis that increments the load to failure.

For any given loading there will be several multipliers on the load that could cause the structure to buckle. Even a single column can buckle several ways (Figure 6.12).

Figure 6.12: Buckling modes

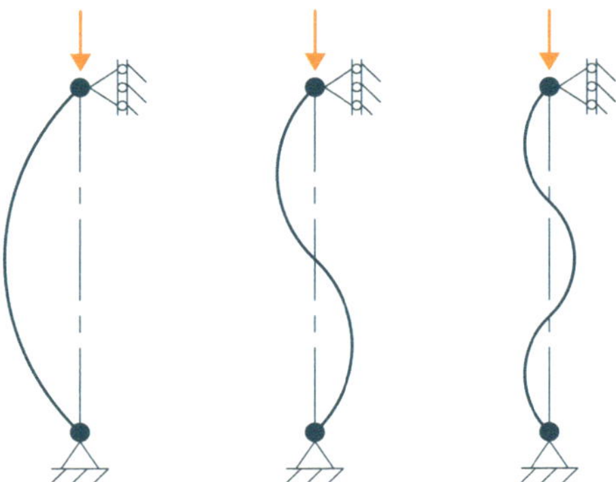

Which buckling mode occurs first will depend on the restraints on the column, and thus which has the lowest critical load, but all are mathematically possible. While finding the critical load on a column by using the Euler formula is simple, how do we find them on a FEA model? Real models have not one load but a whole series of load cases, multiplied and added together.

Applying the critical axial load would prevent a P-delta analysis from solving, as the stiffness and geometric stiffnesses cancel out, leaving us with an uninvertible matrix. This is not a problem here as we can use this knowledge to find these critical load values. We know there will be some multiplier on the initial loads that will cause some parts of the model to lose stiffness:

$$P_{crit} = \lambda_{crit} P$$

Or in matrix terms the overall stiffness matrix K will be equal and opposite to the geometric stiffness matrix K_g (which is based on the applied loads) when multiplied by some unknown factor λ and when the deflected shape $\{u\}$ is the buckling mode for that load:

$$[K]\{u\} + \lambda[K_g]\{u\} = 0$$

$$[[K] + \lambda[K_g]]\{u\} = 0$$

Some of you may recognise this formula as being in the form of an eigenvalue and eigenvector; we can use eigensolver analysis to find our missing values.

6.4.1 Eigenvector analysis

First some background to eigenvectors and eigenvalues. If you multiply a vector (a column matrix) by a matrix then you get a new vector. In other words, the matrix transforms the vector into a new vector (Figure 6.13).

Figure 6.13: Matrix transformation

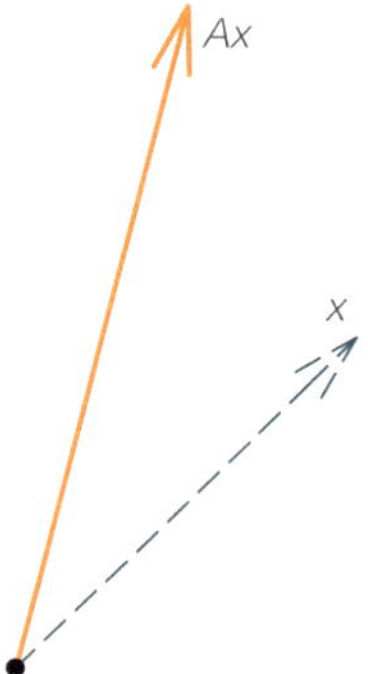

Most transformed vectors point in a different direction to the original one, but some stay pointing in the same direction, with the only difference being a change in their length. The new vector might be longer or shorter than the original, or even point in the opposite (negative) direction.

Figure 6.14: Various eigenvectors

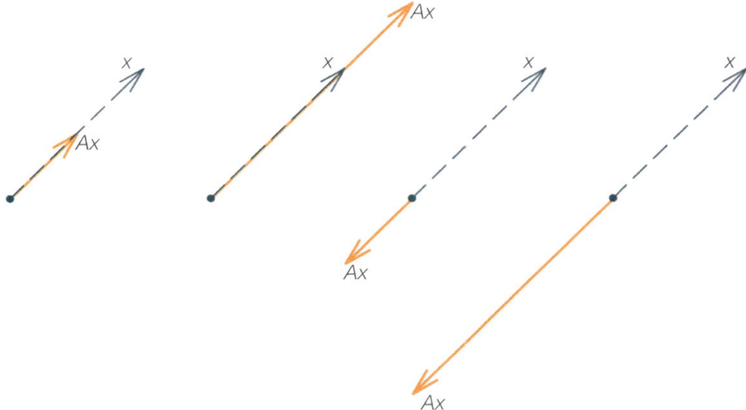

There are only a few of these vectors associated with any matrix, no more than the number of rows in the vector or matrix. As these vectors are particular to the matrix, we call them eigenvectors, from the German word 'eigen', meaning particular or inherent.

As each of these vectors is still pointing in the same direction but with a change of scale, the action of multiplying the vector by the matrix is the same as multiplying the vector by a scalar (i.e. a number). In matrix notation we would write this as:

$[A]\{x\} = \lambda\{x\}$

Or rearranged:

$[[A] - \lambda[I]]\{x\} = 0$

Where $[I]$ is the identity matrix. Multiplying λ and $[I]$ will give a matrix with λ on the diagonal and 0 everywhere else.

Note that there will be just one scalar (an eigenvalue), for each of these eigenvectors, but several eigenvectors might have the same eigenvalue.

One solution to this equation is when the vector $\{x\}$ is all zeros, but that is not only an uninteresting solution, it also breaks the rule about there being only one eigenvalue for each eigenvector. In general, it is difficult to find the other solutions directly, but there are various techniques to estimate their values with enough accuracy for engineering (Appendix C).

The maximum possible number of eigenvectors on our analysis model will be equal to the number of degrees of freedom of that model, which is typically the number of unrestrained nodes multiplied by six for a three-dimensional model.

In a two-dimensional bar example (Figure 6.15) there are 18 degrees of freedom, though there are likely to be fewer eigenvectors. Alternatively, if we used beam elements there would be more eigenvectors as the nodal rotational degrees of freedom would come into play. The eigenvalues associated with the force Fx will, in this instance, mostly be negative as the load is putting the structure into tension. Fx would therefore only buckle this truss if it were reversed; if the load cannot reverse (gravity loads for example) then you can ignore these modes.

Figure 6.15: Bar truss

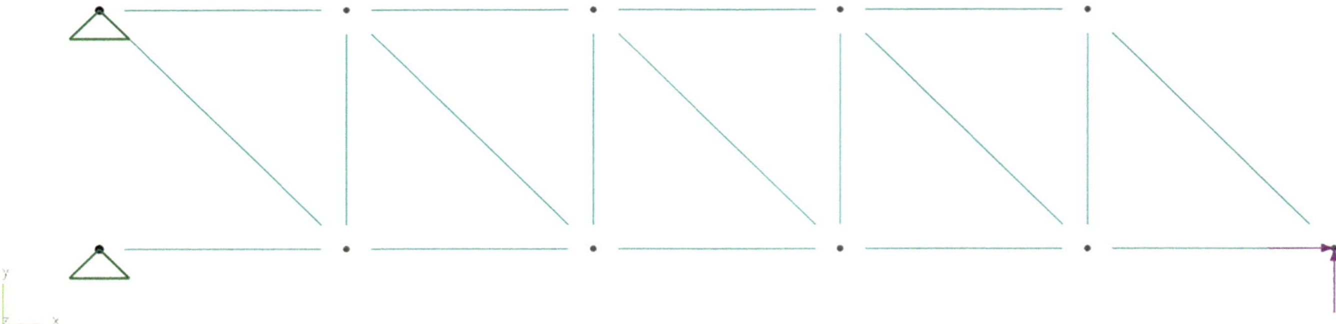

6.4.2 Using buckling load analysis
6.4.2.1 Effective lengths
The formulas used in the design of members against buckling usually depend on the effective length, but modal analyses do not give us that value, only the load at which those parts of the structure buckle. If we revisit Euler's buckling equation, we can rearrange it to give us the effective length factor:

$$P_{crit} = \frac{\pi^2 EI}{(nL)^2} \Rightarrow n = \sqrt{\frac{\pi^2 EI}{P_{crit}L^2}}$$

One thing to watch out for though: while the length of some members, such as columns, are obvious, it is not always easy to work out what the 'lengths' of some parts of the structure are. One option is to look at the bending moments generated by the buckling analysis, as you can use these to find the points of contraflexure for each buckling mode.

A useful application of this calculation is for determining the effective lengths for member design. The standard lengths used by design routines are based on standard structural arrangements, but do not work well for particular cases, such as grid shells and arches for example. They are also difficult to apply to columns in a moment frame, where the effective lengths are dependent on the stiffnesses of the connected members. The 'automatic' lengths may underestimate the effective lengths and underdesign your structure as a result.

6.4.2.2 Load magnification
It is important to note that the λ_{crit} value is not a factor of safety against buckling[74,75]. All structures have the effects of buckling in them to a greater or lesser extent, which affects their ability to carry load. Buckling does this by reducing the structural stiffness and hence amplifying the effects of the other loads.

If we look back at the two-bar column (Fig. 6.11), we can say that the total lateral deflection (δ_t) at the centre point is a combination of the deflection from the lateral load (δ) plus the initial imperfection (δ_0). Starting from the equilibrium equation:

$$F = k\delta + \frac{2P}{L}(\delta + \delta_0)$$

And remembering that:

$$\lambda_{crit} = \frac{P_{crit}}{P} = \frac{-kL}{2P}$$

We can rearrange (Appendix D) to get:

$$\delta = \frac{\lambda_{crit}}{\lambda_{crit} - 1}\frac{F}{k} + \frac{1}{\lambda_{crit} - 1}\delta_0$$

This means that as $\lambda_{crit} \Rightarrow 1$ it magnifies the effects of the deflection from lateral loads and the initial imperfections become more significant on the result (Table 6.1). If the results are to be accurate to within $\pm 10\%$ we need to take buckling effects into account for any λ_{crit} less than 10.

Table 6.1: Buckling load factor magnification

P/P_{crit}	λ_{crit}	F amplification	δ_0 amplification
0	∞	1.0	0
0.1	10	1.1	0.11
0.2	5.0	1.3	0.13
0.3	3.3	1.5	0.14
0.4	2.5	1.7	0.17
0.5	2.0	2.0	0.20
0.6	1.7	2.5	0.25
0.7	1.4	3.3	0.33
0.8	1.3	5.0	0.50
0.9	1.1	10	1.00
1.0	1.0	∞	∞

The magnification due to λ affects the lateral loads and is nonlinear, meaning we cannot analyse each load case individually and add them together afterwards. Instead, we must conduct the buckling analysis using factored load combinations in each buckling analysis. Also, while buckling magnifies the effects of other loads and imperfections, it only does it in those regions that are starting to buckle[†].

There are a number of ways to make use of the buckling analysis results. One is to assume that the initial imperfections are a scaled version of the buckling mode and then run a P-delta or nonlinear analysis. Another is to scale the bending moments from the buckling analysis and add them to the static results. A third approach is to reduce the target utilisation in the section design. In all cases the scaling needs careful engineering judgement led by design codes and guides.

Whatever approach you take, a λ_{crit} value of <10 means you must consider the effects of buckling on the stability of the whole structure and on the individual components. Interpreting which elements in a structural model are contributing most to the buckling mode can be tricky. When reviewing the buckling results look at which elements are deflecting the most, but also look at the elements with the most strain energy density[††].

As mentioned previously, the bending moment diagrams can give you a good indication of what you can take the member effective lengths to be. Alternatively, you can take the buckling capacity from the analysis and back-calculate the effective length value using Euler's formula.

[†] This means that we cannot simply multiply the lateral loads by the factor in Table 6.1 as that would be unrealistic.
[††] Note that you can also use Strain Energy Density to see which elements in a truss contribute most to its deflection (Appendix B).

6.5 Dynamic analysis and the mass matrix

6.5.1 Introduction

Dynamic analysis tends to involve lots of formulas, including differentials over time and eigenvectors, but all texts agree that a good understanding of dynamics is *fundamental*[†] to analysing and designing for dynamics. It is also traditional to draw a diagram of a spring and mass (Figure 6.16).

Figure 6.16: One degree of freedom system

In my experience the best way for an engineer to learn about dynamics is to learn to play a musical instrument (preferably a stringed instrument). I play the guitar and bass, and find they supply useful insight. Other options are available, but I will use the bass guitar to illustrate the principles of dynamic analysis.

Picture an acoustic bass guitar (Figure 6.17). It has strings mounted on a large, hollow wooden body, and a neck which is a wooden cantilever. At the top of the neck the strings are attached to pegs that allow us to tune the strings. Strumming or plucking the strings causes them to produce sound, hopefully of a musical nature.

Figure 6.17: Bass guitar components

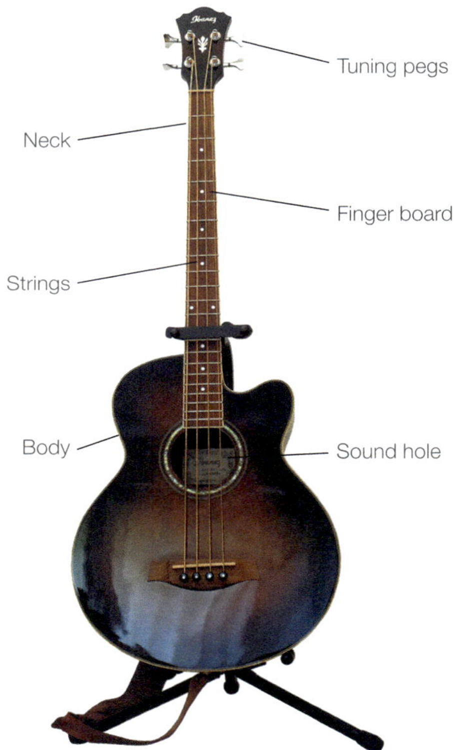

Note that the strings have different thicknesses. The thinner ones produce higher notes and the thicker strings produce lower notes. If the notes are not correct, we can adjust them using the tuning pegs. Turning them one way pulls on the string to add tension, causing the note to rise, and the other way reduces both the tension and note. As it is not practical to play tunes on a guitar by adjusting the string tension, we can also change the note by pressing the string down onto the fingerboard. This makes the string shorter and, consequently, the note higher. You may also notice if you remove a string, the string has negligible stiffness on its own, until you tension it on the instrument.

[†] Excuse the pun, dynamic analysis tends to attract a lot of word play, including the classic "you can tune a guitar, but you can't tuna a fish". I did not say that the jokes were any good…

When you strum a guitar the strings themselves are not loud, which you will find if you strum an unplugged electric guitar. This means that the string vibrations need amplifying, which the electric guitar does using magnetic 'pickups'. The acoustic guitar amplifies the sound because the string vibration causes the guitar front, which is large but made from thin wood, to resonate. If we want to make the music louder we must hit the strings harder.

What can we learn from this in engineering terms? Firstly, we must change some of the terminology: while musicians talk about notes, we talk about frequencies. They are the same thing: for example, the note A above Middle C is 440Hz, where Hz is the number of beats or oscillations per second. The structural vibrations that we are interested in are seen or felt rather than heard: floor structures typically vibrate at about 4–8Hz and we do not deal with much vibration above 20Hz[†].

If you strum or pluck the string in various places along its length, the quality of the note (or 'timbre') changes. It sounds bright and harsh if you play near the end of the string, dull in the middle, and well balanced when you strum over the soundhole. The changes in timbre occur because there are multiple frequencies playing simultaneously on the string. The fundamental frequency of the string is the named note. The other frequencies come from integer divisions of the string length: 1/2, 1/3, 1/4 and so on.

Each of these vibration modes (Figure 6.18) has fixed points called 'nodes', which are not to be confused with the nodes in our analysis models.

Figure 6.18: Vibration modes

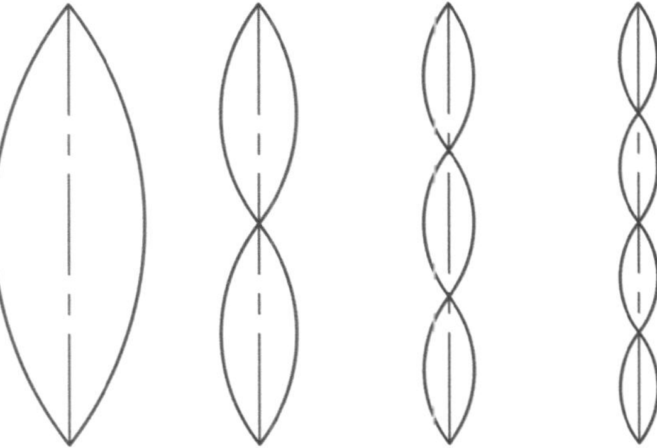

You can only excite these modes if you pluck the string away from these node positions, which is the reason a string plucked in the centre sounds dull: it loses exactly half of the higher modes.

The formula for the fundamental frequency of a string is:

$$f = \sqrt{\frac{T}{4mL^2}}$$

Where:

f = frequency (Hz)
T = tension (N)
m = mass per unit length (kg/m)
L = length (m)

When we compare beams to guitar strings there are some aspects that carry over: mass and length. Beams are not (generally) in tension, instead the stiffness comes from the material's Young's modulus multiplied by the section's second moment of area: its EI.

[†] 20Hz also happens to be the lower limit of human hearing.

The frequency of a beam is:

$$f = \frac{\pi}{2}\sqrt{\frac{EI}{wL^4}}$$

Where:

E = Young's modulus (N/m^2)
I = bending moment of area (m^4)
w = uniform load (kg)
L = length (m)

6.5.2 Modal dynamic analysis

Static analysis assumes static equilibrium, so:

$$f = ku$$

force = stiffness × deflection

But with dynamic analysis the structure is not static but moving. It is still in equilibrium, otherwise something has gone horribly wrong, but in dynamic equilibrium. Calling on Sir Isaac Newton and his legendary apple, he realised that:

$$f = ma$$

force = mass × acceleration

If our structure is in dynamic equilibrium, all the forces must again cancel out and so:

$$ku + ma = 0$$

In other words, mass × acceleration must be equal and opposite to stiffness × deflection at all points. Note also that because the forces have cancelled out, the equation of dynamic equilibrium is independent of any external forces.

While we know (or can calculate) the mass and stiffness of the structure, the deflections and accelerations are unknown. Luckily for us they are not independent. Remember that speed is the rate of change of distance over time, and referring back to Newton and, this time, to his calculus, we can write speed v as $\delta u/\delta t$ or $\dot{u}$. Likewise, acceleration is the rate of change of speed over time, so $a = \ddot{u}$. The acceleration is thus a function of the displacement, the mass, and time.

While there can be any number of displacements and thus accelerations that a structure can experience, what we are interested in are the vibrations that the structure prefers: its natural frequencies. So how do we find these frequencies?

Firstly, we are interested in constant ('steady-state') vibrations and these are sinusoidal. In other words, the lateral movement due to the vibration follows the same pattern as a point on a revolving wheel, which, if moving, would trace out a sine wave. This means that we can replace the frequency term f in our calculations with an angular frequency ω, which is $\omega = 2\pi f$. If the amplitude a of the vibration is the maximum displacement, the displacement u at any time is:

$$u = a\sin(\omega t)$$

Differentiate once to get the velocity ($v = \dot{u}$) and again to get the acceleration ($a = \ddot{u}$):

$$\dot{u} = \omega a\cos(\omega t)$$

$$\ddot{u} = -\omega^2 a\sin(\omega t) = -\omega^2 u$$

Our basic expression of the dynamic equilibrium:

$$ku + m\ddot{u} = 0$$

now becomes:

$$ku - \omega^2 mu = 0$$

Which is another eigenvalue problem. In this instance the eigenvalue λ is ω^2, which we can convert back into a normal frequency. The vector u is the eigenvector, which is the deflection of the structure at that frequency. This deflection is also known as 'modal shape'.

To sum up: as with buckling analysis, we can use eigenvalue analysis (that is, modal analysis) of the stiffness and masses to find the structural natural frequencies and mode shapes. In the same way that we have stiffness and geometric stiffness matrices, we also need a mass matrix[76], which for a bar in a two-dimensional model looks like:

$$[M_e] = \rho AL \begin{bmatrix} \frac{1}{2} & 0 \\ 0 & \frac{1}{2} \end{bmatrix}$$

This version of the mass matrix lumps the mass at the nodes, which will be sufficiently accurate if you have enough nodes along each member.

Previously, I mentioned that adding tension into a system will increase its natural frequency. We can account for this in our vibration analysis by adding the geometric stiffness to the structural stiffness in our modal analysis. You may see this called 'P-delta modal analysis'. We cannot however, combine modal vibration analysis with modal buckling, as a buckled structure has no stiffness and so cannot vibrate.

6.5.3 Dynamic response analyses
Once we know the natural frequencies of our structure (to follow the musical theme earlier: the notes) we can determine how it will react to dynamic loads (how loud will it get and will the strings break?).

6.5.3.1 Harmonic analysis
Structures are most vulnerable to dynamics when the frequency of the loads is close to or matches the natural frequencies of the structure. Tacoma Narrows had a low torsional stiffness, and one of its natural frequencies matched that of the vortex shedding or aeroelastic flutter[†] when the wind speed was 67km/h, causing it to collapse when the wind reached exactly that speed.

London's Millennium Bridge was checked for wind excitation and found to be ok, but on the opening days the large crowds dropped the lateral frequency to about 1Hz, making it susceptible to pedestrian-induced vibration. While vertical pedestrian-induced vibrations were a known risk, hence the instruction for marching troops to break step over some bridges, lateral induced vibrations had not previously been reported as a problem. Also, unlike vertical vibrations, lateral oscillations forced members of the public to walk in step, causing resonance and feedback that amplified the oscillations.

Harmonic analyses can model these situations by applying a force to nodes in the model and then varying them sinusoidally over time. This means that the force is effectively generated by something that is rotating, so like a washing machine with a heavy load attached to one side of the drum, or the pistons in a car engine (if the crank shaft was driving the engine and not the other way around). This varying force then induces accelerations, velocities and displacements in the structure, which can be large if the excitation frequencies coincide with the structural natural frequencies and if there is little damping to absorb the energy.

[†] There is still some debate on which it was.

The Millennium Bridge refurbishment used viscous dampers and tuned-mass dampers to attenuate the oscillations, but all structures have damping from friction in connections or concrete cracks, or from the contents such as furniture, demountable partitions, ceilings and people. The damping is quite low though, ranging from about 0.5–4% of critical damping, which is the amount of damping that brings the structure back to its starting place in the minimum time. The Millennium Bridge now has over 20% critical damping, which prevents the pedestrian vibration problem from returning.

6.5.3.2 Linear time-history analysis

Time-history analyses are similar to harmonic analyses in that they have loads that vary over time. While the harmonic load repeats with a given frequency, the time-history sets explicitly how the load varies with time. The time-history might be a recording of an earthquake, whether simulated or real (Figure 6.19), a stadium crowd jumping to their feet, or wind gusts on a facade.

Figure 6.19: Seismograph recording (El Centro earthquake, California, 1940)

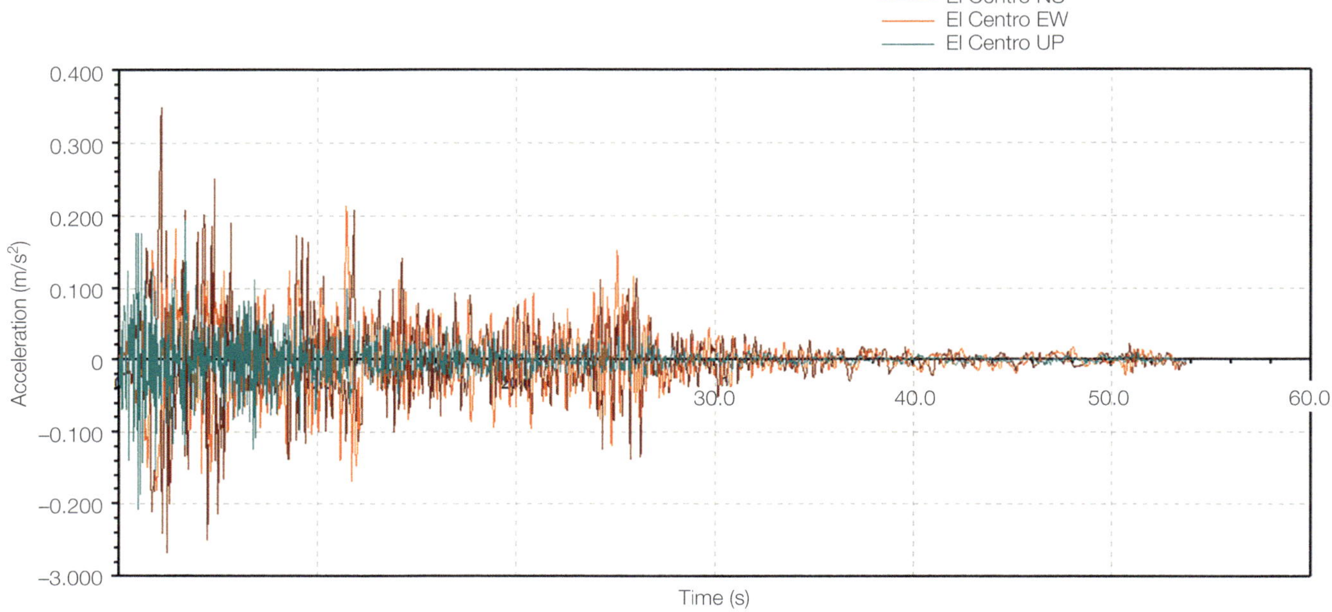

6.5.3.3 Response spectrum analysis

Time-histories can be recordings of specific earthquakes, but when designing a new structure, we want to see how it will respond to a range of future events. If we extract the response spectrums from the El Centro recordings[77], we get the curves shown in Figure 6.20.

If we were to draw an envelope around these curves (Figure 6.21) and compare these to the standard response spectrum in BS EN 1998-1[78] (Figure 6.22) you can see the similarity of the shape. The design response spectrum is an envelope of the historical and potential earthquakes the structure might experience.

Response spectrum analysis takes this curve and generates a series of excitation frequencies and intensities, applies them to the structural model using the modal dynamic analysis results, and outputs the maximum forces and moments experienced by the members.

Figure 6.20: Response spectrums (El Centro earthquake, California, 1940)

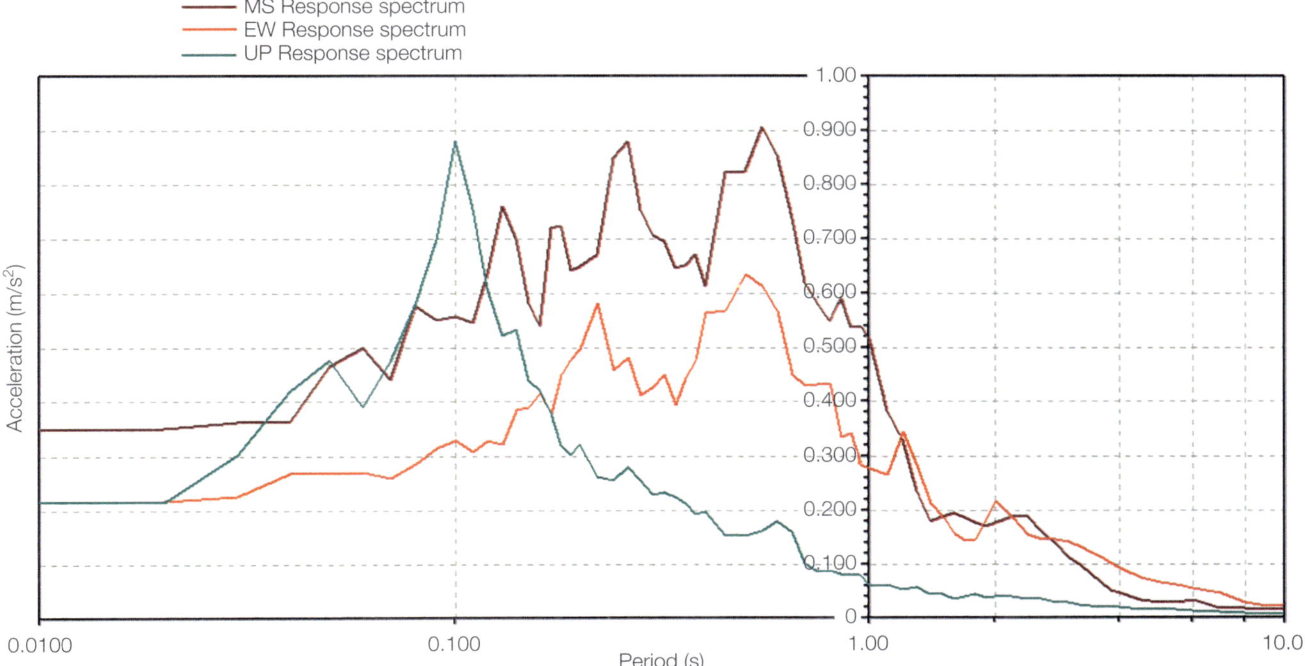

Figure 6.21: Response spectrums including envelope (El Centro earthquake, California, 1940)

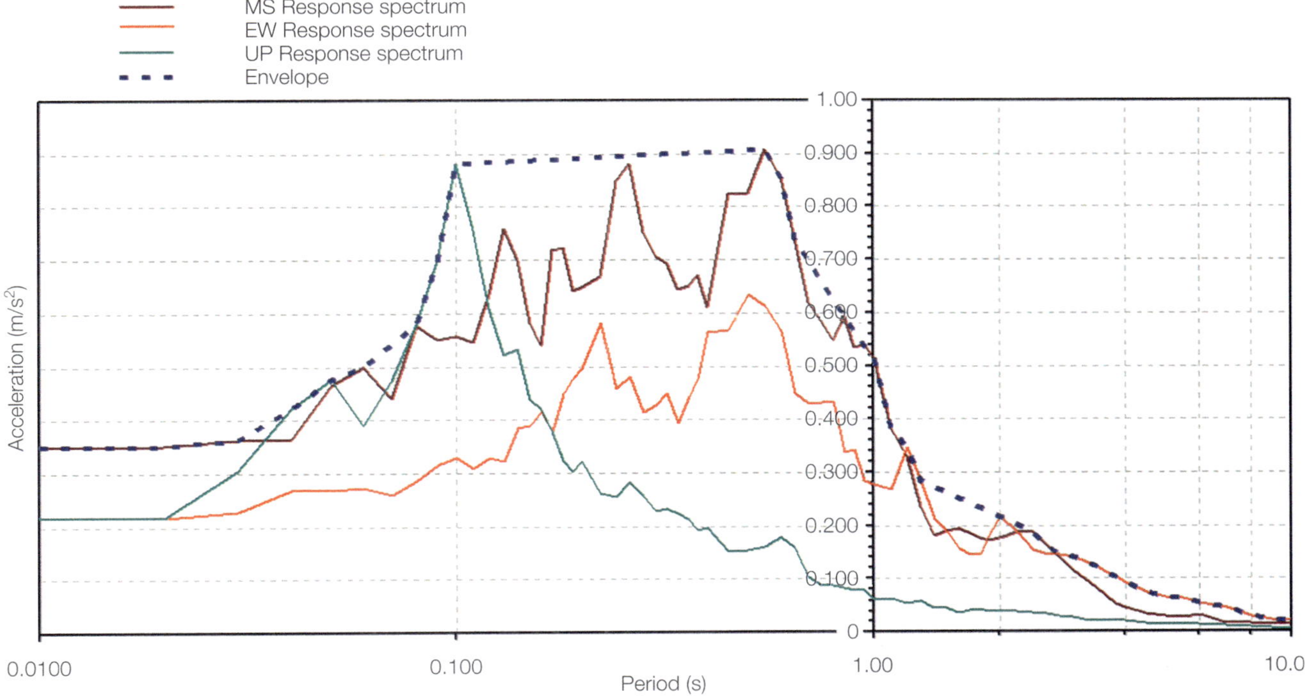

Figure 6.22: Shape of elastic response spectrum (derived/adapted from Figure 3.1 of BS EN 1998-1)

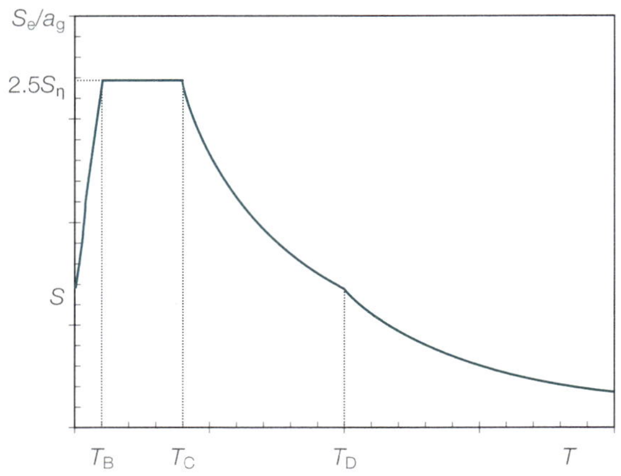

6.6 Nonlinear analysis (with and without matrices)

*"Students... learned that nonlinear systems were usually unsolvable, which was true,
and that they tended to be the exception — which was not true."*
James Gleick[79]

It is true to say that all structures are nonlinear, but also that for many of them the difference between linear and nonlinear is so small that we can ignore it. On the other hand, there are some structures where a linear analysis would produce grossly inaccurate results.

With a linear structural analysis, if we double the load, we also double the deflection, the stresses and the strains. Similarly, if we reverse the loads, we reverse the results likewise. This is not normally the case with nonlinear structures.

Nonlinearity in structures comes in three basic forms:

- Material
- Geometric
- Directional

Material nonlinearity means the stress-strain relationship is not straight. A familiar example is that of steel, which behaves linearly up to yield, at which point it shows large strains for little additional stress, before failing (Figure 6.23). Though this is often simplified to an elastic-plastic curve (Figure 6.24) some materials, e.g. stainless steel, do not have any linear behaviour in their stress-strain relationship (Figure 6.25).

With linear analysis we always assume that the section dimensions stay constant, but under compression an element will increase in area and reduce in tension; the difference in the axial and cross-sectional strains giving us the Poisson's ratio[†] (Figure 6.26). The difference is usually small enough to ignore, but you will have to address it if you work with particularly soft materials.

[†]   Fun fact: the Poisson's ratio of cork is effectively zero, making it ideal for pushing into wine bottles and then, more importantly, pulling out. It is only friction that resists this — a material with a positive Poisson's ratio would swell when pushed, making it much harder to insert or remove. Another fun fact: there are materials with negative Poisson's ratios, referred to as 'Auxetics'.

Figure 6.23: Steel stress-strain curve

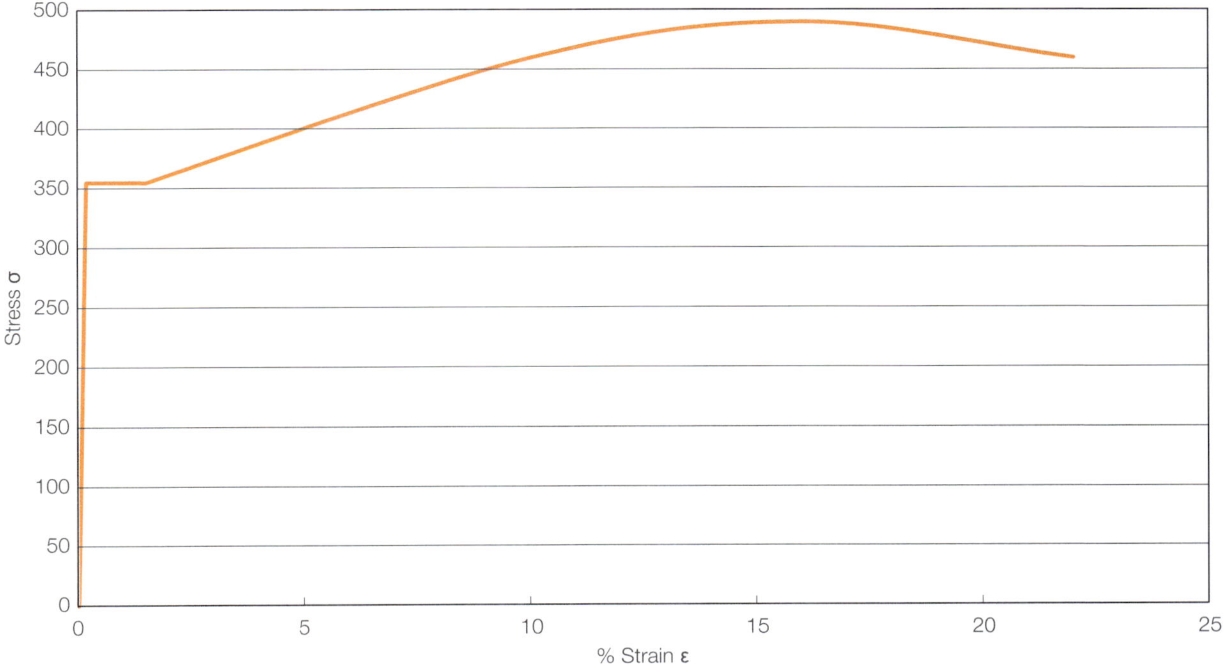

Figure 6.24: Elastic-plastic stress-strain curve

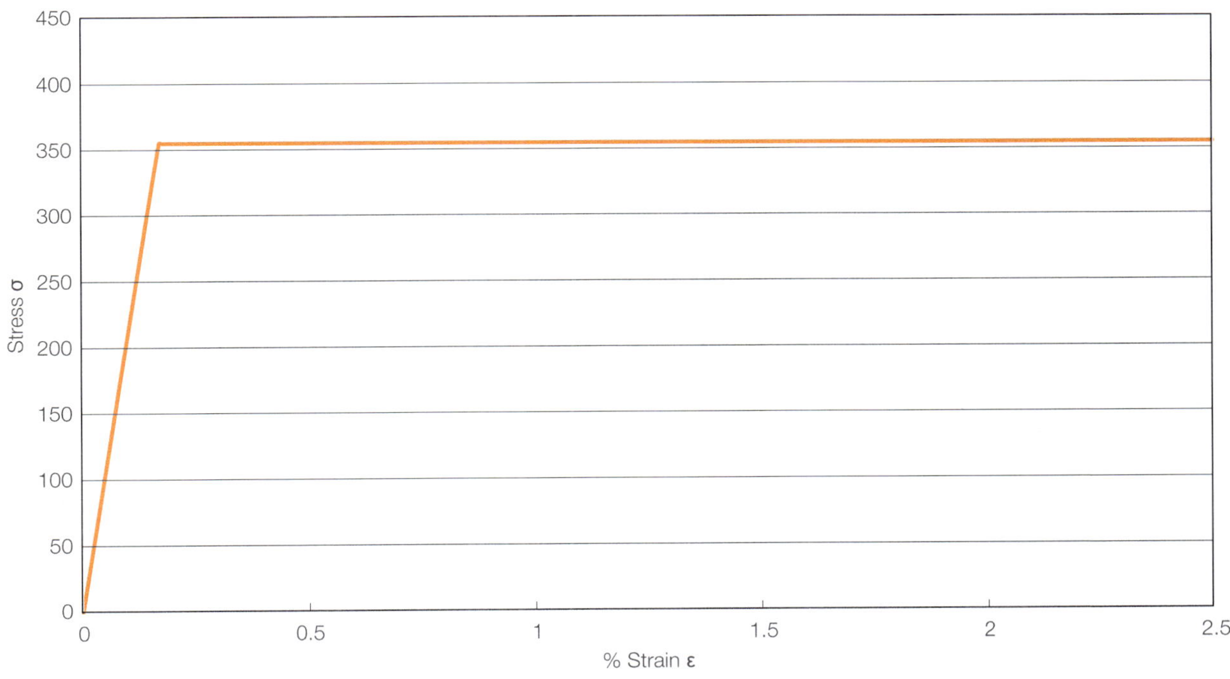

Figure 6.25: Stainless steel stress-strain curve

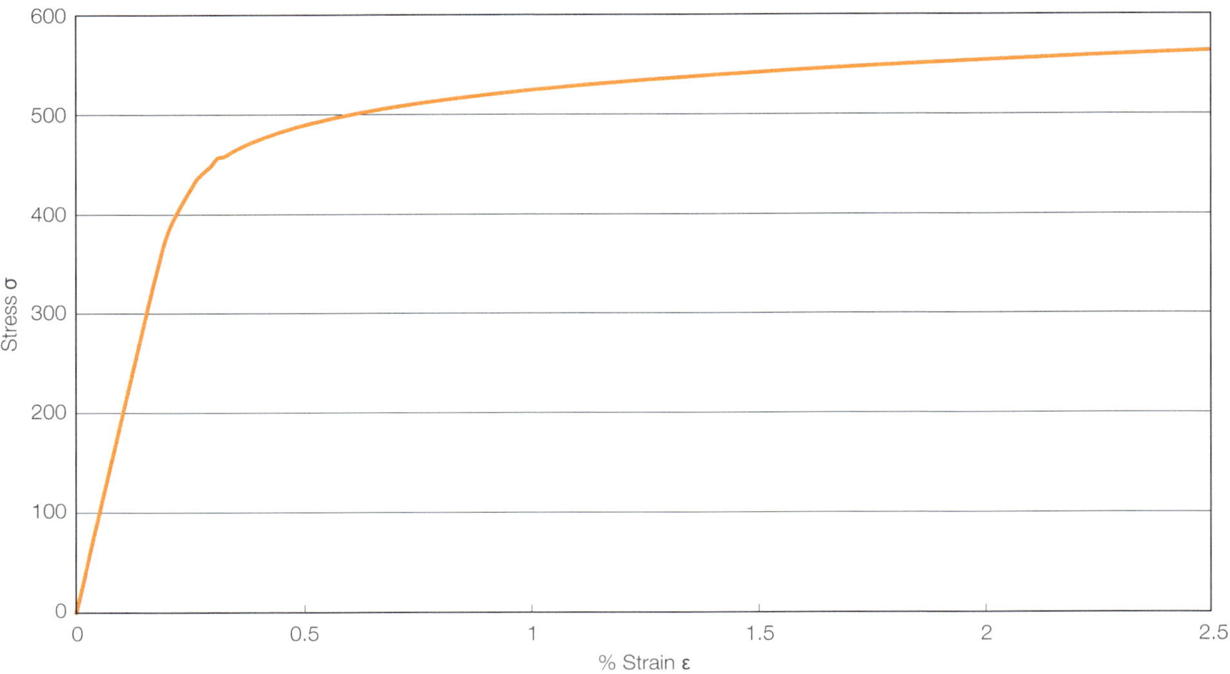

Figure 6.26: Poisson's ratio effects

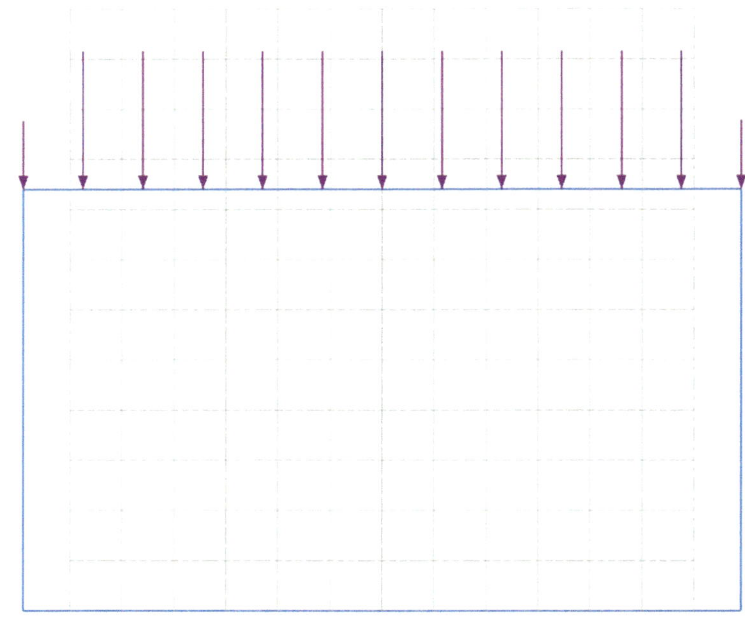

Geometric nonlinearity occurs when a structural response changes because of large deflections. P-delta analysis is a mild form of this, but also includes cable nets and fabric structures where they must deflect significantly to carry the load. Including geometric stiffness effects in the nonlinear analysis means that you might see buckling occurring in the model. If you increment the loads, you may also be able to extract load-deflection relationships from the analysis.

An example of directional nonlinearity is masonry, which is strong in compression but has little/no strength in tension. Similarly, cables have stiffness in tension but buckle if they go into compression.

Forces can also be effectively nonlinear, in that their direction can change during the analysis. Loads that are in local directions to the elements, such as those that are perpendicular to the element, will follow the element as it deflects (Figure 6.27).

Figure 6.27: Nonlinear loading

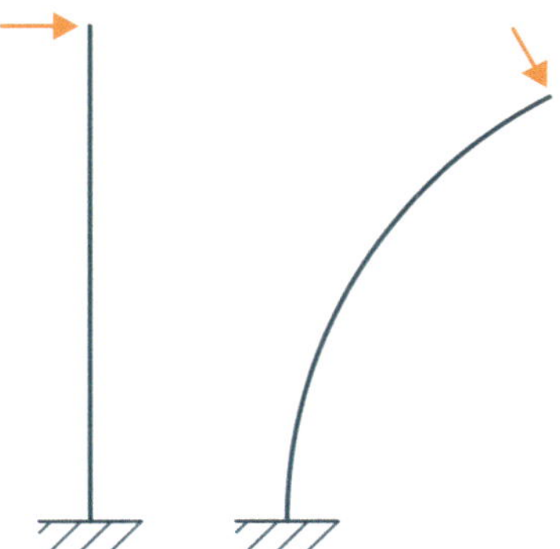

Other examples of nonlinear loads include hydrostatic or buoyancy forces, and aerodynamic loads that change the shape of the structure. I once helped with the FEA analysis of a floating fish farm. To simulate the general buoyancy effects on the rectangular floats, we used compression-only springs but with a linear stiffness, which I set to match the floatation effect of 1kPa/m. This needed adjusting to the element sizes. Supported on these springs the structure would then settle until it generated enough uplift to give us equilibrium. The non-rectangular objects were trickier as the water displacement varies with depth, so they needed a nonlinear spring curve. We could not ascertain the lateral pressures from the water automatically, so had to set those once we knew how deep the structure would float (Figure 6.28).

Figure 6.28: Floating fish farm analysis model

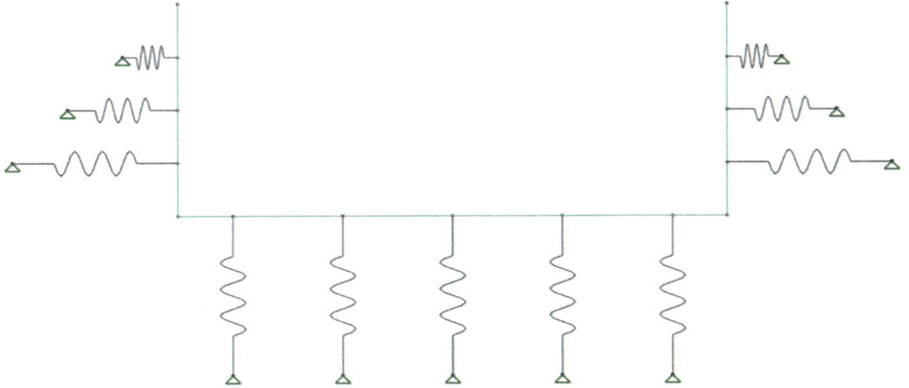

Nonlinear analyses are more complex than linear because there is, usually, no direct way to solve the structure. For a linear analysis we assume the deflections are insignificant, meaning that we can establish the static equilibrium on the undeformed geometry. With nonlinear analysis we do not have that luxury and instead we must find the deformed shape that gives us equilibrium, if there is one.

Consider this cable with a single point load (Figure 6.29). Cables have no stiffness except when in tension: they rely on geometric stiffness. But we cannot conduct a linear analysis, as the stiffness in the vertical direction is zero at the start, and likewise P-delta will not work as it needs an initial linear analysis. What we do know is that it will deflect until the tension builds sufficiently to counterbalance the loads.

Figure 6.29: Forces on cable with single point load

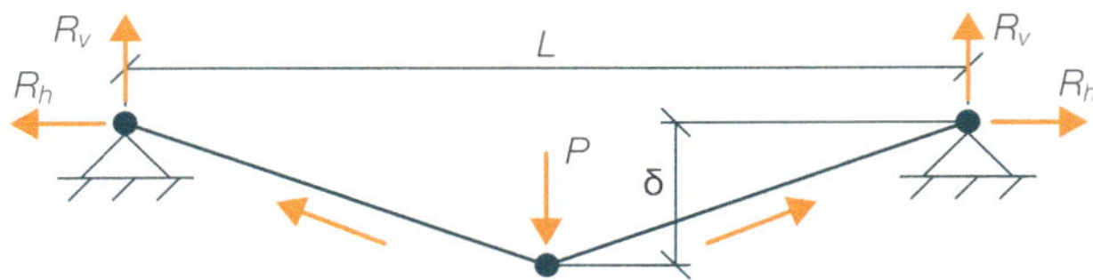

The sum of vertical reactions will be equal to the vertical load:

$$P = 2R_v$$

But the horizontal reactions are another matter: we need to look at the triangle of forces (Figure 6.30).

Figure 6.30: Reaction force triangle

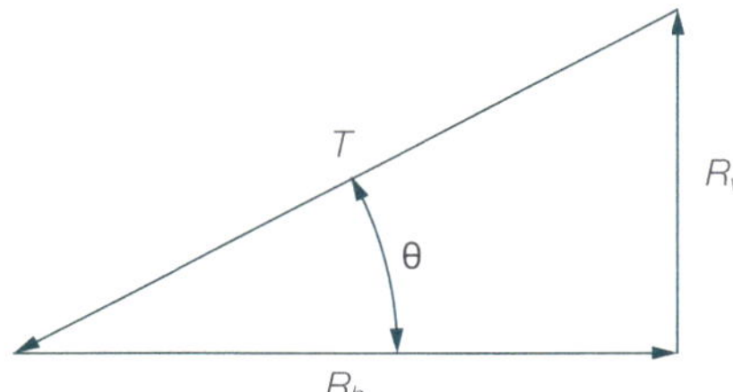

R_v is constant but R_h changes with the angle of the cable, which will depend on the axial stiffness of the cable, the amount of pre-tension on the cable, and the stiffness of the supports. If you know what the deflection is then the maths is quite simple, but it is difficult to work out what the deflection will be for a given combination of loads, prestress and section size. It is not possible to directly calculate the deflections because there is no single answer: the structure can achieve equilibrium for any number of deflections. Instead we must search for the final deflections based on our initial conditions. To put it another way: nonlinear solvers are iterative, approximate, and stop when the answers are sufficiently accurate.

As discussed, linear structures have a linear relationship between force and displacement of a structure (Figure 6.31).

Figure 6.31: Truss force-deflection

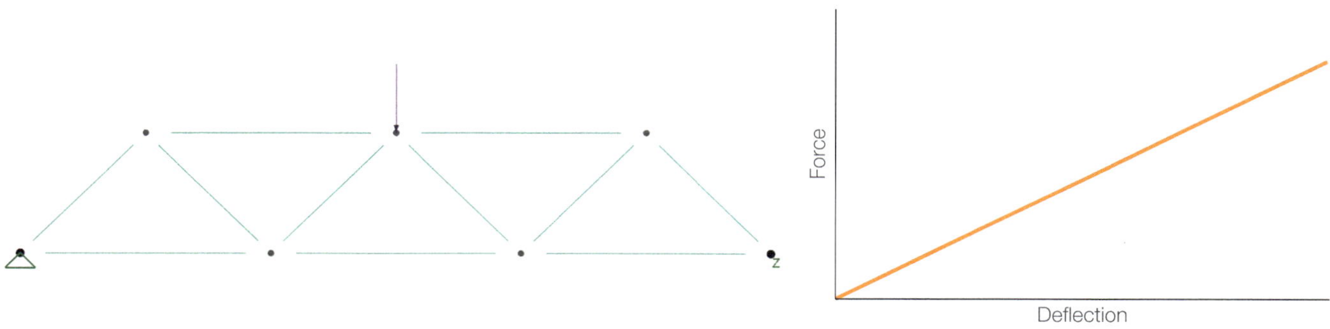

With nonlinear structures however, the relationship (known as the 'equilibrium path'), is much more complex: the real path is multi-dimensional, so it is not always possible to draw the line in just two-dimensions. Note that every point on the equilibrium path represents a stable combination of load and displacement, and the displacement value shown on the equilibrium graphs is not necessarily in a particular direction.

Figure 6.32 illustrates the equilibrium path if the structure softens or 'snaps-through'.

Figure 6.32: Equilibrium path for snap-through

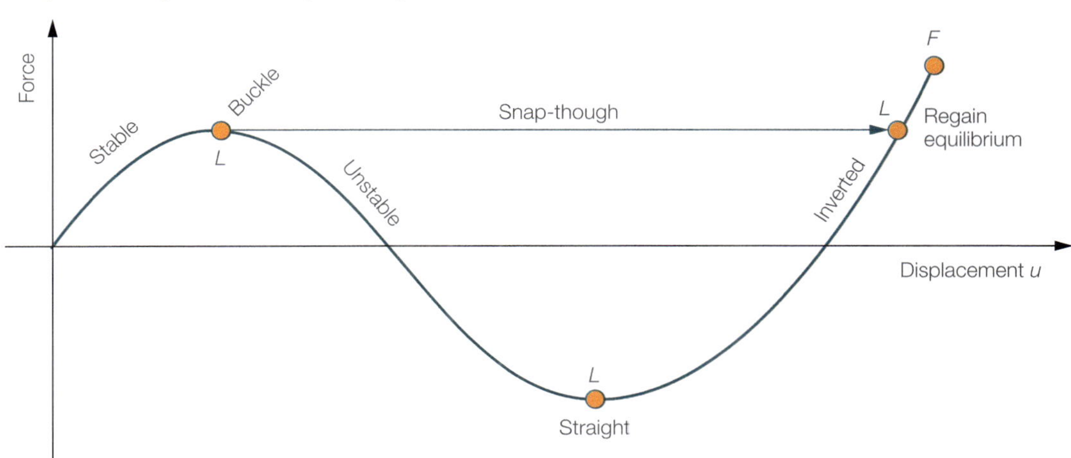

Where:

L = the limit point, at which the tangent is horizontal, meaning that deflection increases with no change in load
F = the failure point, where some or all of the structure breaks

The downward slope on the equilibrium path shows that the load must reduce if equilibrium is to be maintained. If the load stays constant, the structure will deflect until it either finds a new point of equilibrium or collapses.

An overloaded arch or dome is an example of snap-through, as is the pressure-check bump on a jar lid.

While Fig. 6.32 is an idealised path, real equilibrium path shapes are more complex. In the following example, I have taken an arch and traced the equilibrium path by applying a displacement in small steps until the arch buckles, snaps-through, and takes up the load as a tension structure (Figure 6.33).

Figure 6.33: Real load path for arch snap-through

I then plotted the applied displacement against the resulting reaction at the midpoint[†] (Figure 6.34). The graph does not form a smooth line because even with initial imperfections there are various ways the struts can buckle; sometimes symmetrically, sometimes not. As expected, the imperfections, which are present in all real structures, do give a lower buckling load than that of an ideal strut.

Figure 6.34: Arch snap-through deflections

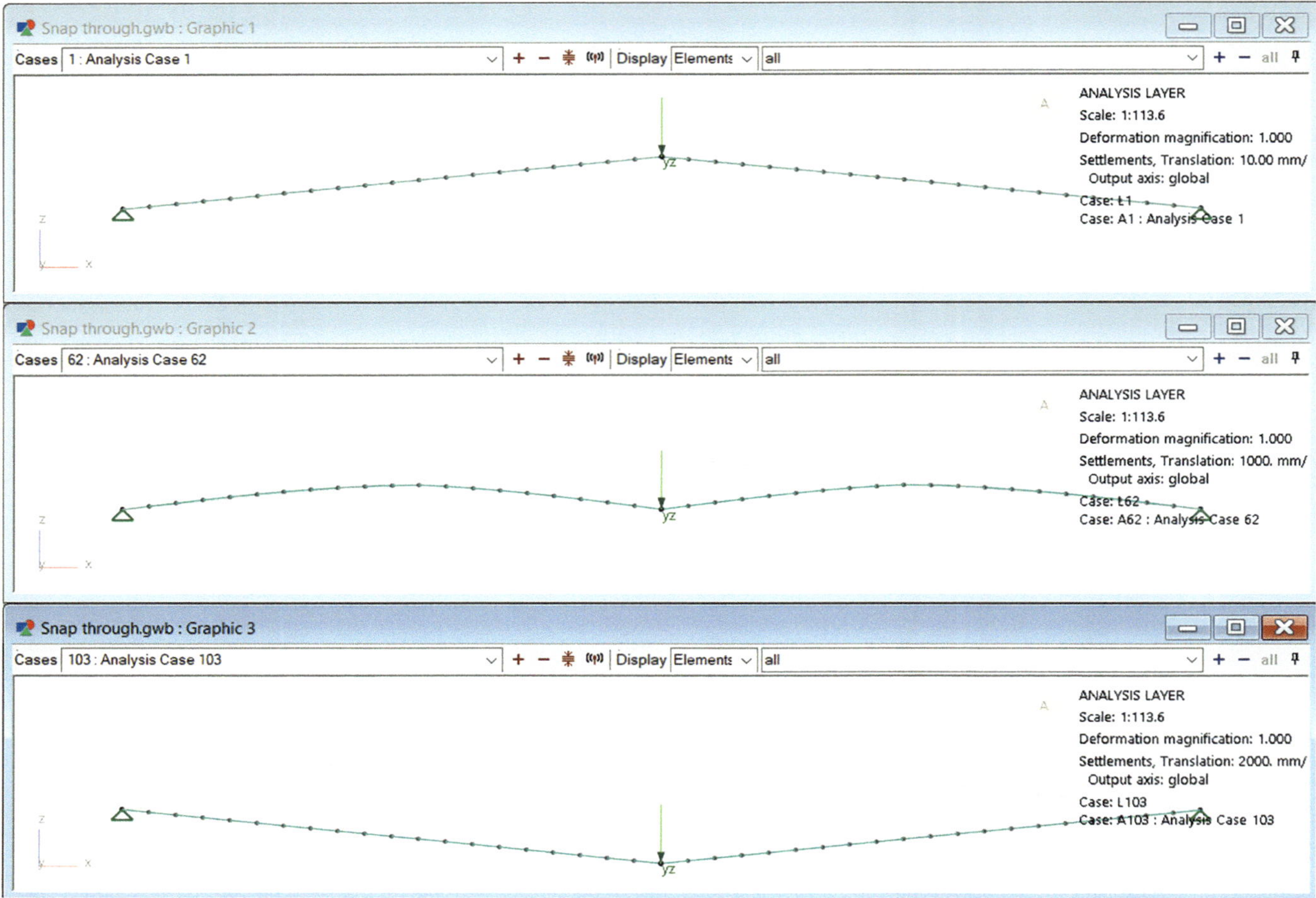

This example assumes that the structure can survive the snap-through. A real structure might suffer catastrophic damage, weakening it to the extent that it cannot carry the load. Such damage is difficult to capture in general FEA terms, but it is possible to explore it with some explicit solvers.

'Snap-back' (Figure 6.35) occurs when there is a sudden jump in deflection (e.g. at the formation of a brittle crack). To maintain equilibrium in these situations, not only must the load reduce but so must the deflection.

[†]   Incrementing the load instead of the displacement would have produced a large jump in the displacement at the point of buckling.

Figure 6.35: Load path for snap-back

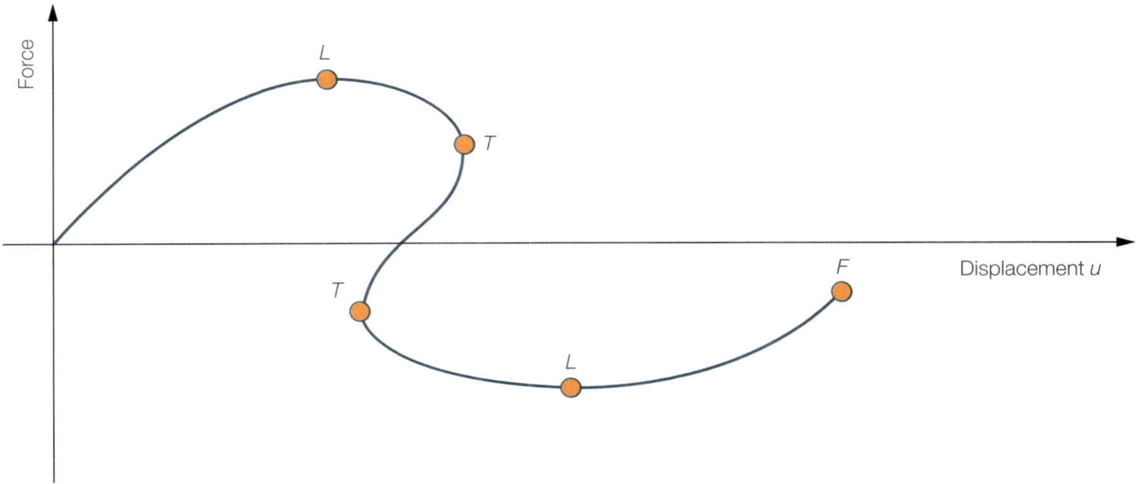

T is a turning point, where the tangent is vertical.

Bifurcation points can occur when the structure could go one way or another, giving us multiple equilibrium paths. Consider a strut buckling: it could go left, right, or stay straight if you are using an analysis that ignores buckling (Figure 6.36). The equilibrium has three options in this instance. Imperfections in the structure or eccentricity in the loads will mean there will be a particular direction in which it will want to buckle.

Figure 6.36: Bifurcating load paths

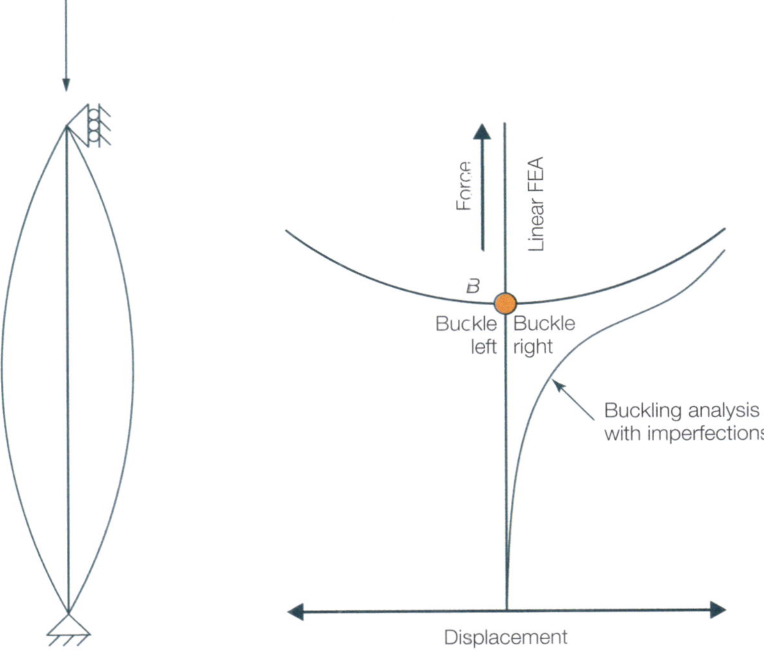

Points failure, limit, tangent, and bifurcation are known as 'critical points', and will complicate the search for the equilibrium path.

6.6.1 Newton-Raphson method

The Newton-Raphson method (N-R), named after Sir Issac Newton and Joseph Raphson, was originally an iterative method for finding the roots of equations by making an initial guess, differentiating to find the slope at that point, projecting that to the baseline, and repeating until the step difference is small enough (Figure 6.37).

Figure 6.37: Newton-Raphson solution for equations

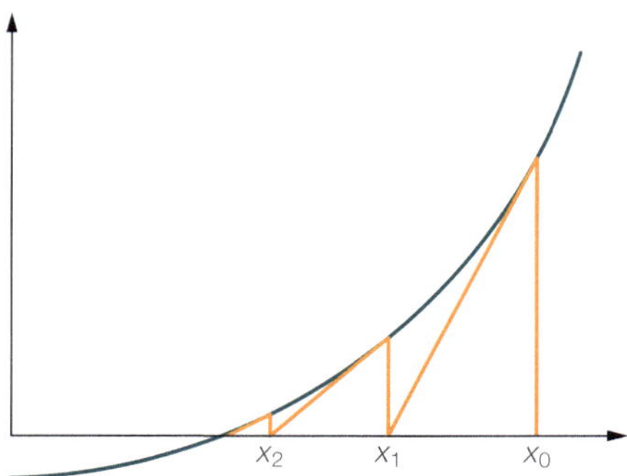

It can have convergence problems though if the slope has any local bumps in it (Figure 6.38). This means that the solution speed, and indeed the answer itself, can be sensitive to the starting point.

Figure 6.38: Divergence with Newton-Raphson method

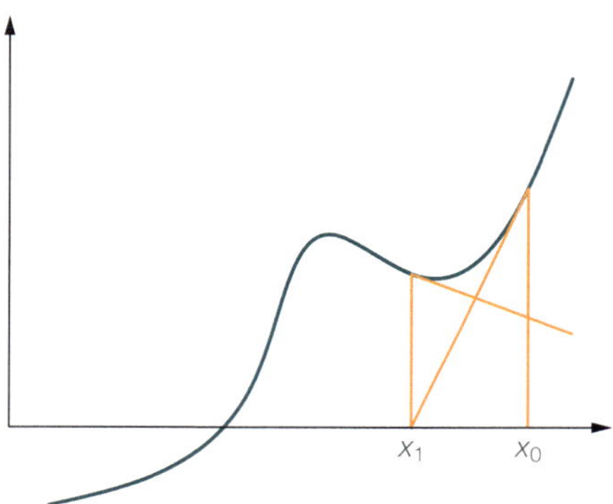

So how does N-R work for structures? It is an extension of the P-delta method, but analyses over multiple iterations, not just two. Like P-delta, N-R creates the geometric stiffness matrix to analyse a second time, but then uses those results to calculate the stiffness again, repeating until it converges. Unlike P-delta, N-R can also take material yielding and other nonlinearities into account, which it achieves by updating the stiffness and geometric stiffness matrices for the newer nodal positions. It can also vary the Young's modulus based on the stress levels to give material nonlinearity. We will call the new stiffness, considering the geometric stiffness and other nonlinear effects $K(u)$, which is the stiffness at deflection u. Updating $K(u)$ at every step is more accurate but can be very time-consuming. Also, if any part of the structure buckles (its stiffness falls to zero) the analysis will fail to converge.

As the equilibrium path is likely to be complex and the load might be beyond the failure point, we need to work our way up the path in steps, which are increments of the final load. Just like P-delta analysis, we analyse to find the geometric stiffness, then combine that with the static stiffness to give us the combined stiffness $K(u)$ and hence the deflections. But as we are including nonlinear effects in the analysis: geometric, material, load, contact etc. the results might not be quite in equilibrium, but have a residual force:

$$F_{residual} = F_{external} - F_{internal}$$

We can then update $K(u)$ and reanalyse. Once $F_{residual}$ equals zero (or is near enough, as it is virtually impossible to get it totally to zero) we then start on the next load iteration (Figure 6.39).

Figure 6.39: Newton-Raphson method for nonlinear load paths

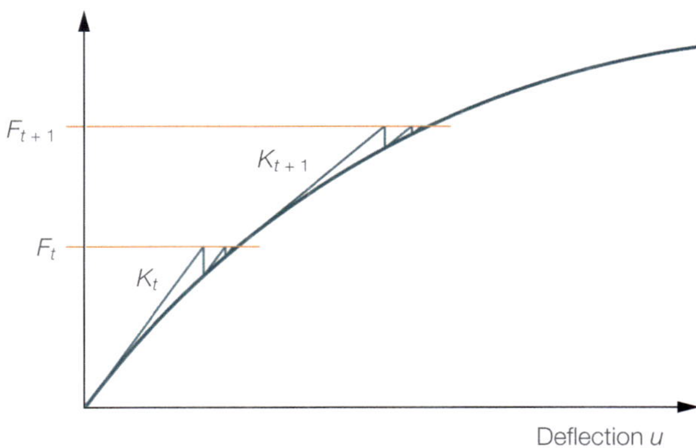

When we use the N-R method to find roots of an equation we differentiate the slope (the rate of change) of the graph to find the tangent, project that onto the zero line, then progress from there. With a structure, we do not have a formula to differentiate, but instead we look at the rate of change of stiffness of $K(u)$ and predict what the deflections will be at the next load step. From there we calculate the new geometric stiffness, etc. and continue until the residual is sufficiently low again. This rate of change of stiffness is called the 'tangent stiffness' (K_t), and we can calculate it just once at the start of the analysis, once per load increment, or at every iteration. In a linear FEA, the tangent stiffness is the stiffness matrix.

Calculating K_t at every iteration will, usually, converge the analysis in the least number of steps, but the time involved in recalculating K_t can be more than that involved in each iteration. Reaching a limit point can cause divergence if we do not update K_t (Figure 6.40), as can updating K_t too often (Figure 6.41): the next point found on the equilibrium path might be behind or so far away that calculating the new geometric equilibrium is impossible.

Figure 6.40: Newton-Raphson method diverging due to softening

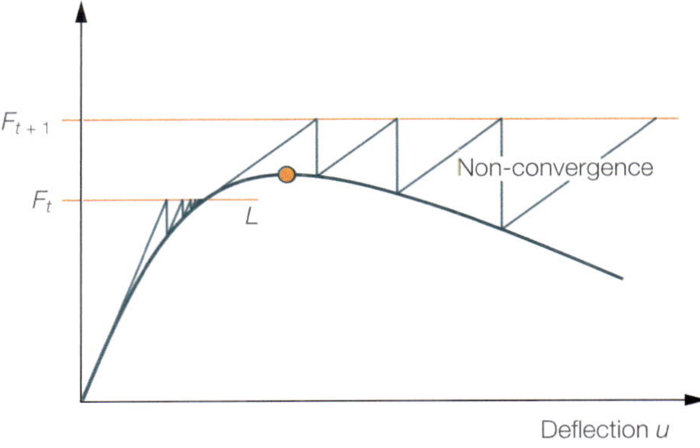

Figure 6.41: Newton-Raphson method diverging due to snap-through

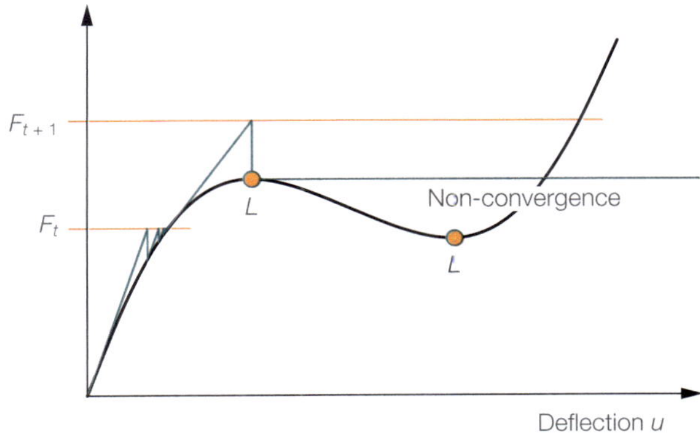

To deal with these divergence problems we can use the arc-length method, where the load step can be automatically adjusted. We do this by intersecting the K_t slope with the next load increment only on the first step. This gives us a distance which we use to define an arc. Subsequent steps intersect with this arc instead, until we reduce the residual forces back to almost zero, when we try for the next load increment (Figure 6.42).

Figure 6.42: Arc-length method to avoid Newton-Raphson divergence

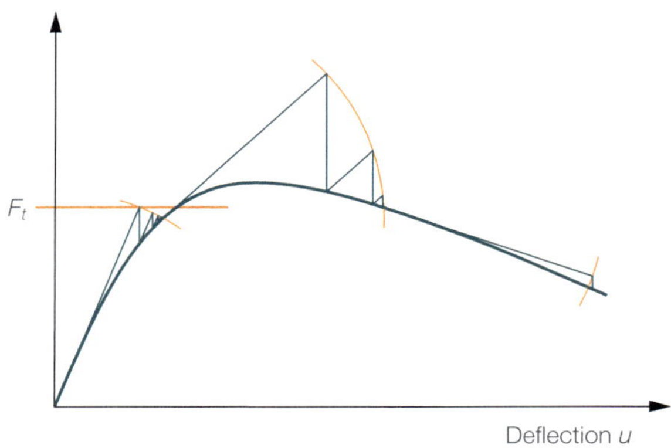

Newton Raphson can be a fast nonlinear solver but can also struggle with large deflections or tangent points in the equilibrium path. It will also fail if there is a mechanism in the structure.

6.6.2 Explicit solvers

There is no matrix.

We now segue from implicit solvers to explicit ones. Implicit methods are time-independent: the results are the same no matter when you examine them as they have already converged on the answer. This includes all the analysis methods we have looked at so far, with the possible exception of the linear time-history analysis. The implicit methods solve the structural problems either directly using matrix methods or by iterating through a series of steps. These matrices can be huge: storage needs a lot of memory and inverting them is computationally expensive. Implicit methods are very suited to linear and slightly nonlinear problems, so for these cases they are still the best approach.

Explicit solvers were once only found in very high-end FEA programs but are now gradually becoming available in mid-range programs. They are used for analysing highly nonlinear problems such as cable and fabric structures, crash simulation, blast loading, seismic nonlinear time-histories, and solid body motion. Other engineering disciplines might use them for heat flux and fire analysis, airbags and ballistics.

One spectacular example of a model verification was organised by Arup in 1984[80,81]. They had used explicit analysis to design and test the 50t flasks for transporting nuclear fuel. Following the computer simulations using explicit solvers and physical drop tests, a flask filled with pressurised water was loaded on to a lorry trailer, which was then tipped over and laid across a train track. A driverless diesel locomotive with four carriages was then crashed into the flask at 160kph (45m/s). The train was destroyed, the flask only scratched, and the analysis results validated (Figure 6.43).

These analyses are for short-term actions, some needing to be measured in milliseconds, so are not looking for static equilibrium but instead measure how the stress flows through the materials. They are also notoriously slow to run, taking hours or days to complete.

Explicit solvers do not use matrices to calculate the nodal deflections, but instead track how the nodes move over time as the structure searches for equilibrium. Explicit solvers use concepts like acceleration, mass, velocity and damping, but can also allow for fracturing, detachment of parts, and contact. Explicit analysis tries to simulate the actual accelerations of the structure, and hence the strains then stresses in the elements. The stresses then act as forces resisting the movement, as they do in real structures. The damping in the materials and friction between the parts might then, eventually, bring the structure into equilibrium, whether static or dynamic, but it is not guaranteed.

Figure 6.43: Operation Smash Hit: full-scale testing on a 50t nuclear transport flask

Explicit solvers need a time-step small enough to maintain accuracy and to prevent divergence of the simulation, but not so small that the simulation takes forever to run. But processing speeds are improving all the time: the first FEA analysis conducted by Arup was for the Sydney Opera House c.1960. The analysis of a side shell rib, with 136 nodes and five load cases, took about three weeks to set up and then four hours to run. When I used FEA on the Surrey University mainframe for my final year project in 1989 the 1,000-element model would take about 24 hours before crashing. Today's computers and programs can analyse such models in a fraction of a second, but explicit analysis runs can still take hours or days to complete.

The explicit analysis sequence is:

1. Apply forces and other boundary conditions.
2. Find element strains → stresses → nodal forces → out-of-balance forces on the nodes.
3. Use the out-of-balance nodal forces and masses to calculate the new nodal accelerations.
4. Calculate the new nodal velocities and displacements at the end of the time-step.
5. Update geometry and repeat from Step 1.

There are several approaches to solving the displacements in Step 4: two popular options are the explicit and implicit integration time-step schemes.

Explicit integration assumes that the acceleration will remain constant over the time-step, so we work out the average velocity of each node and hence its new location:

$$f_t = f_{app} - ku_t$$

$$a_t = \frac{f_t}{m}$$

$$v_{t+\frac{\delta t}{2}} = v_{t-\frac{\delta t}{2}} + a_t\delta t$$

$$u_{t+\delta t} = u_t + v_{t+\frac{\delta t}{2}}\delta t$$

Where:

f_t = residual forces applied at time t
f_{app} = external forces applied at time t
k = structural stiffness
u_t = nodal displacements at time t
v_t = nodal velocities at time t
a_t = nodal accelerations at time t
M = nodal mass
δ_t = time-step

The nodal velocity is the tangent to the graph (Figure 6.44). The larger the time-step, the lower the accuracy of the predicted new displacement; too large and the analysis will diverge and not find a solution. This means that we need an extremely short time-step to remain accurate, which should be less than the element size divided by the speed of sound in its material.

Figure 6.44: Explicit integration with an explicit solver

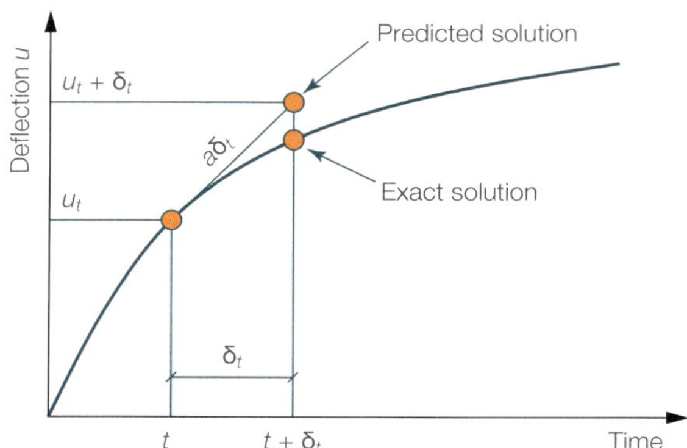

Using an **implicit integration**, we can improve the accuracy by assuming that the acceleration varies linearly over the time-step, so we need to calculate the rate of change of acceleration (or jerk) at time t. From this we can estimate what the acceleration might be at the end of the time-step and thus calculate the velocities and displacements more accurately. This also allows larger time-steps, though it can be difficult to work out what the time-step should be that still maintains enough accuracy.

f_t and a_t as before:

$$f_t = f_{app} - ku_t$$

$$a_t = \frac{f_t}{m}$$

$$v_{t+\delta t} = v_t + \tfrac{1}{2}(a_t + a_{t+\delta t})\delta t$$

$$u_{t+\delta t} = u_t + \tfrac{1}{2}(v_t + v_{t+\delta t})\delta t$$

$$= u_t + v_t\delta t + \tfrac{1}{4}(a_t + a_{t+\delta t})\delta t^2$$

If we combine these equations and substitute for $u_{t+\delta t}$ we get:

$$a_{t+\delta t} = \frac{f_{app} - k\left(u_t + v_t\delta t + \tfrac{1}{4}a_t\delta t\right)}{m + \tfrac{1}{4}k\delta t^2}$$

Each implicit integration time-step is computationally more demanding than an explicit integration time-step, but you need less of them (Figure 6.45).

Figure 6.45: Implicit integration with an explicit solver

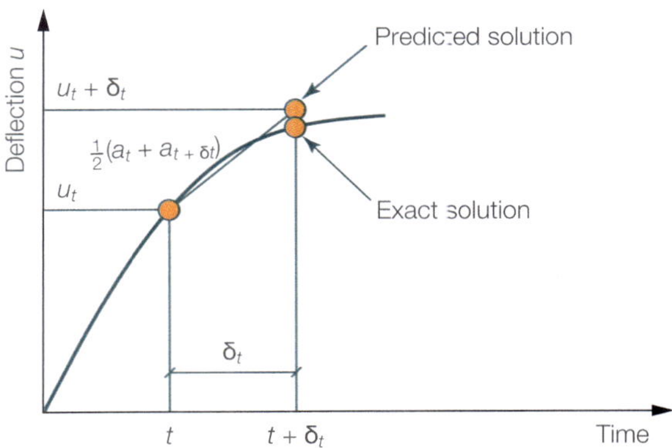

You can reduce the computation time if you know that certain aspects are linear. For example, you might use a matrix method to determine the stresses, strains and deformed geometry under variable and permanent actions, then use the explicit solver from that starting point to model the resulting response to an earthquake.

It is important to remember that an explicit method analysis simulates the structural response to just one load case. We can use a response spectrum for an implicit seismic analysis, but an explicit analysis must be for a particular set of acceleration time-histories. While we might analyse the structure to see how it would have performed in an historical earthquake, for a robust design we must check it against the events that might happen in the building's lifetime. Luckily, there are ways to generate artificial time-histories from a given response spectrum, but for confidence the building will need checking against approximately 12 combinations of these before we can be reasonably sure it will stand.

Another drawback is that to model behaviours like crushing and spalling of reinforced concrete you need to model all the reinforcement and relevant details. This will be very time-consuming to build unless you can automate it, plus you will not know what reinforcement is required until you detail design the building. Detail design requires an initial analysis, indicating that the nonlinear time history is a check on the design and detail of the structure. A failure in the structure at this point will mean a redesign.

6.6.2.1 Dynamic relaxation
The dynamic relaxation method is a form of explicit solver that has proven effective for a range of large-deflection nonlinear static structural problems. While the general explicit solvers are a simulation of the structural response to static and dynamic loads, accelerating them into their final states, the dynamic relaxation method instead works with fictitious masses over an arbitrary period to find the static equilibrium of structures despite large deflections.

The advantage is that we can solve nonlinear static problems in a single load step and deal with snap-through and other large displacement problems, including mechanisms, that Newton-Raphson would struggle with. While it does not use load-steps, it does use time-steps, so can be slower than Newton-Raphson for simpler nonlinear problems. Like all nonlinear methods, it can have convergence problems, but not necessarily for the same reasons.

Alistair Day invented dynamic relaxation in 1958 while working at Arup[82,83]. He had seen the approach being successfully used to model water flow and realised that he could apply it to nonlinear structural analysis. One of the drivers was that the method, unlike matrix methods, needed a much smaller memory footprint, which was an important consideration at the time.

The structural model is idealised, as usual, as a series of nodes, but each has a mass. This mass can be based on the supporting elements' stiffnesses or be arbitrary. The loads on each node are summed and then applied, causing the nodes to accelerate. The node's acceleration and velocity are then used to calculate the node's new position after a short time-step. If the node is restrained, its movement in that direction is ignored. The nodal positions are now updated.

Looking at each element in turn, we now calculate the element strains and hence internal forces using whatever stress-strain relationship is desired, including nonlinear and hysteresis. These internal and external forces are then summed and applied to the nodes, ready for the next time-step. So rather than calculating all the internal forces simultaneously, like the matrix methods, dynamic relaxation causes the forces to cascade across the structure until equilibrium is found.

One problem with the dynamic relaxation method for static analysis is that, if unchecked, the structure would just bounce backwards and forwards with all the structure's natural frequencies. One way to bring it to rest is to apply a viscous damping to all the movements by reducing each node's velocity by a percentage at each time-step. Another is to use 'artificial' damping, which detects when nodes have passed their equilibrium points and then removes their velocity completely. To see how artificial damping works, consider the range of movement of a cable with a point load (Figure 6.46).

Figure 6.46: Dynamic relaxation method searching for equilibrium

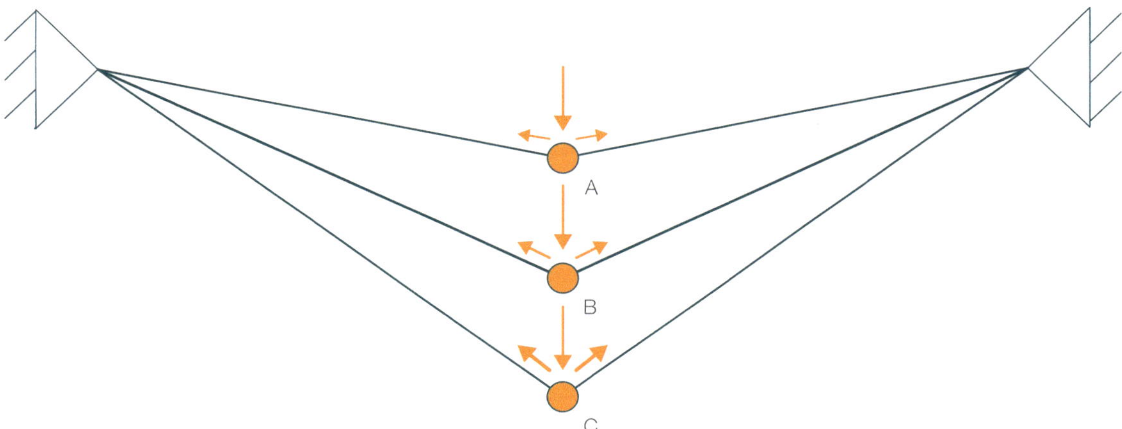

At Point A, the minimum deflection, the internal forces are smaller than the external applied force, so the net force causes a downwards acceleration: acceleration is maximum and velocity minimum. It is similar at Point C, the point of maximum deflection, but here the internal forces are much larger than the external: again, acceleration is maximum and velocity minimum. At Point B the internal and external forces are balanced so acceleration is minimum (ideally zero) but velocity is maximum.

In a real structure of masses connected by springs we would be able to see the extents of movement and acceleration, but reality uses an infinitely short time-step[†]; we need to choose a time-step such that the analysis finishes in the shortest calculation time with sufficient accuracy.

If we start the node at its highest point, the first few steps will see the velocity build and the acceleration reduce. As soon as we see the acceleration has changed sign, we know we have passed equilibrium in the last time-step, so we must be close. Zeroing the velocity at this stage means that the node will start to accelerate upwards, until it passes the equilibrium point again, this time at a much slower velocity. We repeat until the sum of the residual forces is small enough, at which point we stop the analysis.

While the time-step length and fictitious mass size are arbitrary, the time-step does need to be able to track the nodes as they travel within their bounds. If the time-step is too large or the mass too small, we will over-predict the distance the nodes will move, and the next time-step will see it further from equilibrium not closer, causing the analysis to diverge. If the time-step is too small or the masses too large, it will take an excessive time to converge. Unlike this simple example, real structures have a wide range of stiffnesses: larger stiffnesses exert larger forces, and hence accelerations, for a given displacement. If we are not careful, some nodes might crawl to their final positions while others speed way beyond their equilibrium points. The overly fast nodes will prevent convergence and need reining in.

For this reason, we might give nodes connected to stiffer elements a higher fictitious mass. The ideal is for all nodes to accelerate at the same rate and for the damping to be the critical value to bring all the nodes to their equilibrium positions in the shortest time. The time-step can also be adjusted, preferably automatically, so that fast nodes are kept under control.

Other options to reduce the convergence problems caused by a wide range of element stiffnesses include replacing the excessively stiff elements with link elements and replacing stiff supporting structures with restraints. For example, if you have a fabric roof supported on heavy steel or concrete members you might analyse the fabric substructure first and then apply the reactions as loads on the primary frame.

Dynamic relaxation deals with large displacement problems like snap-through and form-finding very well. It can also deal with individual elements buckling if the rest of the structure can carry the load, but will not converge if the load is too large for the structure as a whole. It is more sensitive to stiffness ranges than matrix methods but can analyse problems that Newton-Raphson cannot.

6.6.3 Meshfree and particle methods

One of the perennial problems in finite element analysis is the creation of the mesh. The FEA method assumes that the elements are square or equilateral, yet they are often far from these ideal shapes. In addition, while pairs of nodes in a beam model are joined by a single element, in a mesh they are connected by two or more, which can be different shapes or sizes. While we can calculate the displacements of the nodes, the calculated stresses in the adjacent elements are unlikely to be the same.

Meshfree (or 'meshless') methods, as the name suggests, do not use a mesh of elements, but instead all the nodes interact with whichever are their nearest neighbours. This allows the analysis of problems where formal relationships between nodes is not a possibility, such as simulating granular or viscous flow, or computational fluid dynamics (CFD) problems that involve 'sloshing'. Meshless methods can also be useful for large displacement problems where elements would become distorted and thus lose accuracy or even fragment. They are also good for problems involving rapidly changing variables, such as shockwave, fracture, and cracking problems, by automatically adding nodes as required. The alternative would be a fine mesh throughout the model, which is both computationally expensive and mostly redundant.

It is possible to mix FEA and meshless methods in a single model, keeping FEA for the parts with low deformation. The two parts will need to be coupled together to ensure continuity in displacements and stress.

Transport and fire engineers also use meshless methods for modelling the movement of traffic and people. Here, instead of nodes, the models have 'agents', which have rules of movement and interaction programmed into them.

[†] Or nearly so, but Planck Time at 5.39×10^{-44} seconds as a time-step is too incremental to be useful to a structural engineer. It will be a long time indeed before our simulations can approach the computational power of a steel frame determining its own state.

The result is a simulation for how long it will take to evacuate a building or how a new road junction is likely to affect congestion.

While meshless methods remove some problems, they add others, such as increased computation time, as you need to figure out which are the closest nodes for every iteration. It is also not easy to apply boundary conditions to meshless methods. They are not yet common in structural analysis.

6.7 Conclusion

There are a variety of analysis methods for you to choose from: static or dynamic, linear or nonlinear, implicit or explicit. It is up to you to choose the most appropriate one for your particular problem.

7 Resolving problems in FEA models

"To err is human but to really foul things up you need a computer."
Paul R. Ehrlich (attributed)

On 17 January 1978, 5,000 basketball fans were watching the University of Connecticut beat the University of Massachusetts Amherst in the Hartford Civic Center Arena, Connecticut, USA. At 04:15 the following morning, an accumulation of snow brought the 1,400t roof crashing down onto the (thankfully) empty seats (Figure 7.1). Six hours earlier and very few of those sports fans would have survived[84,85].

On 21 August 1991, the original gravity base for the Sleipner A offshore platform was being towed towards the gas field in the North Sea when a deep, bang-like sound was heard. 18 minutes later, the massive structure had disintegrated at the bottom of the Gandsfjorden[86–88]. Figure 7.2a shows the replacement platform, complete with the 40,000t topside structure that includes accommodation for 160 people and drilling equipment. Figure 7.2b is a gravity base of similar design and scale to the one that sank.

At 12:51 on 22 February 2011, a magnitude 6.3 earthquake struck Christchurch, New Zealand. About 15 seconds later the Canterbury Television (CTV) Building collapsed (Figure 7.3). 115 people died as a result[89].

What links these structural disasters, and no doubt many more, is major failings in the use of FEA. Yes, there were other serious design flaws in all these structures, but they were designed using bad FEA models plus a lack of understanding with regard to what the models were not telling them.

Figure 7.1: Collapse of Hartford Civic Center Arena, Connecticut, USA, 1978

Figure 7.2a: Redesigned Sleipner A, North Sea

Figure 7.2b: Unsubmerged offshore platform gravity base

Figure 7.3: CTV Building wreckage

Over the last few decades the FEA method has become the pre-eminent way to analyse statically indeterminate structures. Previously, engineers needed to use specific algorithms and calculus to analyse particular structural shapes, which meant the analysis was highly inaccurate if the structure deviated from the specific forms. Today, models of almost any shape or problem can be solved using linear and nonlinear solvers.

The FEA method is covered on most university courses these days, but in my experience the education tends to concentrate on the maths of the stiffness method, which will not help at all if the model itself is wrong. "Garbage in, Garbage out" as the software industry saying goes. The most important thing for practicing engineers today, is not to write FEA software but to know how to build good models and to have an understanding of what makes them good.

There may be myriad ways to build good FEA models, but I have found there are also some very common ways to build bad ones. I too, have spent many years building FEA models of dubious quality, as well as helping other engineers sort out problems with theirs.

So how do we build bad models?

7.1 Finite element approximation

The answer is to 16 decimal places, so it must be accurate...

It is easy to forget that the model you have created is just that: a model. It is not the real structure but a gross simplification of it. To reiterate: *All* **models are wrong, but some are more useful than others**.

The art of engineering design is to derive answers that are 'accurate enough' — and 'how accurate is accurate enough?' can depend on the contract stage. 'Within 50%' might be ok for scheme, but for the final design, if 100% is impossible, do we need to be 90%, 95% or 99% accurate? And how do we know how accurate we are? These are tough, possibly unanswerable, questions, but something we must keep in mind while designing.

The stiffness method used in most static linear analyses is finding the solution to a large number of simultaneous equations describing how the nodes in the model move. If the structure is statically indeterminate there is no definite answer and the algorithm is designed to get as close as possible. The equilibrium of the model is also based on the undeformed structure, so the larger the deflections the lower the accuracy.

Nonlinear static analyses, on the other hand, are trying to converge on the solution, where equilibrium is based on the final, deformed structure. How close they get depends on several factors, especially how tight the convergence criteria are and how much time you have. As for nonlinear dynamic problems, well that way chaos lies — literally[†].

Your model will also contain several simplifications and approximations that reduce accuracy, which is right and proper if they are within appropriate limits. While mechanical engineers model individual components in fine detail (or at least, finer detail), we may be modelling an entire bridge or building. We cannot afford to model every single bolt and weld in such a situation; in fact, we would be quite foolish to do so in a design model. Not only are we often building the analysis model to give us the information to detail the connections, but any changes to the structure will be unnecessarily time-consuming.

We may, and should, have a reasonable idea of what section sizes we are probably going to use, but we cannot be certain until we crunch the numbers and work out exactly what we need. Only then might we take a closer look at details like connections. But first we model whole beams or columns with just one or more 1D elements.

[†] An online search for double pendulum videos can be quite rewarding. You may be able to model a double pendulum with an explicit nonlinear time-history solver, or just write your own script and see what happens.

Even this simplistic approach can lead to models with thousands of elements. Indeed, I have seen an airport structural model with over 100,000 beam elements. Meshing of slabs and walls rapidly increases the number. While a bay may require only four beams, the bay slab mesh is likely to require two dozen or more two-dimensional shell elements. And three-dimensional solids are likely to require at least an order of magnitude more elements than the 2D mesh.

In many respects, our model is a map of the structure, containing the significant aspects, but in an abstract fashion. And just as a map simplifies and brings clarity to a landscape, our models do the same to the behaviour of the real structure. This simplification takes thought and effort: it is simple to make something difficult and difficult to make something simple.

7.1.1 Ill-conditioning
The various FEA algorithms are trying to get close to the answer, but the model, and the way you model, can throw them off.

7.1.1.1 Rounding errors
The first thing to remember is computers model the real world using floating-point numbers of limited precision. It is true the limited precision is now 64 bits, which is binary bits, or about 16 decimal places in our more familiar base 10 number system. These floating-point numbers can only approximate to real numbers (like Pi and the square root of 2), and real structures don't even use numbers as they are just physical objects subject to forces and other actions. We are approximating the real structure and subjecting it to approximate loads using numbers that approximate to the real values. And yes, the computer may give us answers to 16 decimal places, but how many are justified?

To illustrate, let's look at how integers relate to real numbers. In Figure 7.4 you can see how each integer (on the lower part of the graph) maps to a spread of real numbers (the upper line): the integer 1 might represent any number between 0.5000000 and 1.4999999. As one of my colleagues joked to me: "one plus one equals three for some large values of one".

Figure 7.4: Integer vs. real numbers

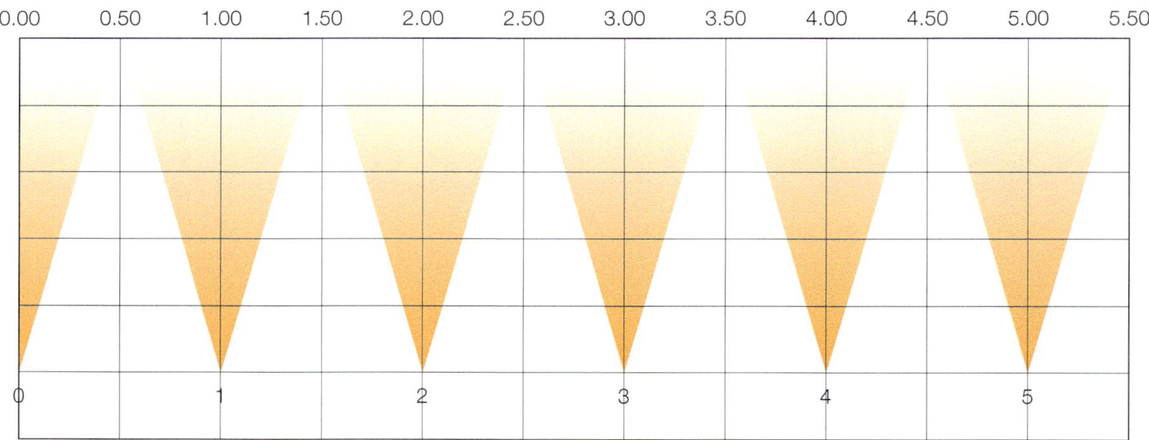

Likewise, floating-point numbers (lower line in Figure 7.5) relate to a spread of real numbers: rather than 1 we now have 1E+0, which represents numbers from 0.5000000 to 4.9999999. Here, for simplicity, I will just take floats with one significant place. Computers work to more significant places, but the principle is the same. This means the model coordinates are also going to affect the analysis results. If the model is a long way from the origin then the coordinates are large, and each floating-point number is going to approximate to a larger spread of real numbers.

Figure 7.5: Float vs. real numbers

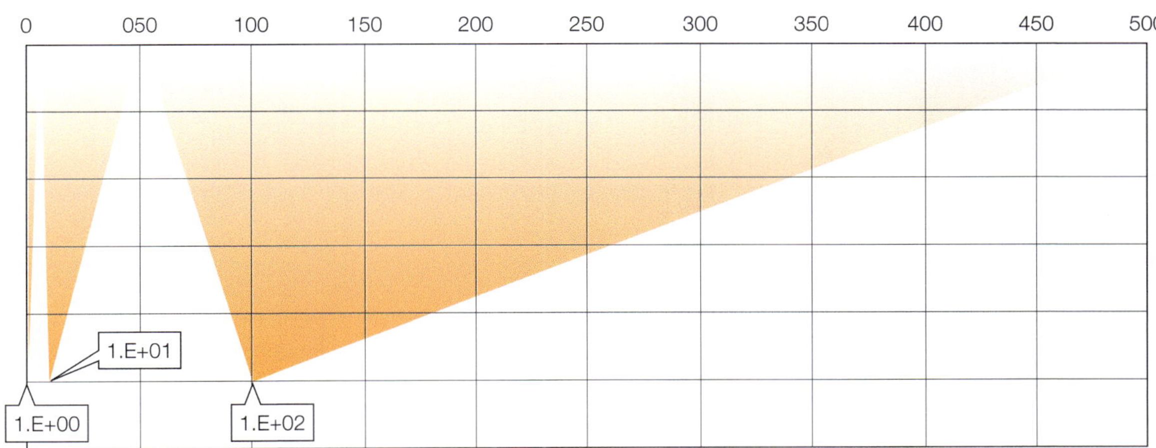

7.1.1.2 High/low stiffness

Possibly the best way to reduce the accuracy of your model is to have an extremely wide range of element stiffnesses. These stiffnesses are a function of the elements' properties (materials, cross-sectional area, etc.) and the element length or size (the smaller the element the stiffer it is). In other words, very short elements with large cross-sections can reduce the accuracy of the calculation, even though they might be intended to increase the realism of the model.

Even in the simple model in Figure 7.6, the short beam is over 15,000× stiffer than the long beam. Good FE programs recognise this and allow you to add elements such as links, which are effectively infinitely stiff but avoid problems because they constrain two or more nodes to move together.

Figure 7.6: Problematic cantilever

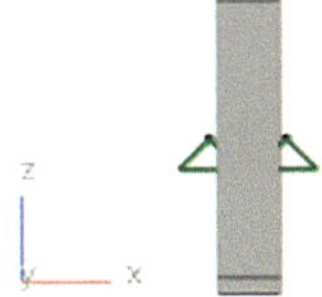

Many ill-conditioning problems are also caused by extremely low or zero stiffness. Freely-spinning beams, trusses unrestrained out of plane, and materials without sensible properties are bad moves. If you need these low stiffnesses, consider using element types with less degrees of freedom, such as replacing beams with minuscule bending stiffness with bars, which do not even consider bending moments. Alternatively, can you leave the unstiff elements out of the model?

Some programs have a model stability diagnostic tool[55] designed to find the locations of low or zero stiffness as well as the locations of highest stiffness. If you don't have access to this, a modal analysis can find your locations of low stiffness, though note this will fail if you have any parts with no stiffness at all. Another option is to use a nonlinear explicit solver, which can deal with mechanisms more effectively, and see what parts of the model move too much.

7.1.1.3 Total loads and reactions

So how do you know how ill-conditioned your model is? Some programs publish a 'condition number', which is a mathematical assessment of the reliability of the matrix solution[†]. Another is to look at the total loads and reactions, and see if they cancel out.

[†]   See the program's help file or documentation for interpretation of the condition number.

You should be able to export, or at least calculate, the total loads and total reactions from any program and see what the difference is between them. In the real world, all actions have an equal and opposite reaction, but in the digital world we can only get close. Or not close at all if the model is ill-conditioned, or if a nonlinear solver has not sufficiently converged. Check these numbers — as missing loads and/or reactions are also a big clue to problems.

7.1.2 Warnings and errors

Speaking of problems, you may get warnings and errors when you run your analysis. READ THEM. I have been asked so many times why some analysis has gone wrong, and I have worked out the problem just by reading the warning messages from the solver.

Some warnings are just warnings because you have modelled something unusual. You are the engineer and sometimes you do need to build unusual models. Take responsibility, read the warnings, and decide whether you need to act on them or not. You probably should act on 'severe warnings'; while 'errors' will stop the analysis, so you have no choice but to resolve them.

We do not know what warning messages the software gave, if any, when modelling the Hartford Civic Center. We do know there were warnings on site, such as the deflections being twice what was predicted. This should have sounded alarm bells: if the deflections are doubled, what else is double the prediction? One thing we do know, is the effective lengths of some major members were twice what was assumed in the design.

7.2 Getting the units wrong

As I mentioned, I do a lot of software support for customers and fellow employees, and the most common errors I see are the misuse of units, missapplying signs, and mistakes in orders of magnitude, whether in dimensions, material properties or loads. The mistake is easily made and takes very little effort. And if you put incorrect values in to your model you will get the incorrect answers out.

One of the easiest ways to misuse units is to get confused between metres and millimetres. In either case your model or parts of your model will be too large or too small by a factor of 1,000. Likewise, confusing newtons and kilonewtons will put your forces out by a factor of 1,000. Combine the mistakes and the error might be wrong by a factor of a million or more. Also, pay close attention to the section size units. You are probably safe with steel beams from the software catalogues, but user-defined plate girders, concrete beams and slabs are classic items to scale incorrectly.

So how might you easily check the model scaling? Simply display the sections on the analysis elements. If the beams and columns remain as lines, either the model is too big or the sections too small. If the lines completely vanish it's probably the opposite and your viewpoint is now inside the beams.

To check the loads, aside from a glance at the model data, run a static linear analysis and look at the deflections. If the results are in micrometres or kilometres then you've probably made a mistake.

If it's self-weight deflection, check you have applied the right materials: a colleague once modelled a building with a 200mm steel slab and then wondered why it didn't vibrate as much as he expected. Another material error nearly caused a bridge design to fail a CAT 3 check: one model used the long-term concrete Young's modulus and the other used the short-term. The change in stiffness resulted in very different moment distribution in the models. Inclusion or exclusion of shear stiffness in analysis will likewise change the moments in a continuous structure.

Another thing to check is the deflection direction. Building modelling programs are dedicated to those structural types and thus typically expect gravity loads to be positive. On the other hand, with FEA programs you usually define the loads with respect to the model's global axes. This means vertical loads, such as those generated by gravity, must be negative. The global Z axis goes up, so gravity loads must have a negative sign. It's an easy mistake to make and an easy one to check: just switch on the load graphics and make sure the arrows are going in the right direction.

It's not just gravity loads that can be applied in the wrong direction. I was told of a 12m high, 12m diameter reinforced concrete water tank that failed when first filled, with cracks in the concrete big enough to put your hand into. The circular tank had been modelled in 3D, loaded, analysed, then designed using those results. The problem was that the water pressures had been applied pointing in not out, putting the concrete walls into compression. Rather than being detailed to withstand the hoop stress, the walls instead only had minimum rebar, which was wholly inadequate when the water pressure acted in the correct direction[†].

Applied moment graphics can be harder to check, so examine them closely. Your software will use either the left- or the right-hand rule for moment directions, where your thumb is the rotational axis and your fingers thus point in the direction of positive moment or rotation. Make sure you know which it is.

A quick sum to calculate the total loads, which you compare to the model results, is also very useful.

7.3 Exercising restraint

7.3.1 No restraint
All models need boundary conditions, which in our case are usually restraints. Restraints are fixed points of zero translation and/or rotation that prevent the structure accelerating off over the horizon or towards the centre of the Earth. The restraints usually model where the structure is connected to the rest of world, such as the foundations, or can represent another larger structure the model is connected to, such as the column positions in a floor model.

Appropriate boundary conditions can be very difficult to determine. When I was at university, I knew a PhD student looking at the vibration characteristics of the International Space Station. The space station can have no external restraints: it's floating in space! This means you would need a special restraint that requires the overall displacement to be zero, rather than restraining individual nodes.

Aircraft have a similar problem, but they can deal with it by creating sub-models. The aeroplane structure is usually broken down into wing and fuselage models. The wings cantilever from the fuselage connection and the fuselage is supported at the wing locations. These are obviously particular cases and not ones that we should normally expect with our bridges and buildings.

If there are no net forces in a direction, say horizontally, does this mean you do not need restraints in those directions? The answer is you **do** need them, otherwise your model will have a zero stiffness in that direction. While the structure may not need the restraint, the matrix solution does, otherwise a divide by zero will occur. Also, the real world has friction, so there will be restraint even if it is not used by the construction. These additional restraints may be horizontal ones for vertically-loaded structures, or out of plane for plane frames and beams. Essentially do not give your model the opportunity to either slide or rotate as a whole in any direction.

Forgetting restraints is a classic mistake, though one that the FEA programs are usually good at spotting and informing you of before actually running the analysis.

7.3.2 Too much restraint
When you are modelling a statically determinate structure such as a pin-ended steel frame, you can model the supports as pinned or sprung without changing the force/moment results, but there are some structures where you do have to be careful regarding the supports. One obvious example is the portal frame, and it's sibling, the arch. Both rely on the horizontal reaction to function correctly so in both cases you must include that restraint in your model.

A more subtle problem is caused by putting too much restraint onto the model. I am sure we all remember first year structures classes where the lecturer was keen to emphasise that a beam model should have a pin restraint at one end and a roller at the other. Or to put it another way, have one and only one restraint in the X direction. As an undergraduate I was a little confused by the emphasis on this. If you have a simply supported beam with a nonlinear analysis, then the sag will generate pull on the supports. But with linear analysis, it doesn't make any difference to the result if the beam is pinned at one end or both.

[†]   The contractor's foreman had reported that the tank was under-reinforced, but the site resident engineer insisted that it was correct. Always listen to those with experience as they might be right.

The reason for this, though I did not realise it at the time, is because we are assuming the beam is supported on the neutral axis. But what happens if it is not? It is true, that making one end a pin or a roller makes no difference to the result if the line represents the neutral axis of the beam, but once we move the support away from the neutral axis, the support conditions become a lot more important. This is easiest to see with trusses as they are generally deeper than solid beams and thus the effect is more pronounced. Take the truss bridge in Figure 7.7 as an example, where the supports are on the bottom surface. One version has pins at both ends and one has a roller release.

Figure 7.7: Two similar trusses?

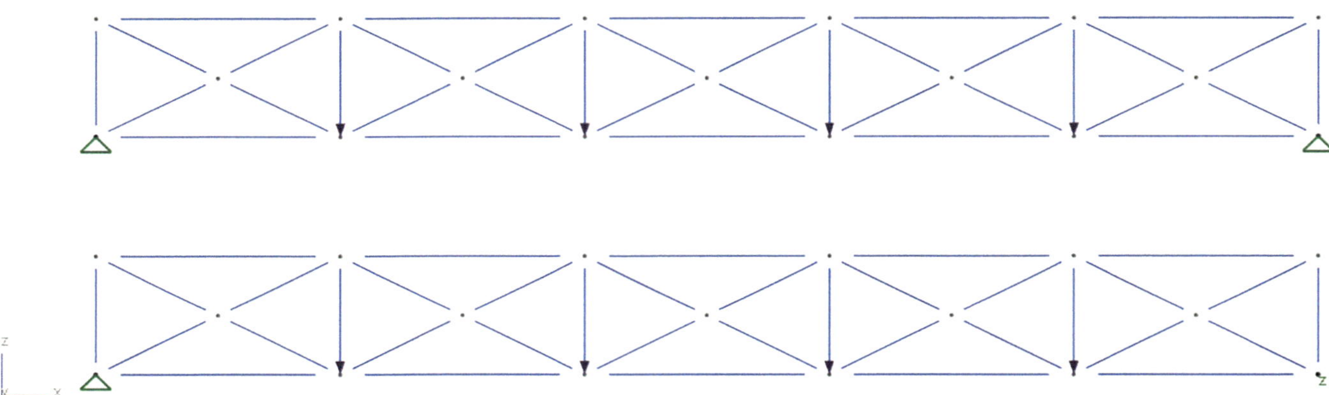

You will note the distribution of axial loads in Figure 7.8 is very different in the two cases, as are the reactions. In the double-pinned example the bridge is actually acting as an arch (remember that compression structures are more efficient than bending ones, so the structure will want to work in the most efficient way — in compression). The bottom chord will have virtually no tension in it and thus is especially vulnerable to under-design.

Figure 7.8: Simply supported truss. Or is it an arch?

If you do analyse and design your model with this double-pinned support arrangement, the real structure will want to generate those reactions. To do this it will push the supporting structure until either the reactions are generated, or it is acting as a beam. The risk is that either the supports or the bottom elements will be overloaded and fail. In the truss with the roller support we can see that the axial forces are balanced and thus the truss is acting like a beam, as we would expect.

If we have assumed the supports to be infinitely stiff there can be no movement at the support, otherwise the analysis will not mirror reality. The problem is that zero movement is impossible as any real support will have a finite stiffness: whenever there is a force there is a stress, and wherever a stress there is a strain, and therefore a movement. This means the final distribution of forces in the truss will be somewhere between the two analyses. To analyse this truss with pinned supports really requires you to replace the horizontal constraints of zero movement with an appropriate spring stiffness (Figure 7.9). This stiffness can be very tricky to establish as it is dependent on

Figure 7.9: Truss with lateral support stiffness

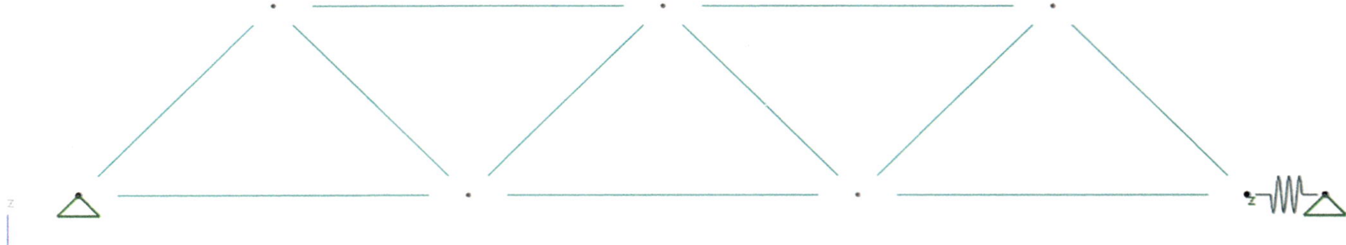

the stiffness of both the supporting structure and the ground. This spring turns the truss into an indeterminate structure as the stiffnesses of the truss elements will now affect the final distribution of the element forces. Bridge engineers like to avoid situations like this by putting a bearing under one end.

The right-hand support, whether sprung or on a roller, will move outwards under gravity loads as the supports are not on the neutral axis of the truss, which is ok if the supports allow that movement to occur.

In the truss model the vertical stiffnesses didn't matter because it was just one span. But what if there was more than one? Two or more spans make the continuous beam statically indeterminate and thus sensitive to the support stiffness. Here we have one example, with a UDL and the bending moment is as you might expect (Figure 7.10a).

But if we replace the zero-displacement supports with spring supports you will note that the bending moments are quite different (Figure 7.10b). This redistribution of the moments occurs not just because of yielding in the beam, but also through differential support movement.

Figure 7.10: Effects of support stiffness on moment distribution

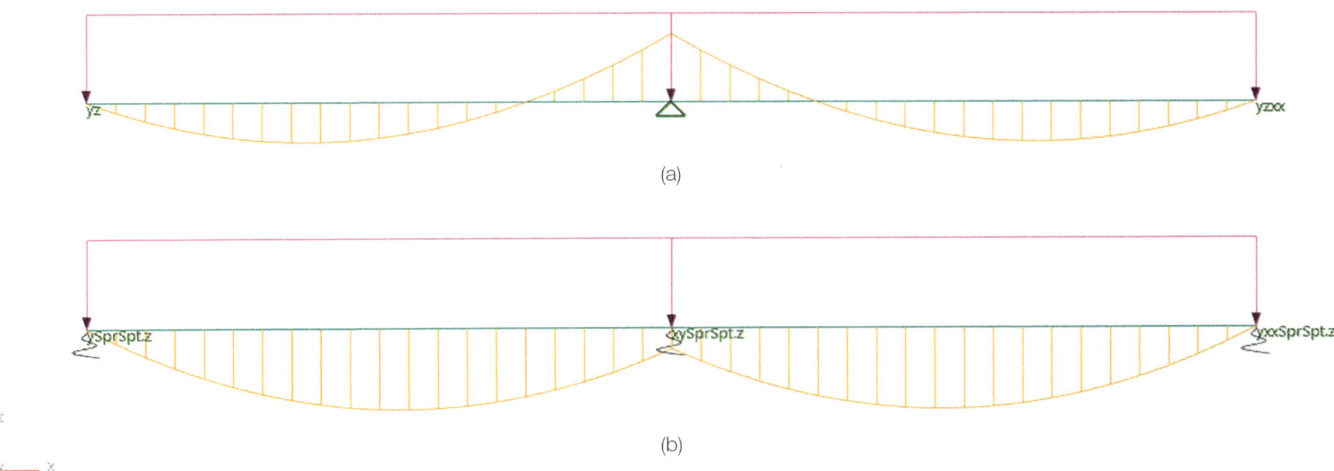

It is important to determine what the support stiffness may be for statically indeterminate structures. We cannot be certain what they are, especially in advance, so you are best determining what the range of stiffnesses might be, analyse the structure with first one extreme value and then the other, and checking what difference they make to the forces and moments. If the supports are considerably stiffer than the structure, the forces will be little changed, and we can say that it is not sensitive to the support stiffnesses. Very soft supports will have a large effect on the forces, as well as on the deflections: the influence of the central supports will diminish. You will get a similar behaviour on the upper storeys of tall concrete structures, when the column axial stiffnesses begin to influence the beam and slab results.

I first discovered this problem when, as a graduate, I needed to model a concrete reservoir (Figure 7.11) and find the moments and forces in the wall and base. As usual, I modelled all the restraints along the base as pins as I knew that it was continuously supported by the soil.

Figure 7.11: Fully restrained reservoir cross-section

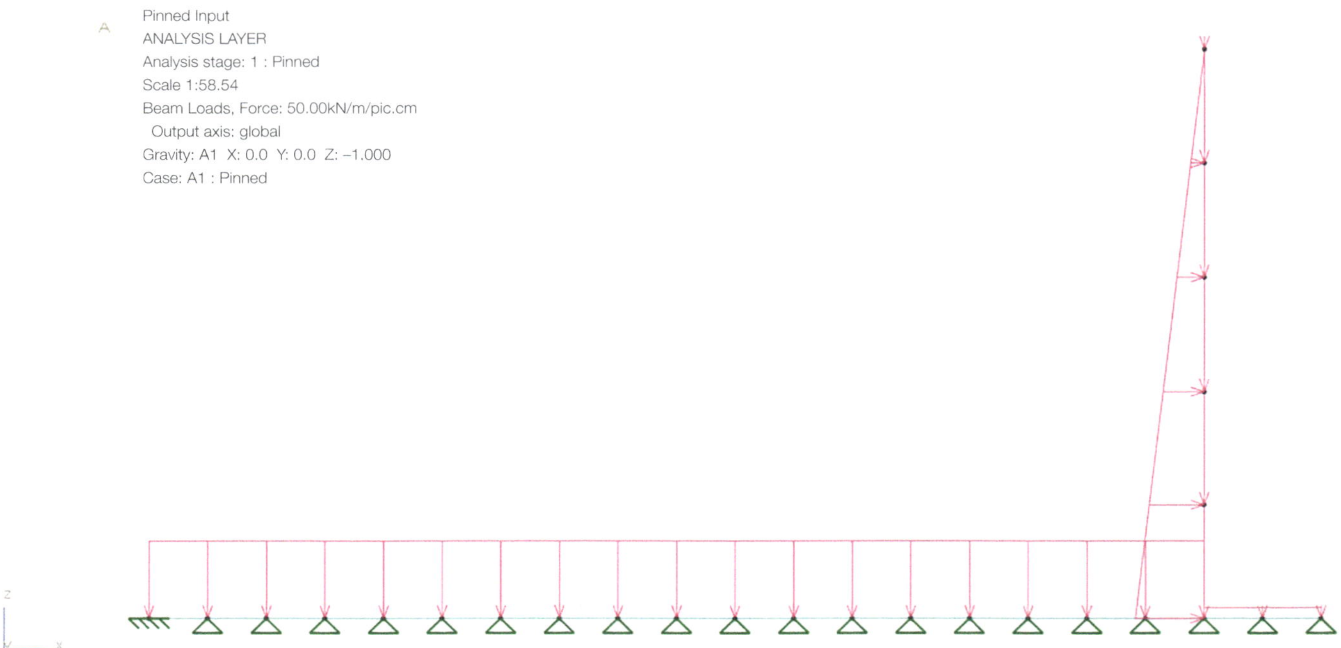

The wall moment was as expected but not in the base. I was expecting a smooth moment distribution but instead got zig-zag moments and counterintuitive reactions (Figure 7.12).

Figure 7.12: Fully restrained reservoir results

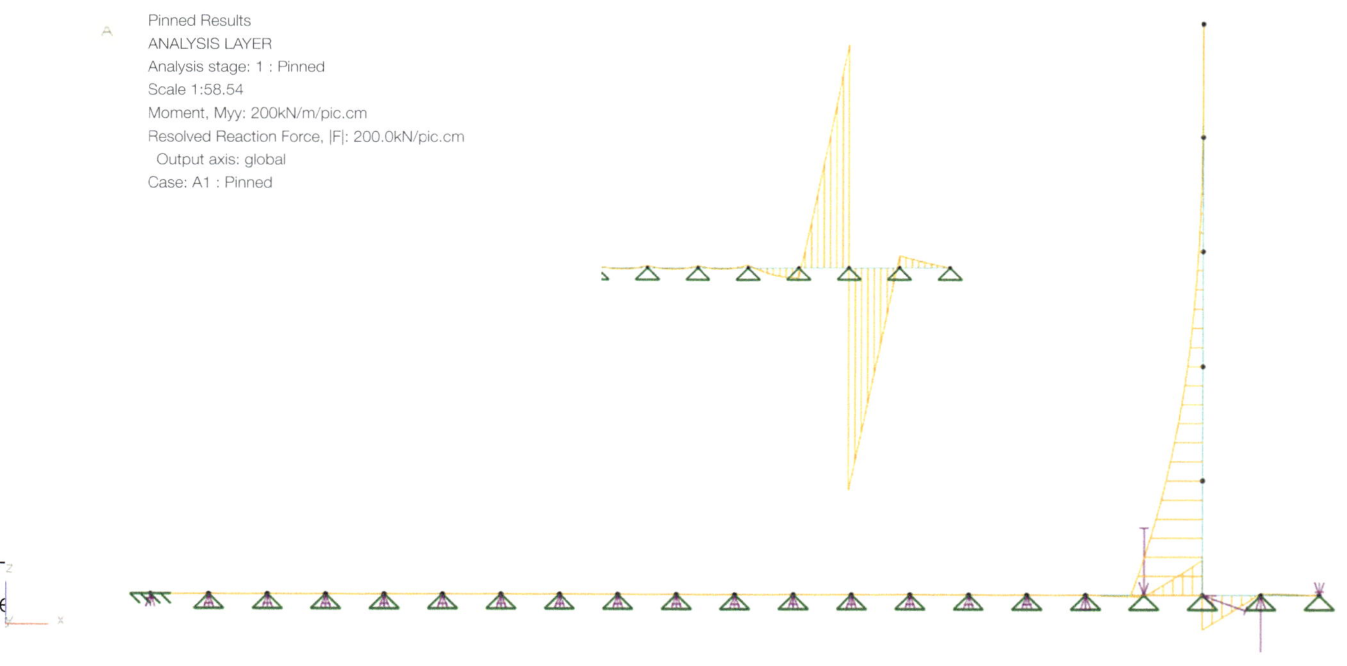

Figure 7.13: Spring-supported reservoir results

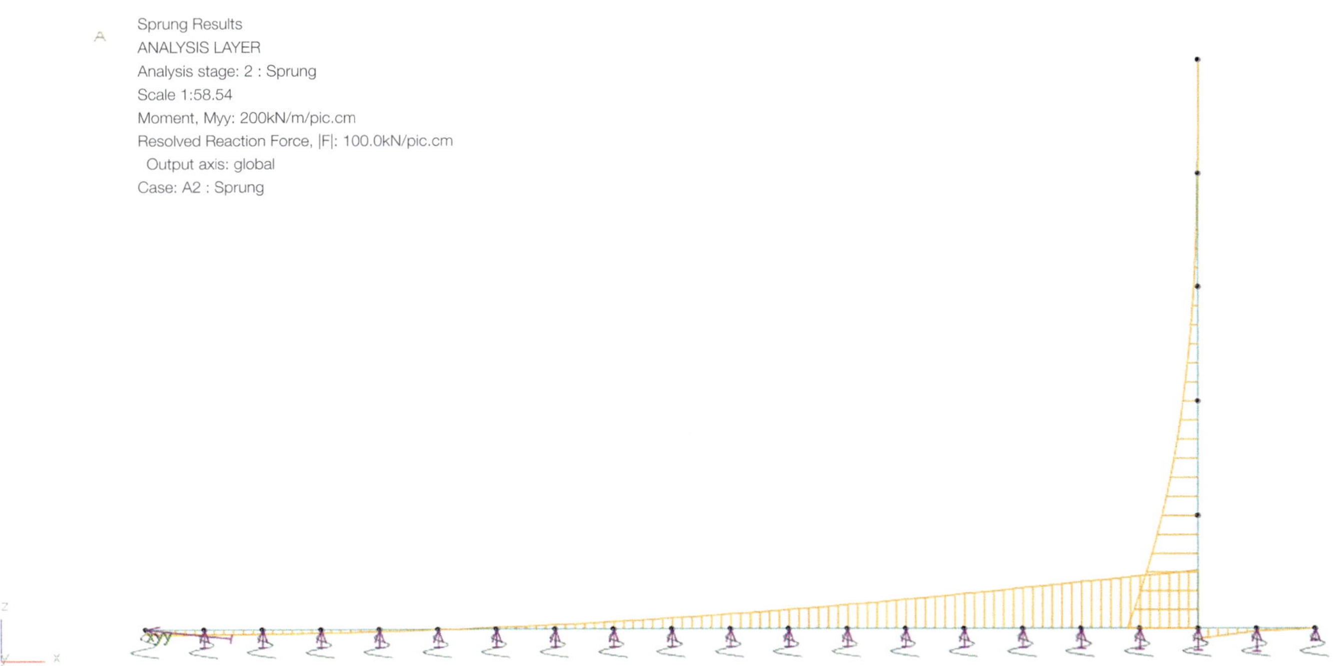

Today, having more sophisticated FEA software I might go further. With the springs, I have just assumed a constant soil stiffness under the base, but in reality there will be an interaction between the soil stiffness and that of the structure. This gives higher stiffnesses around the edge of the foundation and lower stiffnesses in the middle. We can model this using a two-dimensional plane-strain analysis (Figure 7.14) or, to account for the corner stiffening effects, model it in three-dimensions.

Figure 7.14: Plane-strain reservoir results

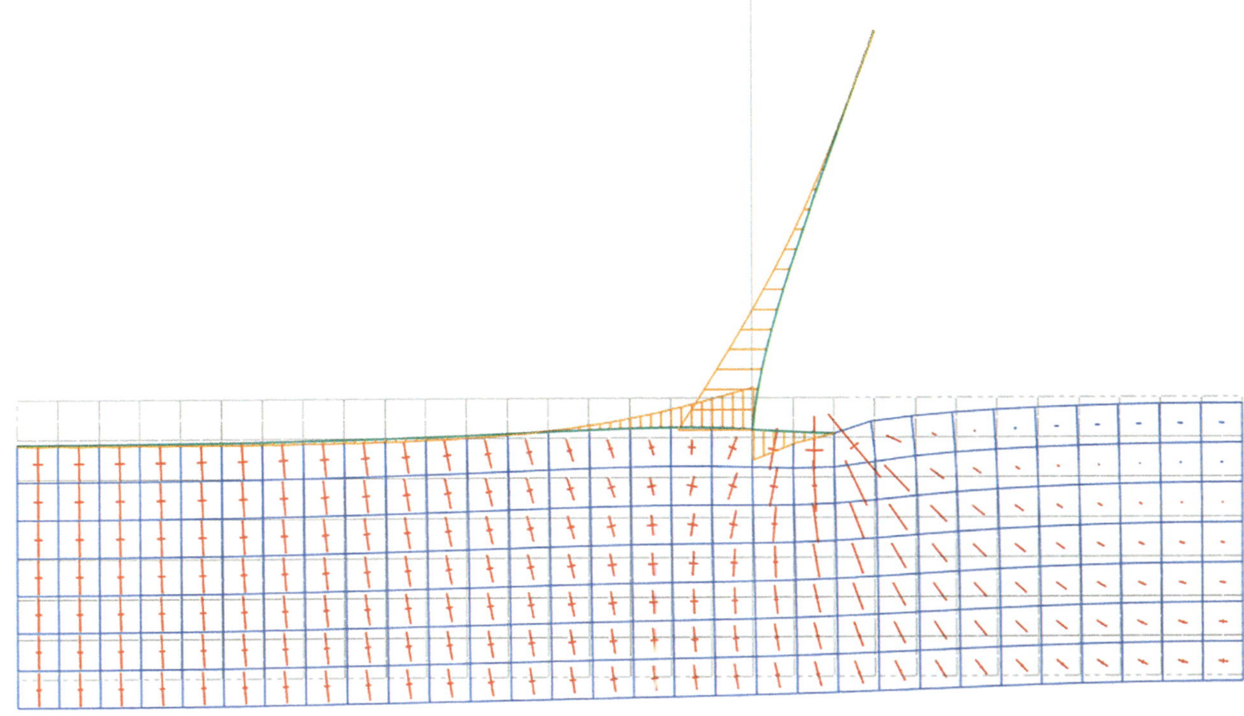

7.3.3 Confusing 'restraints' with 'releases'

I have seen a few problem models where the user has confused 'restraints' with 'releases'. It may be to do with them both being referred to as 'pinned' and 'fixed' or that both begin with the letter R. One example that springs to mind is: "I have put pinned restraints on the ends of all my beams. Why am I getting no axial loads in my columns?". The answer was that they should have used pinned **releases** instead. Restraints are the points where the structure joins to the rest of the world (foundations, primary beams, division on a line of symmetry, etc.) and can generate reaction. Releases allow the elements to replicate the behaviour of the physical connections within the structure (bolted fin plates, web cleats, etc.).

Over-releasing elements can cause other problems. When using beam elements, don't release all the end moments at a node, otherwise you will have nodes that have no rotational stiffness and the analysis can fail. Similarly, because primary members will be broken down into multiple elements, do not release the end moments on all the beam elements as that will also break the analysis[†].

Beware of mixing trusses with rigid constraints or diaphragms: if the truss chord is included in the floor diaphragm that chord will carry zero load, while the diaphragm will carry all the load.

Do use bar elements in trusses where you can. Bars only have an axial degree of freedom, meaning they cannot carry moment and need no end moment releases. As there is no moment degree of freedom, the rotation of the end nodes are not even addressed in the analysis.

Do not use bar elements if the internal nodes are laterally unrestrained. It's fine if we analyse trusses with bar elements in two dimensions as the nodes are restrained out of plane, but if we look at the trusses element from the Hartford Civic Center (Figure 7.15), we can see that while the top chord is restrained by the diagonals internally, it is unrestrained horizontally at the edges[††].

Figure 7.15: Hartford Civic Center FEA model (recreation)

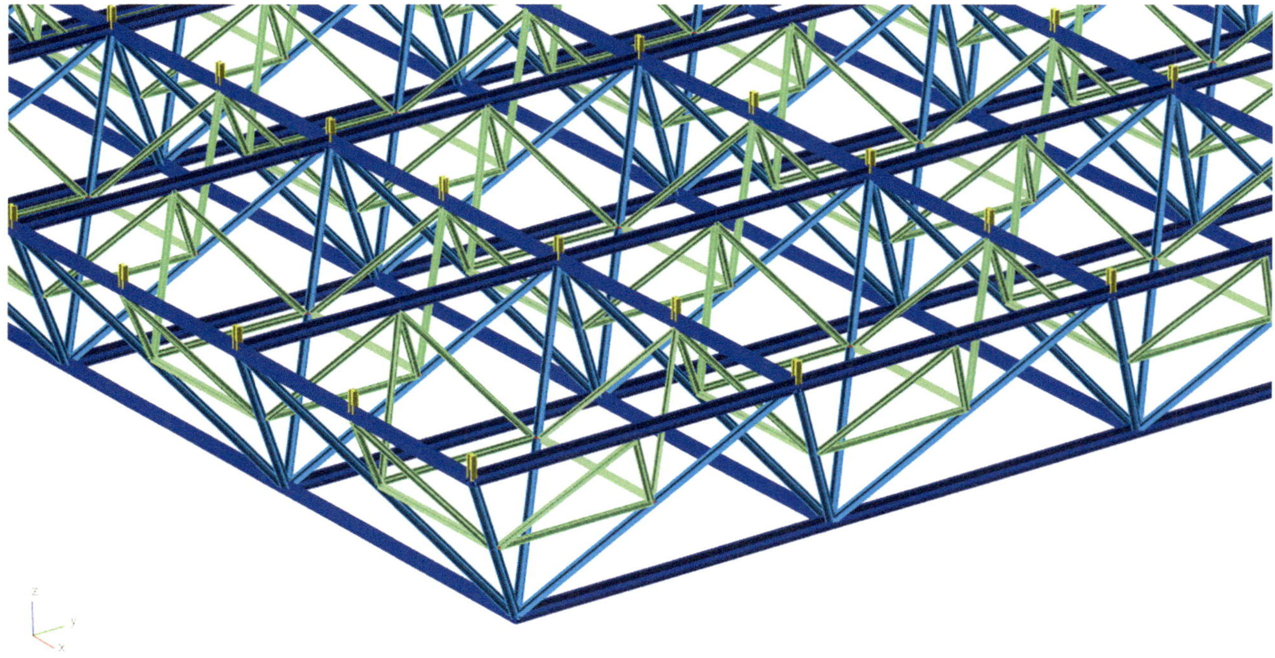

There is another release that can catch you out: not actually being connected at all! For FEA to work, the elements must share nodes, but you might have elements that pass through the nodes, or possibly 'coincident nodes' may occur. Coincident nodes may arise when you move an element, action an import, edit coordinates in the table, and so on. The downside is that while the nodes appear as one in the graphics, they will separate during the analysis. A good program will be able to detect coincident nodes for you and allow you to merge them together. Otherwise you will need to display the node numbers and look for the bad text — or run an analysis and spot the parts that are deflecting far more than they should.

A similar problem to coincident nodes is that of coincident elements. Here, the nodes might be shared, but there are multiple elements all contributing stiffness, self-weight, and so on. Occasionally they can be a useful thing, as sometimes you may need, say, a damper and a spring element working together; other programs might not allow them. Again, a good program can detect coincident elements and resolve them, otherwise you need to look carefully at the element number display.

7.4 Forgetting torsion

Torsion is something that structural engineers are often careless about. Yes, we are all aware of lateral torsional buckling, but the other types are too often forgotten. And you can get into serious trouble if you ignore it.

Here is an example, based on a real case. The design engineer in question was tasked with the structural design of a masonry house, which needed a piled foundation. I understand that they used an FEA (Figure 7.16) to load up the beams, extract the moments and shear forces, and design for those. Just as any of us might. But that careful design contained a fatal oversight. During the construction of the masonry walls, the suspended ground beams started to fail. Thankfully, this failure was slow, due to the nature of masonry construction, and no one was hurt.

Figure 7.16: Piled ground beams

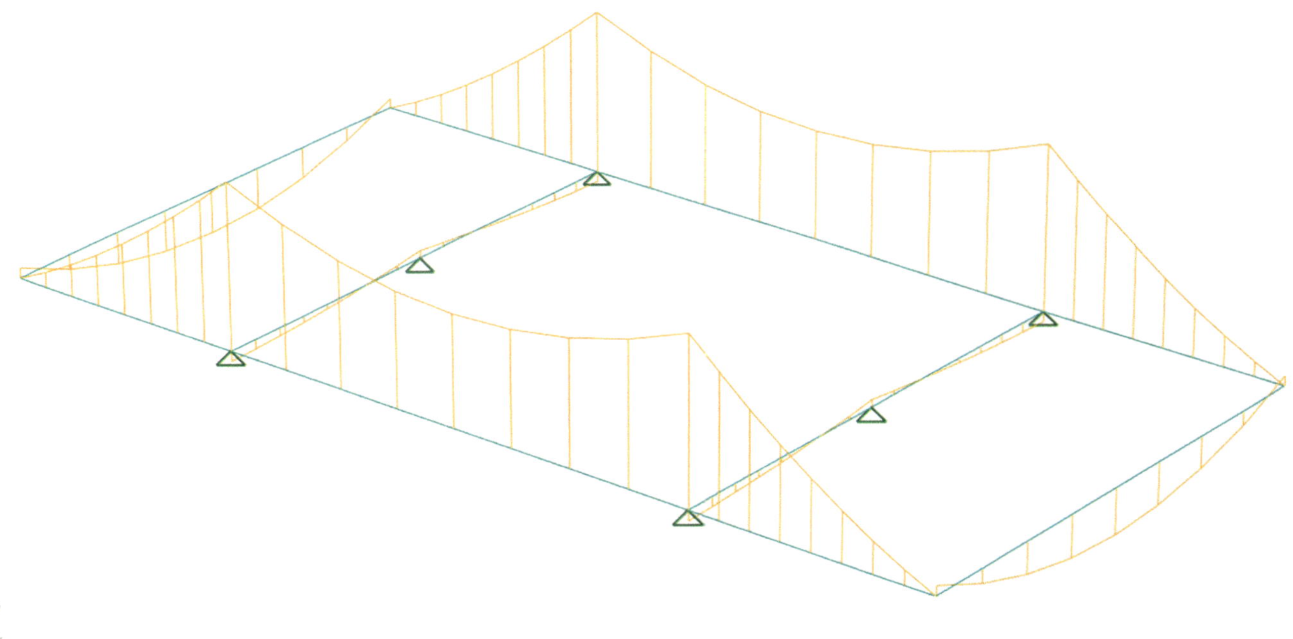

If you look at the results on this frame (Figure 7.17) you may note that there is something curious about the moments in this corner: they do not match up. Forces and moments must be preserved, as for every action there is an equal and opposite reaction. So where had the missing moment gone?

Figure 7.17: Missing moment

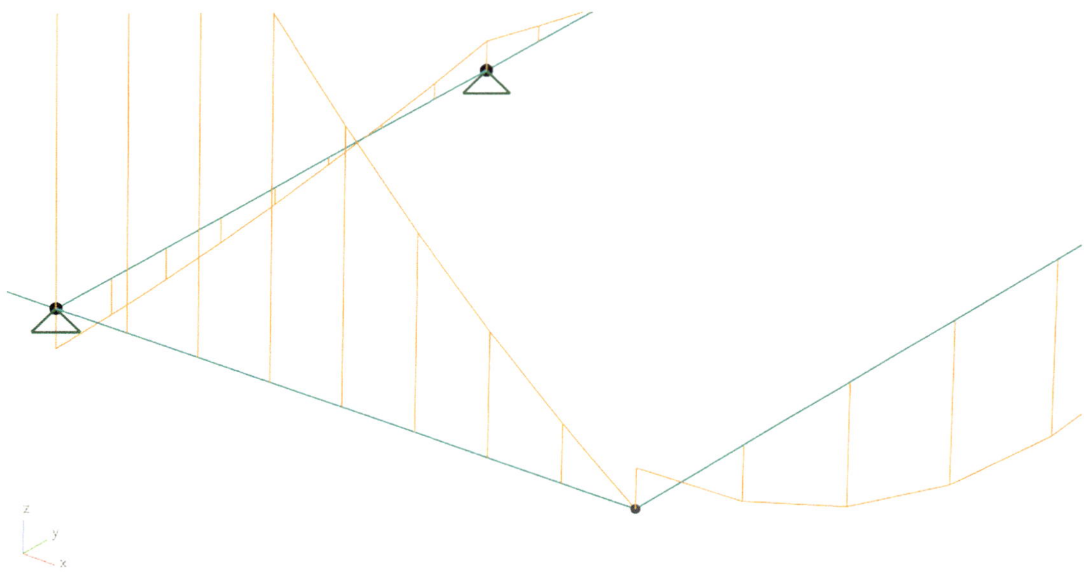

The answer is: into torsion. While there were sufficient links in the beams to deal with the shear forces, they were insufficient to resist the addition of torsion, and yielded. The lesson is that you must design for all the applied forces and moments, not just the common ones.

You will get torsion in concrete flat slabs, so be very careful if you are extracting the moments to do the rebar design. Most in-built reinforcement design routines will use the Wood-Armer moments, which are a combination of the principal direction moments (Mx or My) combined with the twisting moment (Mxy), but you might need to switch that aspect on[90].

Some structures need consideration of torsion in their design, in others you want to avoid it where you can. For example, when using open steel sections such as I beams. Unless you have explicit off-centreline loads, such as masonry shelf angles, you would not expect torsion in a simple steel frame, though it can still happen, so always check for it.

While not directly related to FEA, the designers of the Hartford Civic Center decided to use angles in a cruciform section, which are highly vulnerable to torsion, and in this case, axial torsional buckling. We might check for buckling in the major and minor axis, but how often are checks made for buckling about the third axis?

7.5 Misusing offsets

7.5.1 Bad offsets
Unlike 3D finite elements, where the structural shape is modelled explicitly, by default, all 1D and 2D finite elements are defined about their centroid. So, if you apply a lateral load to one you get a moment, an axial load produces an axial force, and so on. This means that when modelling floor beams, we often need to take a few liberties with the overall geometry to get satisfactory results. This is also where the requirements of FEA differ to CAD, blurring the integration of BIM.

Let's take an example of steel beams supporting a concrete floor. In the real structure all the top surfaces of the beams are aligned at the same level to enable the supporting of the slab. But the FEA beams are all modelled at their centroids, so does that mean we should work out what size all the beams are first, carefully work out what level all the centroids will be, and then model them at those levels?

If your beams are non-composite and have pinned connections the answer, thankfully, is no. It will make no difference to the behaviour of the beam elements if you change their level a little, as they are all statically determinate. This approach also avoids the problem of the level differences requiring very short, and hence excessively stiff, column elements within the connection. Remember that we need to avoid very short elements as they reduce the accuracy of the finite element approximation.

But what level should they be at then? Is top of steel ok? It is ok for the beams but not for the columns, as it will overestimate the ground to first floor length. But in its favour, it does make the checking easier, and if you are manually calculating the column effective lengths then it shouldn't make a significant difference to the result. But if you are automating the column buckling, or any other analysis where the effective level of the beams is critical, then you need to either use a more reasonable beam level or carefully adjust the foundation level.

But this section is about offsets: what are offsets?

Offsets are rigid links (infinitely stiff elements) that connect the support nodes to the element nodes. Many programs allow you to create these as part of the element properties, but with others you may need to explicitly create link elements and model the nodal locations appropriately. This means you can put all the nodes at the top of steel level, with all the beams offset down so their top flanges align. The result will then look just like the CAD model. A win-win scenario? Unfortunately not, as unless you are modelling, say, composite action with concrete shell floors, the result will be less than helpful.

Let us compare two otherwise identical beams: one modelled on the centroid and one offset by half the depth (Figure 7.18). You will note there is a huge horizontal reaction, which will have an accompanying axial load in the beam. If we had used a roller support, we would have a horizontal deflection instead.

Figure 7.18: Effects of offsets on beams

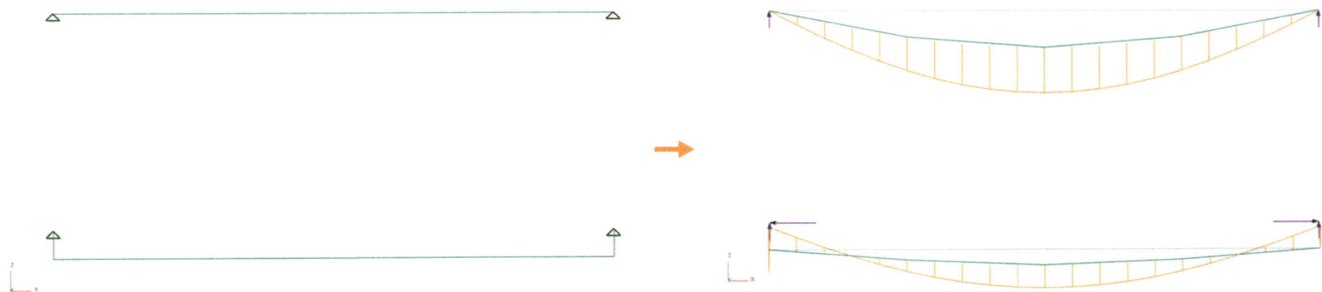

On the other hand, if you do want to model a composite beam (Figure 7.19), then you do get the composite axial forces in the beam and slab — just as you would expect. You will get some discontinuities in the moments and forces at the connection points though. The ideal is a fully connected beam/slab, but that would require far more elements than is appropriate in a large frame. You may also note that modelling partial interaction between the beam and slab is tricky in FEA and will require some careful thought.

Figure 7.19: Composite beam and slab model

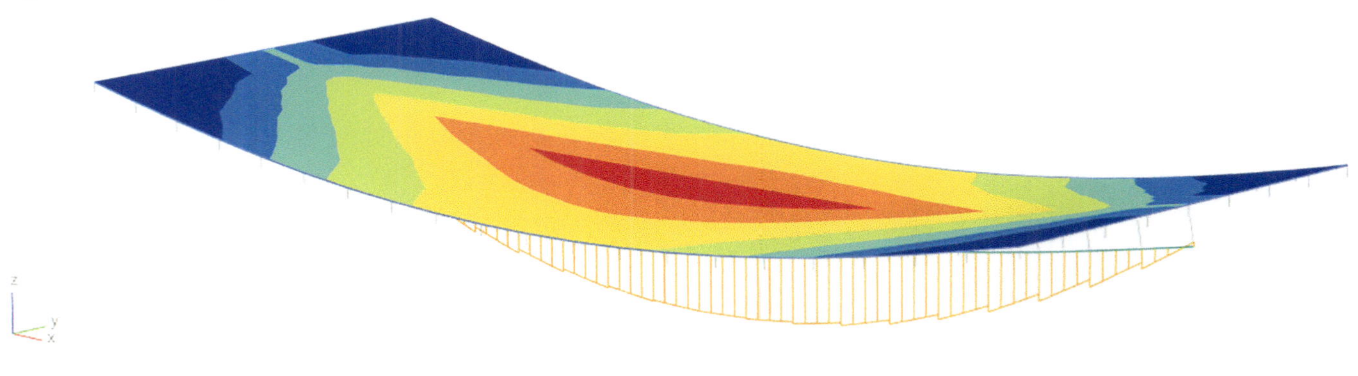

Modelling an entire floor in this way can present problems. The slab will be automatically continuous with the adjacent bay. This is ok for a floor vibration model but not for the ULS condition if the concrete is designed to only span between the beams. In these cases you might either model isolated parts of slab attached to the beams, adjust the beams to include a block of concrete (or equivalent steel) above the top flange[†], or just ignore this aspect in your model and use the forces and moments in a separate composite beam design program.

In summation: while the CAD model is there to represent the physical reality of the final structure, the FEA model needs to instead model its behaviour. Don't get confused and try to make the FEA model look right at the expense of its function.

7.5.2 Good offsets

When should you use offsets? As mentioned, you can use offsets for composite beams, but they are also useful when modelling the behaviour of connections. One of the problems with the simplicity of 1D element models is that they do not automatically generate eccentric loads on the columns[91]. The steel design codes ask that the beam loads on the columns include an offset of half the column depth plus a factor, to allow for the actual point of load transfer in the bolted connections. This eccentricity will generate a moment on the column that needs designing for. I would expect a dedicated steel column design to automatically include these eccentric moments but FEA programs find this difficult as they are dealing with arbitrary elements[††].

So what end offset should you use?

- If no offset, the beams will be slightly over-designed as you are using too long a length, but the columns will be under-designed as they lack the eccentric moments
- Conversely, if you use the code-based offsets the columns will be just right (if you appropriately patch the loads, which is not easy in a three-dimensional model), but the beams will be under-designed
- You can compromise and offset the beams to the column face and get the beams right, but slightly under-design the columns

Or make two separate models, one to design the beams and one the columns. But make sure that you are not losing any column axial load by shortening the beams!

I will leave that decision to you. But before you decide, let us return to the Hartford Civic Center. Here the space frame was modelled and analysed using cutting-edge technology (as FEA was in the early 1970s) and the engineer even negotiated some extra fee to cover it. As was, and possibly still is, standard practice, the truss elements were all modelled to the intersections, but if we compare the design details to what was built (Figure 7.20), we will see that they were not the same. The members are offset from each other, resulting in significant additional, and undesigned for, moments and torsions on the cruciform sections, which we have already noted are not strong in that regard.

As you can see from the connection details, the as-built connections are not only less restrained than as-designed, but the offsets will generate large moments and torsions in the elements that just didn't have the capacity.

Gathering as much original data as I could, I modelled the Hartford roof with and without the offsets[92]. My answers were by no means definitive as I didn't have accurate section sizes or details, but I did find that the as-built first buckling mode occurred at just 75% of the capacity designed layout. Also, because of the 3D nature of the structure, all the connections are twisted. In other words, they would have had to twist the cruciform members to make them fit together. Neither situation would have helped the structural capacity.

To offset or not to offset? That is the question. Offsets are very powerful items in models, and you must use them wisely.

[†] Use the modular ratio to calculate the width of a steel slab equivalent to the concrete one.
[††] Just because the element is vertical, does that make it a column? There are axial forces, but where did they come from? Should all shear forces from connected elements generate eccentric moments or just some?

Figure 7.20: Hartford Civic Center truss connections

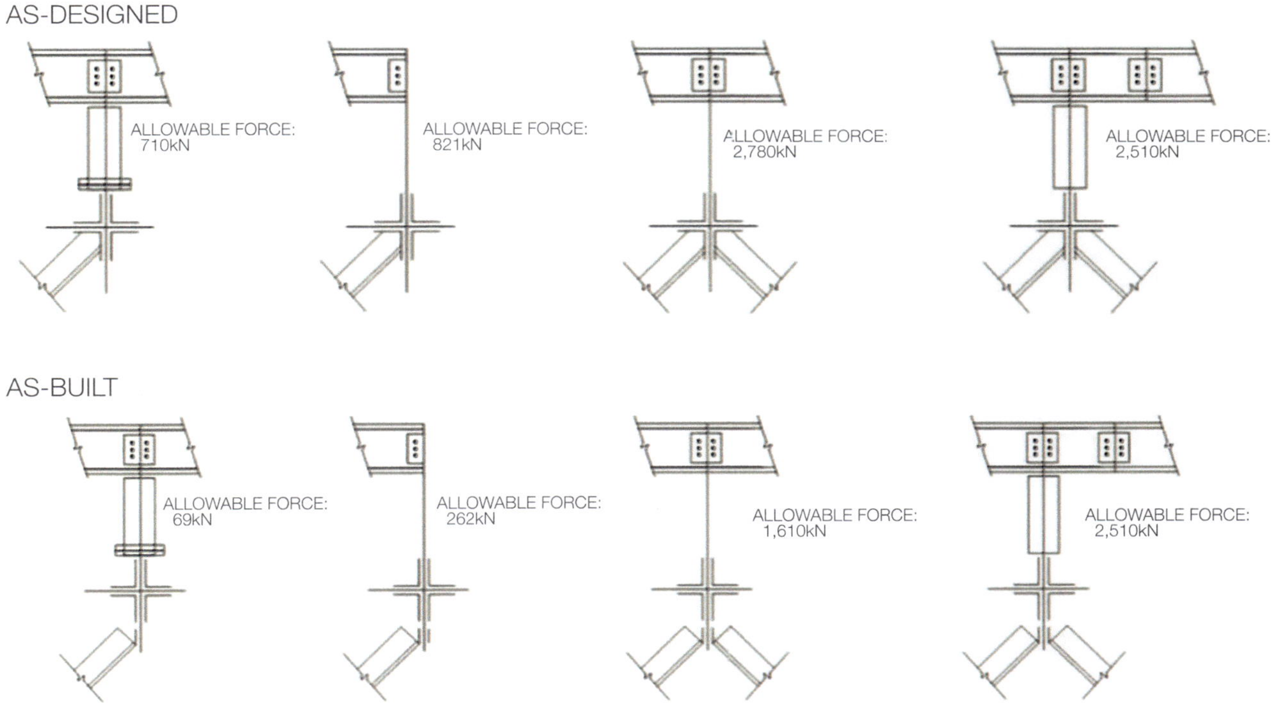

7.6 Meshing too coarsely or too finely

Another fine mesh...

Now we come onto another fine item in the art of engineering modelling: meshing.

When I was an undergraduate learning about finite elements for the first time, I was quite confused by 2D and 3D elements. I could understand that a 1D element represented a beam or a column as it fitted in well with our theories of beams that we had been studying since the first year. But with the meshing elements I thought that we were still modelling individual pieces of structure. In this instance, squares of steel or concrete joined at the corners. Surely that would create enormous stress concentrations at the nodes?

I had fallen into the trap of thinking that FEA is all about the elements, when in fact it is all about the nodes. FEA models work out how the nodes move under load, and thus how the elements strain and therefore stress. With 2D and 3D elements we are modelling a structure continuum. We make our analyses at nodal points in that continuum and calculate the stiffnesses between those nodes using the elements. This means that we were not actually looking at pieces of structure, but rather at points in the structure connected to all adjacent points with the most appropriate shapes (usually squares and triangles).

To get a reasonably accurate analysis of the real structure we need to decide how many points within it we need to include. These points must include the problem boundaries, such as edges and restraints, connections to other structural elements, loading locations, areas where the forces or moments are changing rapidly, and locations where we want to know the results. The spacing between these points then gives us our mesh density.

The point about locations where we want to know the results is a key aspect to the Sleipner A disaster. If we look at their model (Figure 7.21), it appears very promising. It is a large, complex structure subject to very large loads. It was designed to withstand over 80m head of water: 800kPa. The designers wisely took the symmetry into account in their model, meaning they only modelled a quarter of the structure. They also used 3D brick elements throughout for minimum abstraction on the stress calculations. The model was, n my opinion, well positioned to calculate the overall structural behaviour. The problem was that they also used it to examine the detailed behaviour.

Figure 7.21: Sleipner A model and details

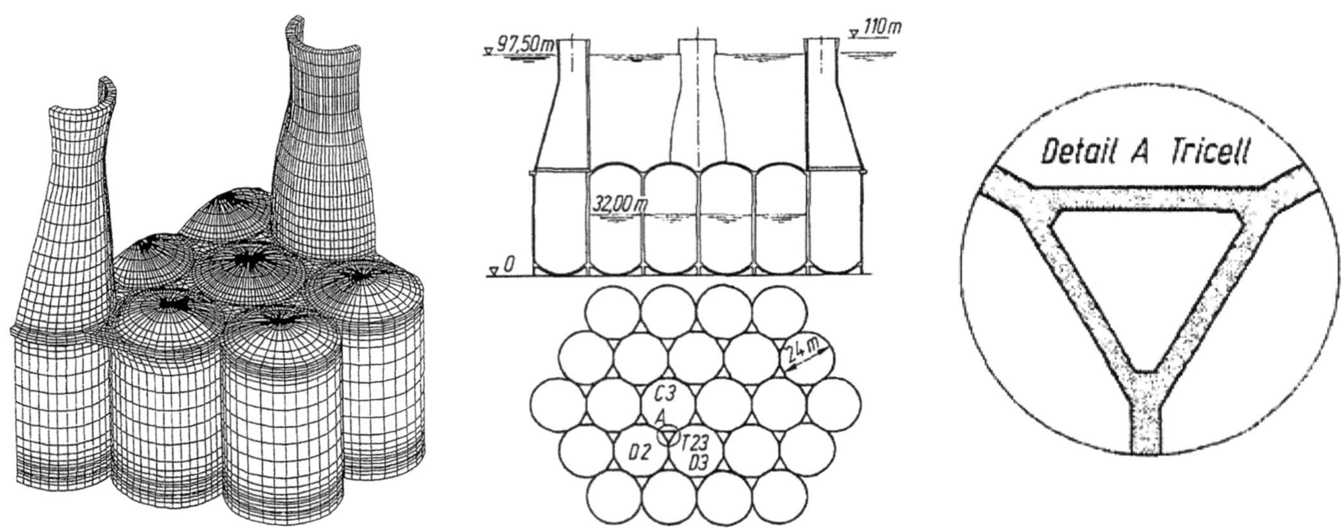

The basic form, when we look on plan, is a series of air-filled vessels that give buoyancy during transportation, enabling it to float, and the gaps in between that were open to the sea. In the service condition the vessels would be partially filled — first with sea water to weight it to the bottom (it is a gravity base design) and then with oil when storage was required. This means that the walls were subject to the highest forces during transit.

If we look at their finite element mesh at this critical location (Figure 7.22), we can see that the mesh density is rather low. Remember that we need nodes to calculate values at locations, though modern FEA programs often include additional points such as at the centre. In this instance the lack of detail underestimated the shear force by nearly 50%.

Figure 7.22: Sleipner A FEA detail

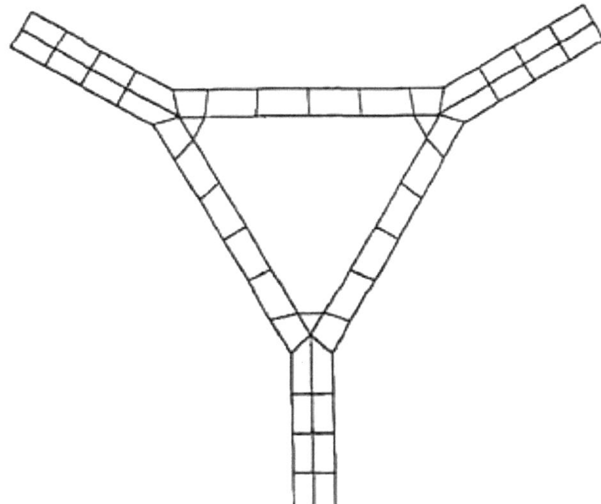

Nevertheless, disaster may not have struck if the designers hadn't detailed the concrete badly as well (Figure 7.23). The T-headed bars were quite inadequate for holding the concrete together under the shear load. I have heard that these areas had cracked badly on a previous structure, which means that the warning signs were already there, but overlooked.

Figure 7.23: Sleipner A reinforcement details and failure mode

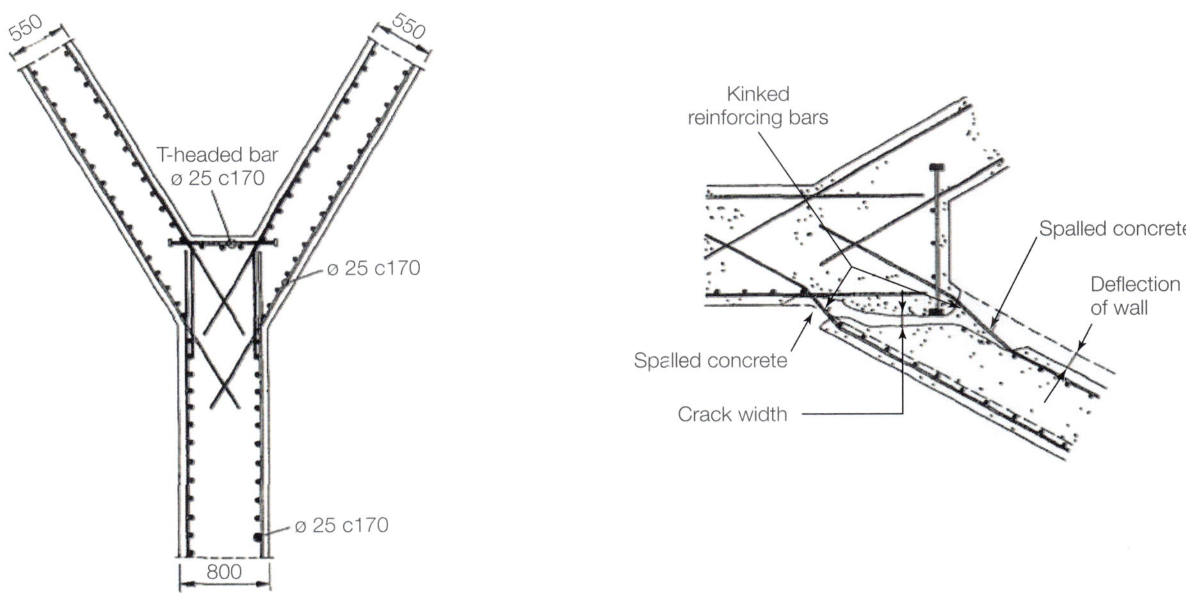

If we look at the average shear forces where the walls meet (Figure 7.24) we can see that even with a modern program, which calculates additional points, it will still underestimate the shear stress if you don't have enough elements. As a rule, with finite elements, you should try halving the size of all the elements to see if the increase in detail results in a change of result or increased accuracy.

Figure 7.24: Sleipner A shear stress analysis

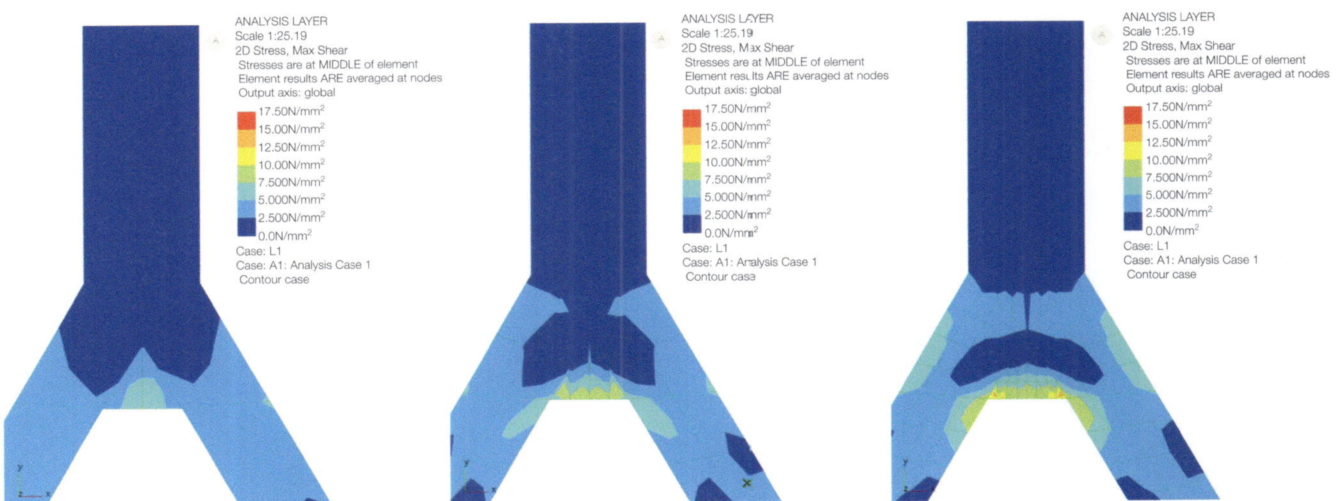

But increase the mesh density here and we see a different problem: the shear force in the corner keeps increasing as the mesh size reduces: there is a stress concentration. The danger is now that the more we refine the mesh, the more the stress increases. In situations like this, known as 'singularities', refining the mesh too far will produce unrealistic results, and care must be taken to keep things sensible.

This can also happen when you have 1D elements connecting to 2D element meshes, such as column to flat slab connections.

So: use too coarse a mesh and the result is underestimating the stresses. But, make the mesh too fine and you can overestimate the stresses. Did I mention that engineering is an art?

7.6.1 Other mesh errors

You can introduce errors to the mesh in many other ways too:

- Discontinuities in the mesh — unless it is on an edge, all nodes must be surrounded and connected by elements (Figure 7.25)

Figure 7.25: Bad (left) and good (right) mesh size transitions

- Changing element formulation within a mesh (i.e. from Quad 4 to Quad 8) as the mid-side node is only connected on one side (Figure 7.26)

Figure 7.26: Mixing Quad 4 and Quad 8 elements

- 1D–2D connections that do not have all required degrees of freedom:
  If the 2D elements do not have a 'drilling' moment degree of freedom the column will be free to spin (Figure 7.27)

Figure 7.27: Flat slab/column connection

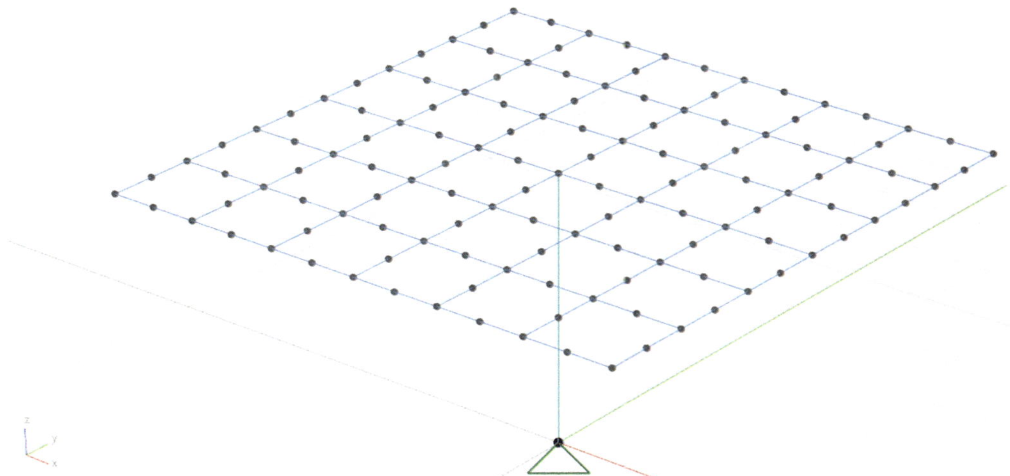

The shape of the elements will have an effect on the result. The algorithms assume that the elements are square or equilateral. The further the elements are from this, either from being elongated or having corners that are too acute or obtuse, the less accurate the results. The solution can include a finer mesh (so that the meshing engine does not need to produce such elements), splitting elements with obtuse corners or merging those with acute ones.

7.7 Second-order effects

7.7.1 Linear when it should be nonlinear
Linear analyses assume deflections are small enough so that you can find load equilibrium with the undeformed structure. Or to be more exact, although all structures are nonlinear, the structural behaviour is virtually the same whether you run a linear analysis or a nonlinear one.

With a nonlinear static analysis, you are looking for the equilibrium in the deformed position, so it is the appropriate choice for a structure that has large deflections.

Take this beam with a central point load (Figure 7.28) as an example. You can see the deflected shape looks sensible, as does the bending moment, and there is no axial load. But there is something in the small print: the automatic deflection magnification is only 2.5. Normally FEA deflections must be magnified to make them visible. This suggests that the actual deflection is extremely large.

Figure 7.28: Beam results

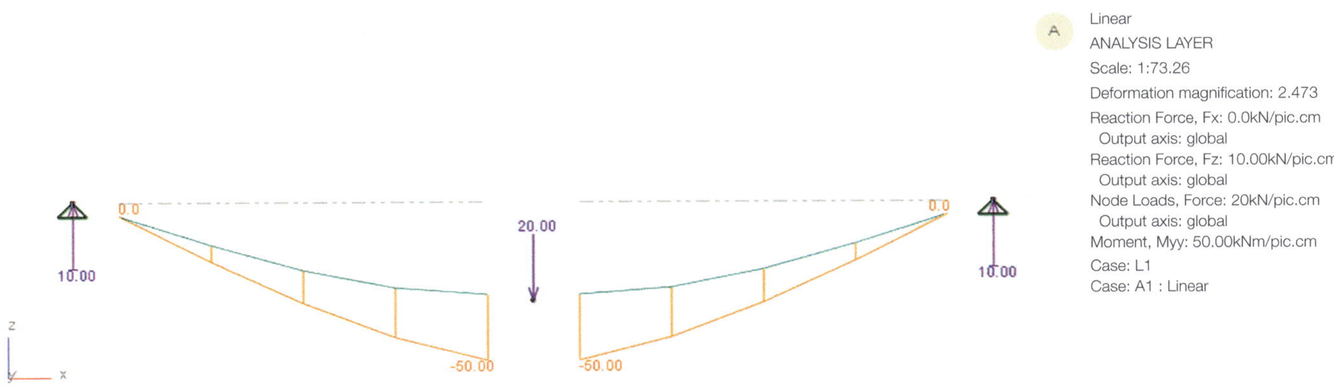

Looking deeper at the values, such as the stresses (Figure 7.29) we see that they are also excessive. This beam should have yielded, but it hasn't because we ran a linear analysis, which cannot consider such behaviours.

Figure 7.29: Beam linear stresses

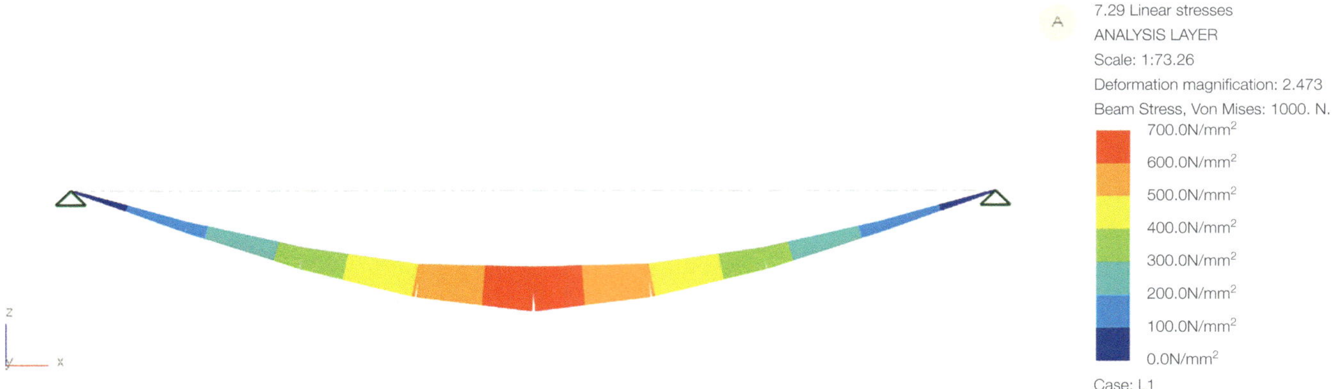

If we rerun the model with a nonlinear analysis we get much better results, including an axial load that was absent previously (Figure 7.30) and considerably lower stresses. You may also note that the deflected shape here is V-shaped rather than U-shaped, which is to be expected where a plastic hinge has formed and axial load dominates.

Figure 7.30: Beam axial force

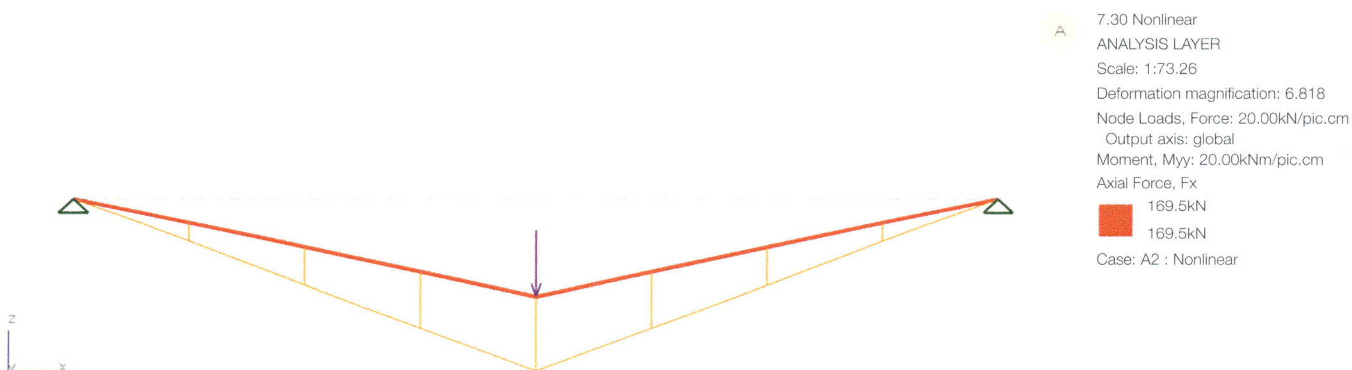

7.7.1.1 Nonlinear load combinations

An error I see all too frequently is caused by engineers thinking that the only difference between a linear and a nonlinear static analysis is the solver. It is not — as model elements, materials, and properties can be nonlinear.

Compare a building to a bridge: buildings resist loads in all directions and, if they have a steel or concrete frame, you might conceivably pick them up and turn them upside down (assuming they are strong enough). Indeed, some buildings have survived liquefaction of the supporting soil and, though rotated were left almost completely intact (Figure 7.31).

This is not the case with nonlinear analysis. Here, if you double the load, the deflection might be less than double, or more, or even fail the structure. If you negate a load in a nonlinear analysis, the load path can be completely different. And some loads will only work if you combine them with gravity loads which, let's face it, are normally there all the time. For example, nonlinear structures such as suspension or cable-stayed bridges (Figure 7.32) only work if the primary loads are downwards, with any uplift being resisted by gravity.

Figure 7.31: Soil liquefaction: Niigata earthquake, Japan, 1964

Figure 7.32: Queensferry Crossing, Edinburgh — Fife, Scotland

Recently I had a call from an engineer who was designing a basement subject to floatation. It was being held down by ground anchors, modelled with tension-only tie elements, with beams representing the basement walls and floors. As a two-dimensional representation of a three-dimensional structure it was a very appropriate model. The problem was that the water uplift load case was analysing, but the gravity load case was not. This was because the uplift was resisted by the ties, but there was nothing to resist a downward movement. Combining the water uplift with the self-weight in a single analysis task cured the problem and allowed it to analyse. Note that the engineer was running a linear analysis, but the model included nonlinear elements, making the analysis also nonlinear (directionally). If the engineer had run a nonlinear analysis the gravity load would have worked, but the basement would have travelled downwards a long way until the ties reversed and took up the load.

The lesson is, if you have a nonlinear analysis or a nonlinear model, you must combine and factor the loads in the analysis task itself. You cannot add or multiply the results in the postprocessor. In fact, the only situation where this can be appropriate is the envelope, where you extract the maximum and minimum values.

7.7.2 Buckling effects

Buckling under pressure...

Linear analysis ignores P-delta and buckling effects, which is why there are separate analysis options for those. If we compare the linear and nonlinear results for this slender column (Figure 7.33) we can see that the resulting moments and deflections are wildly different for the same loads under a linear and nonlinear analysis. If your structure is vulnerable to second-order effects and you only run a linear analysis, you are at risk of under-designing the structure[†].

Figure 7.33: Column under linear and nonlinear analysis

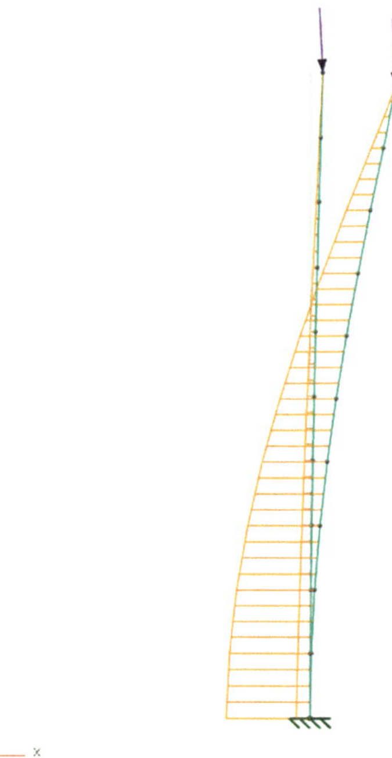

Another thing to remember is that one of the typical characteristics of FEA models is that they are built perfectly straight, and the invariable characteristic of real structures is that they are not. This means, for example, that FEA columns are vertical with no kinks while real ones are built to tolerances that weaken the structure. Analysing the sway sensitivity using notional horizontal loads goes some way to assess how vulnerable the structure is to second-order effects. The problem with this test is that it only looks at large-scale p-delta effects, not local buckling risks. If you want to predict what the buckling capacity might be, it is worth extracting the modal buckling deformations of the structure (as they are probably the worst cases), normalising them to the tolerance limits, and then running a nonlinear incremental analysis on them to find the failure load.

7.7.2.1 Hartford Civic Center

Returning to Hartford, the engineers chose to use cruciform sections built up using angles, and (presumably) chose cruciform sections to make the connections easier. All three of these configurations (Figure 7.34) use the same amount of steel and have the same cross-sectional area, but have very different bending and buckling stiffnesses.

[†] While not exactly the same, I heard a story from the first consultancy I worked at. One of the partners had designed a flagpole for a major client, which had a single pole then a cantilever at the top, from which the flag hung. He checked the deflection of the cantilever and that of the pole, both by hand, but completely forgot that the pole deflection would magnify that of the cantilever.

Figure 7.34: Three ways to join four angles

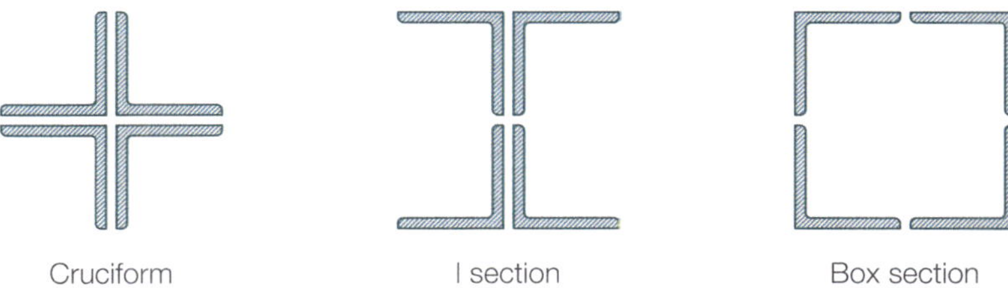

The box is best all round. The I section is strong in the major direction, poor in the minor direction, and poor in torsion. The cruciform section is poor in the major, minor and diagonal directions, and terrible under torsion failure. This, combined with the asymmetric connections, exacerbated the torsional instability and led to the torsional buckling failure of the roof.

7.7.3 Meshing

Mesh density can have a significant impact on problems like buckling analyses. Remember that, like all FEA, the analysis is carried out at the node points, so you need them in there to get the answers. While you can now model a beam or column for a static analysis with a single 1D element (it wasn't always the case), buckling analyses can be very sensitive to the number of nodes (Figure 7.35). For example, if you do not include mid nodes then the eigenvectors are looking for rotation of the end nodes. With an extra node they can find the translation of the middle. This node reduces the reported buckling load factor by nearly 20%, but there is a trivial improvement after that (Figure 7.36).

Figure 7.35: Strut buckling

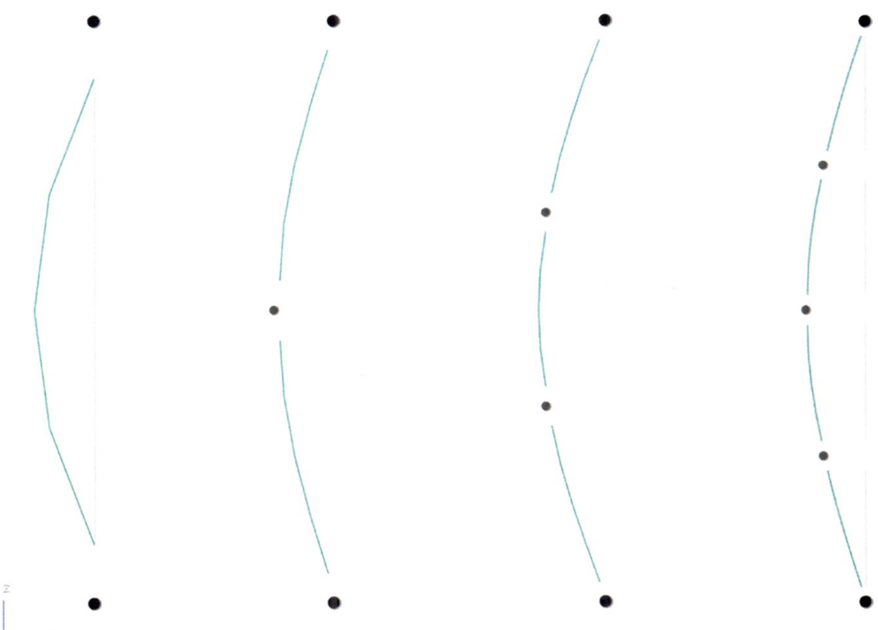

Figure 7.36: Effects of 1D meshing on buckling factor

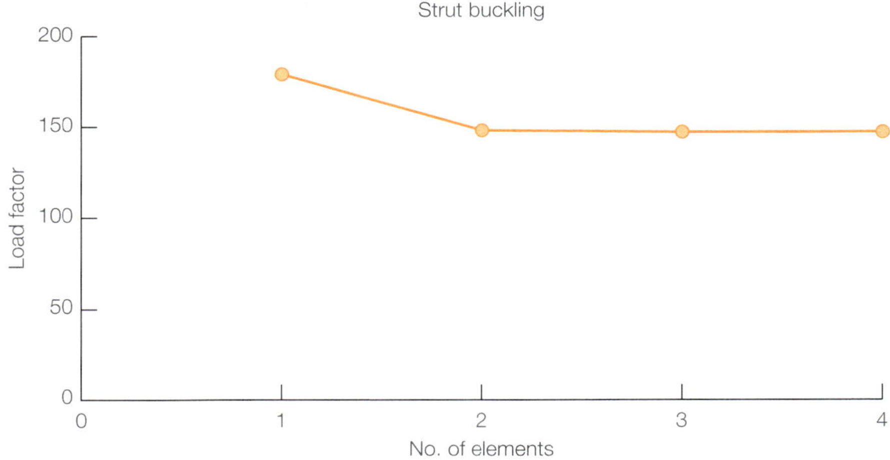

Element subdivision becomes even more important with 2D element buckling. For example, Figure 7.37 is a hyperbolic structure that is not unlike a cooling tower. With a crude mesh the buckling modal shapes are driven by that mesh. Increasing the mesh density reduces the calculated buckling load factor and makes it more mesh independent.

Figure 7.37: 2D mesh buckling

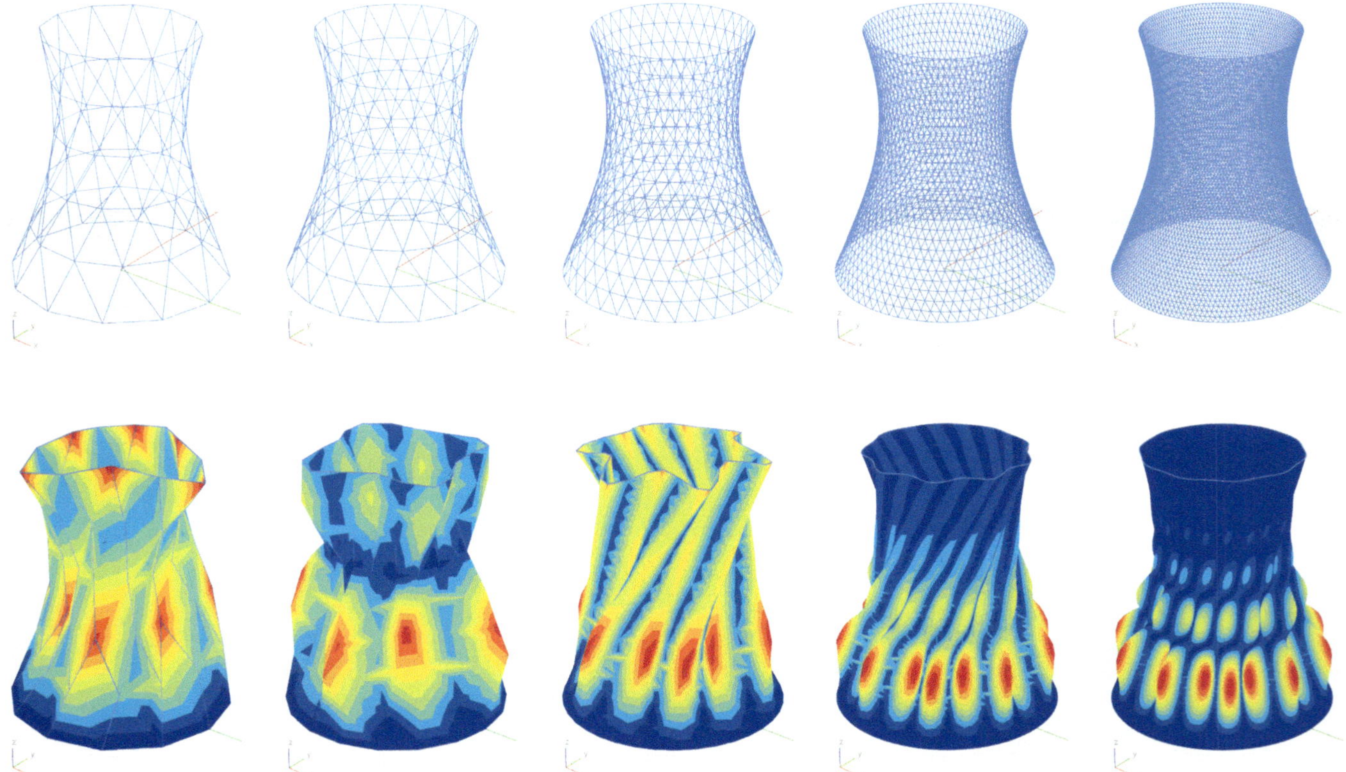

If we look at how the buckling result converges as we reduce the element size, we can see that if the model is insufficiently detailed, the result is going to be entirely wrong (Figure 7.38). Remember: keep halving the mesh size until there is no meaningful improvement in the result.

Figure 7.38: Effect of 2D meshing on buckling factor

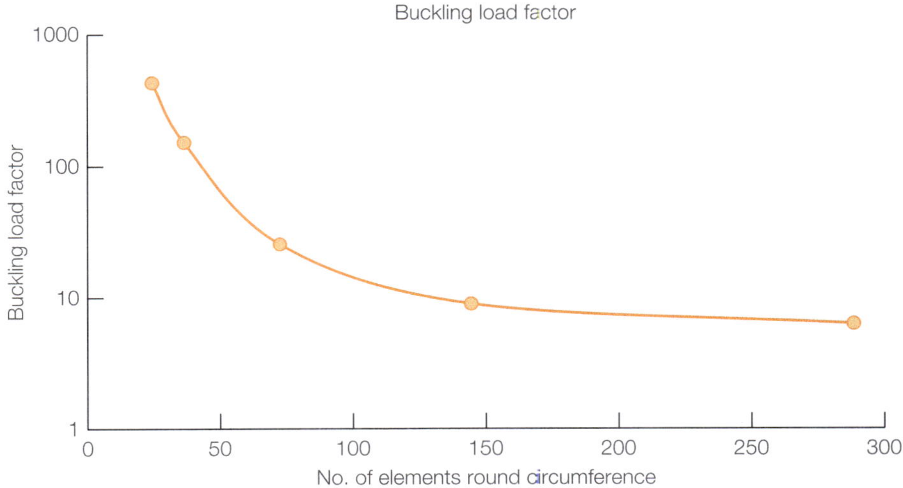

7.7.4 Linear vs. nonlinear buckling

Some programs will give you both linear (modal) and nonlinear buckling. Modal buckling will check all possible buckling scenarios, while nonlinear will check only the first. But do not be surprised when the nonlinear result is less than the modal — this is quite normal. As a general rule, the nonlinear buckling capacity might approach the linear buckling capacity, but it will not reach it (Figure 7.39).

Figure 7.39: Nonlinear buckling curve

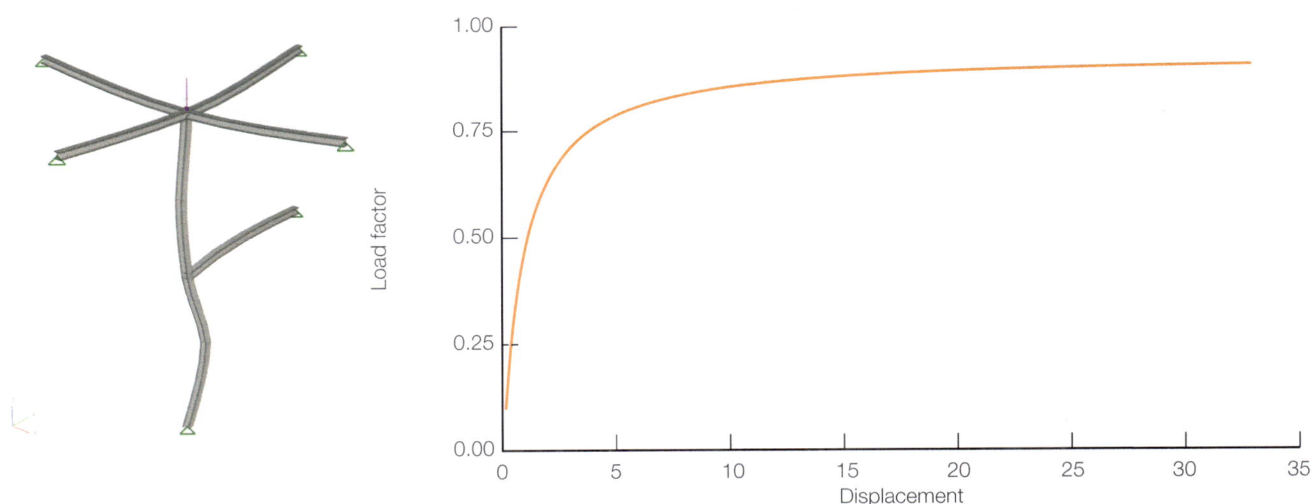

7.7.5 Design

One of the areas engineers can get caught out regarding buckling is with steel design: some programs will use only the results from the analysis, and some will automatically calculate buckling effects in addition to those results. If your program does not automatically calculate buckling effects and you do not manually include them, you will under-design your struts — and if your analysis results do include buckling, say from a nonlinear analysis, and your program adds buckling curves as part of the member design, you will over-design. It is vital that you understand exactly what your program does and does not include.

7.8 Validate the model

Now we come to the part that academics call 'validation and verification', and the rest of us call 'checking'. Validation is **"the process of determining the degree to which a calculation method is an accurate representation of the real world from the perspective of the intended uses of the calculation method"**[93]. This essentially means: is the model realistic[94]? This breaks down into two essential parts: the model and the software.

7.8.1 Valid models

How do you validate your model? Well, much of what has been covered already in this book concerns validating your model: does the model accurately (or sufficiently accurately) represent the reality that the structure will experience, and will it produce the results that you need? Cover those points and you are a considerable way towards validating your model.

7.8.1.1 CTV Building, Christchurch

Let us look at the Canterbury Television (CTV) Building mentioned at the start of this chapter.

The CTV Building was designed in 1986 and had a reinforced concrete frame of *in situ* columns and precast beams with infill masonry panels. It was stabilised by a stiff stair and lift core on the north side and a single shear wall to the south; both of which were off the building centreline (Figure 7.40). This asymmetric stability system introduced dangerous torsions into the building when subjected to seismic loads. Eye witnesses said the building twisted like clothes in a dryer before it collapsed like a stack of pancakes. The only part to remain standing was the north shear core[89,95] (Fig. 7.3).

Figure 7.40: CTV Building — typical upper floor structure

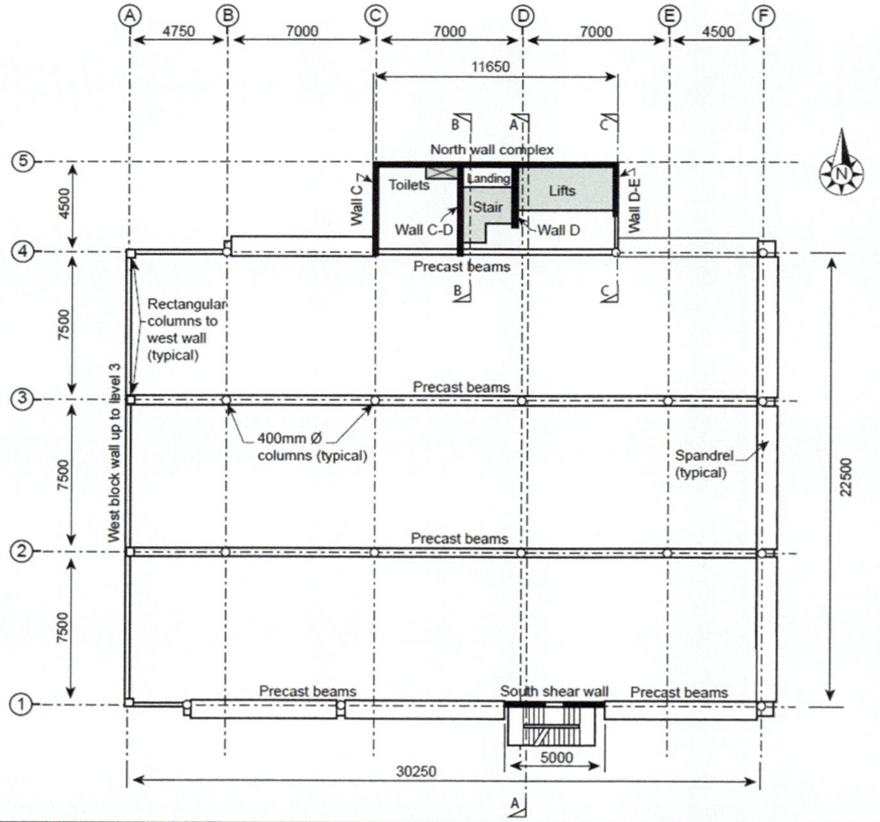

The subsequent investigation into the collapse concluded that a critical column failed due to excessive inter-storey drift. A progressive collapse rapidly followed, exacerbated by poor detailing and insufficient connection into the shear walls. The irregular plan and masonry wall placement was also highlighted as increasing the inter-storey drifts. The building had been badly designed, badly detailed, and the structural drawings were signed off unchecked by the principal engineer on the project.

There were also major problems with the analysis. A 3D model analysis had been carried out, but by an engineer who had little experience with either seismic design or the limitations of the program. The crucial limitation in this instance was that the software only gave a single deflection readout at the centre of mass of each floor. When torsional modes dominate, the deflections at the building corners are going to be considerably higher than those in the centre, but the frame was designed to these lower values. If the engineer had realised both the information missing from the analysis model results and the significance of that information, the result 25 years later might have been very different for those 115 occupants.

7.8.2 Software validation

Can the software be invalid? Yes, if the software is not suitable for your problem. All software is designed to perform specific tasks. If you use a program to do something it is not designed to do you are in danger of it being invalid. Yet this isn't always the case — and this is where engineering judgement comes in.

For example, as a recent graduate I had to design several multi-span timber beams. I had learned how to solve such beams at university, but it was a lot of work. We had an FEA computer in the office (yes, just the one! — the idea of one computer per desk was many years in the future). Anyway, setting up an FEA model was also a lot of work, not least because it didn't have a graphical user interface. You had to create the model using a text file and didn't know what it looked like until you pushed it out onto the pen plotter[†].

What I did have was a multi-span concrete beam design program. I realised that I could use it to carry out the multi-span beam analysis to give me the moments; I just prevented self-weight inclusion and ignored the design output.

Was what I did valid? A lot depends on exactly how sophisticated the analysis was 30 years ago. These days I do get the occasional message from students, and some lecturers, who have been verifying FEA multi-span beams vs. hand calculations, and wondering why the bending moments differ. The reason for this is that modern FEA programs can include the shear stiffness of the beam, which affects the moment redistribution. Hand calculations are unlikely to do this.

I was happy about the validity of my hacking at the time, but today I would have to look closer to be sure. On the plus side, none of my beams failed, so I either did it right or the factors of safety cushioned my mistake.

A riskier approach is when you make greater use of a program and are not aware of its limitations and assumptions, or at least don't compensate for them. Examples of this include:

- Floor diaphragm systems that don't account for the openings around a stability core
- Flat slab design programs that do not include punching shear (Figure 7.41)
- Steel design programs that assume full top flange restraint where the actual design does not have it
- Using a 2D portal program and forgetting out-of-plane stability

---

[†] A type of desktop drawing robot.

Figure 7.41: Flat slab punching shear failure

7.9 Verify the model

"Analysis models can be slow to input but fast to run, encouraging users to run every conceivable load case and combination rather than thinking about it and just running the critical cases. This means that checking is harder (so more likely to let mistakes through) and can hide that the engineer does not actually know what results they are expecting. This is a dangerous situation."
Tristram Hope[96]

The principal aim of FEA is to model the structure so that you understand it in a quantitative sense. In other words, what are the values of the bending moments, shear forces, stresses and deflections? There are many unknowns in structural modelling so it can never be 100% accurate, but must be near enough to be both safe and economical.

Note that I said, understand "in a quantitative sense" as you need to already have a qualitative understanding of the structure. In other words, you need to have an expectation of how the structure will deflect and generate bending moments, etc. before you run the analysis. That way you can identify irregularities and work out why they have been generated. This is usually due to one of three reasons:

1. Your model is wrong.
2. Your assumptions are wrong.
3. The software is wrong.

And it is not often number 3!

Of course, there are times when it is acceptable to run models where you don't understand the structure, e.g. when experimenting to gain understanding. But it is a very risky thing to do on a live project.

All models, designs and calculations must be checked by the engineer who creates them, but also by another engineer, who might spot what you have missed. It is a very good idea to record not just what you have done, but also what you have assumed. If you think that a load case or situation is non-critical then record that so that they can be checked as well — you might just have missed something important.

This part of the checking is called 'verification', which has been defined as: **"the process of determining that a calculation method implementation accurately represents the developer's conceptual description of the calculation method and the solution of the calculation method"**[93].

In other words, questioning whether the model has given the right answers. As mentioned, the best way to do this is to know what the answers should be before analysing the model. This can seem strange at first, because the whole point of using FEA is to find the answers to complex problems that we cannot solve any other way.

Except that we can! Even if we cannot solve the whole structure in detail, we can often simplify it down to a point where we can solve it, at least approximately. Then our approximate model should give approximately the same answer as the complex model. And if it doesn't? Then we know we have made an error somewhere, either in the simple or the complex model, and need to determine where it is.

How might we create a simpler model? One way is to reduce a dimension of the structure.

For example, if we are modelling a truss (Figure 7.42) we can reduce it down to a single beam and use:

- $wl^2/8d$ to calculate the maximum chord forces where d is the truss depth
- $wl/2\sin\theta$ for the diagonal shear force where θ is the angle from the horizontal
- $5wl^4/384EI$ for the bending deflection where I is the truss stiffness
  - Use the parallel axis theory to calculate the 2nd moment of area of each chord $I_z = I_x + Ar^2$ where r is the distance from the truss neutral axis to that of the chord. See Appendix F for more standard formulas

Figure 7.42: Truss

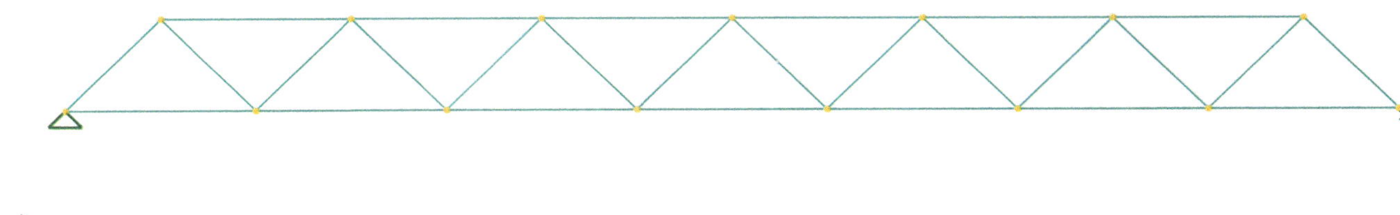

Consider the Hartford Civic Center, which was a two-way spanning space frame. If we assume that the roof was a two-way spanning slab then we could use the standard formulas in the concrete codes to estimate the bending moments and shear forces, and hence the maximum space frame forces.

Or the Sleipner A oil rig: each tricell wall was essentially a fixed-ended beam with a UDL:

- End moment $M = wl^2/12$ or $M = wl^2/24$ at the centre
- End shear $S = wl/2$

A one line calculation would have given the correct shear force on the wall and instantly shown that the model was under-reporting shear, perhaps saving the structure.

If you are running a static model, also run a dynamic analysis and look at the natural frequencies. Bridges might be about 1Hz, floors from 4–10Hz, and so on. If the frequencies are outside typical values, then the stiffness, material or loads might be inaccurate.

You can quickly predict what the frequencies should be, at least for beams: $f = 18/\sqrt{\delta}$ where δ is the static deflection under the load concerned (permanent + 10% variable for example). Likewise, the lateral frequency for a building is often close to $f = 10/n$ where n is the number of storeys.

Similarly, if you are creating a dynamic model, run a static analysis and look at the deflections.

Another verification technique is to sketch out the expected deflected shape and bending moments. These are often known as 'Brohn diagrams' after Dr David Brohn who popularised them through his book *Understanding Structural Analysis*[97] and in the Arup graduate recruiting process. See Appendix I for some examples for you to try.

In many respects they are quite straight forward: fixed supports cannot move; pinned supports can rotate but not move. The bending moment is on the convex side of any deflection. Use supposition for statically indeterminate structures, and so on. I often use them as ice-breakers in my GSA training courses, and it always amazes me how difficult engineers find them. I do believe that in this increasingly digital world of black boxes we need to practise more simple checks.

7.9.1 Envelope results

Watch out for envelope effects. The exact form will vary from program to program, depending on how each treats the permutations that lead to the max./min. results.

For example, some programs may use all the extreme values of moments and forces and not look at the individual permutations. This might be conservative, or it might not, depending on how it takes the signs into account.

For example, in Table 7.1 we have two reactions, P1 and P2:

Table 7.1: Reactions

Case	Fx	Fy	|F|	Case	Fx	Fy	|F|
	[kN]	[kN]	[kN]		[kN]	[kN]	[kN]
P1	5	−7	**8.6**	Max.	5	2	5.4
P2	−1	2	2.2	Min.	−1	−7	**7.1**

If you designed for just the maximum and minimum loads in the axis directions, you will miss the actual extreme value. Just because we find it convenient to resolve the forces down to particular directions, it does not mean that those directions are the critical ones.

If the member or connection design is based on an envelope of the results, is it over-designing? The maximum moment may be in one case and the maximum axial in another, but to combine them is overly conservative. It might under-design if it only designs the maximum and minimum forces and moments with the other coexistence forces, as the worst case might have large loads in multiple directions, but without any of them individually being an extreme value. The detailed design should check for every load permutation in an envelope, otherwise it may not be designing for the correct values.

7.9.2 Software errors

> *"A common mistake that people make when trying to design something completely foolproof is to underestimate the ingenuity of complete fools."*
> Douglas Adams[98]

It is good to remember that all computer software is written by humans[†], just like you and me. They are diligent, hardworking, highly intelligent, and often highly qualified. Professional engineering software is tested repeatedly, with standard models and outlying edge cases. Like all engineers we try to think of ways that the software design can fail, and ensure that it doesn't. These programs must work across a range of hardware (within limits) and whatever operating system updates are released (within reason) and whatever data you put into it (with exceptions). Algorithms are tested. Components are tested. Libraries are tested. The whole program is tested. It is sent out to beta testers to see if there is anything that we have missed. And all bugs that are found are squashed and updates issued.

[†] Though research to get programs to write other programs is underway.

But bugs there are. Major programs can contain millions of lines of code. If you have written your own code you will know how difficult it is to write, and even harder to understand six months later. But these people are dedicated experts who strive for 100% every time, even though 100% is an impossible ideal. In other words, engineering software enables you to design structures with a speed and efficiency not otherwise possible, but be mindful that there may be mistakes in the code.

7.9.3 Checklist for model-checking

Stage 1
Look and think:

- Is the displaced shape correct?
- Does the bending moment or stress distribution look sensible?
- Look for discontinuities in the results
- If there is a problem in the model that you cannot find, breaking the model down into parts will help you locate it

Stage 2
Check it another way:

- Undertake a separate hand calculation, to a reasonable degree of approximation, to see if the answers are similar
- Build a simplified FEA model of the structure to check if the answers are similar

Stage 3
Build the model again, in detail, in another program to see how close the answers are[†]. If you think that something is wrong, check for the following:

- Ill-conditioning and excessively long or short elements
- Incorrect units in the dimensions, loads or materials
- Use of incorrect materials
- Bad restraints
- Bad releases
- Torsion and other ignored factors
- Meshing that is too large or too small
- Second-order effects

In other words: your model must be complete not complicated, simple not oversimple, a realistic representation and an accurate approximation. This is not an easy task but an engineering challenge.

7.10 Conclusion

Maths is a core engineering tool, and computers are incredibly good at calculating at lightning speed. The result of this is that engineering software has revolutionised our industry, with the FEA method the cornerstone of that revolution. FEA allows us to analyse structures that, in the past, were only possible if you knew the correct equations for the form, or could derive them. Now we are only limited by our imagination, the capabilities of the software, and the extent of our own competence.

FEA gives us great power and enables us to get things spectacularly wrong. The Hartford Civic Center engineers produced an invalid, unrealistic model. The Sleipner A engineers did not verify their model results. The CTV Building engineers used invalid software for their analysis. We must be constantly on guard against our own mistakes and those of others; the consequences are too great. That said, the rewards for success are also great, as we can now achieve structures undreamed of by our ancestors.

[†]   Also known as a Cat 3 check.

7.11 Further reading

Ohio University. *FEA Good Modelling Practices: Issues and examples*.
Available at: www.ohio.edu/mechanical/design/Resources/FEA_Examples.pdf [Accessed: October 2020]

MacLeod, I.A. *TN 133 Guidelines for checking computer analysis of building structures*. London: CIRIA, 1988

MacLeod, I.A. 'A strategy for the use of computers in structural engineering'. *The Structural Engineer*. 73(21), November 1995, pp.366–370

MacLeod, I.A. 'Structural engineering competence in the computer era'. *The Structural Engineer*. 85(3), February 2007, pp.35–39

MacLeod, I.A. *Modern Structural Analysis*. London: Thomas Telford Limited, 2010

MacLeod, I.A. and Weir, A. *Essential Knowledge Text No.14: Principles for computer analysis of structures*. London: IStructE Ltd, 2016

Mann, A. and May, I. 'The interpretation of computer analysis'. *The Structural Engineer*. 84(7), April 2006, pp.29–32

Rogers, C. 'What is the need for verification, validation and quality assurance of computer-aided calculation?' *The Structural Engineer*. 94(3), March 2016, pp.48–51

8 Optimisation

"It is the least I can do; a quantity that I specialise in."
Major Denis Bloodnok, The Goon Show

8.1 Why optimise?

Optimisation is a key skill for engineers and is the reason clients employ us: we know how to achieve more with less[†]. Our cave-dwelling ancestors could easily bridge a 1m gap, but to bridge a 1,000m gap is considerably harder; it requires imagination, exploration of the options, and fine-tuning of the chosen solution. As engineers we first want to make things that work. Next, we want to make things that work better: ideally make things work as well as they possibly can, while remembering that perfection is impossible and that good enough is good enough.

As engineers we are always optimising — from deciding on the frame layout to picking the beam size for the job — but sometimes it gets complicated. In this chapter we will look at several types of structural optimisation (topology, shape and size) and techniques that directly relate to them. In the following chapter we will explore more general-purpose optimisation methods, which can be applied to structures or other problems.

The easiest option is for the engineer to over-design everything. Making things twice or three times as large as they need be is an easy and fast calculation, keeping the fees low. So, what's wrong with that? Over-design is not acceptable. When the client engages us to design a structure, an aspect of that commission is to ensure that the structure will be efficient and cost-effective to build. If the structure is grossly over-designed, the other members of the design team will recognise this. You may have saved the client fees, but you will have cost them more overall and they will not be happy. As engineers we should strive to deliver optimised designs. These will lead to further commissions.

There are also our professional responsibilities: we are responsible to the public to produce safe designs, but also have the responsibility to create structures that are material-efficient and sustainable. Building materials, processing, transportation and construction all use energy and natural resources, plus produce greenhouse gases. Buildings and construction leave behind a carbon footprint and are responsible for around 40% of energy-related CO_2 emissions[99]. It is our duty to minimise our impact on the environment.

We all carry out small-scale optimisation in engineering design. We know that the beam must carry a load, so we find the lightest section or least rebar to carry it. This is a natural and instinctive action. But larger-scale or indeterminate problems can get complicated: stiffen a beam and it will attract more load, possibly requiring an even bigger beam, which will be heavier, and so on…

"Success is going from failure to failure without loss of enthusiasm."
Winston Churchill (attributed)

Likewise, design is going from failure to failure without loss of enthusiasm; eventually you will find a design that works for everyone. Design is an iterative process with multiple solutions. Questions at school and on engineering undergraduate courses tend to have just one answer, but engineering practice is not like that. Instead there are several parties that require their needs to be met. Often these needs cannot all be satisfied, which means that all design is a compromise. This can be frustrating when you consider only your own responsibilities, but remember that the purpose of a building is not to provide a structure but is instead the other way around. We are servants of the end goal, as are all the other consultants and contractors. At this point we need the computer to help us sort through all the possibilities so that we can choose the answer we think best.

[†] So why is this chapter so long?

8.2 What is optimisation?

It is, in simple terms, the minimisation of one or more criteria (e.g. weight or cost) while satisfying constraints, e.g. every part of the bridge or building must be strong enough to carry the load. This optimisation can be as simple as choosing the lightest beam section, or as complex as choosing between the different layout options offered by a steel or concrete frame. It can consider the shape of a beam or the depth of a truss. It might derive the minimum cost structure, or it may consider the lifetime energy usage of the entire building from construction to demolition. It might optimise the design of a bridge or it might optimise its location. It is achieving the goal with less.

At some point in any discourse on optimisation we must show 'the formula'[†]. It is:

minimise[††] $f(\boldsymbol{x})$

subject to $g_i(\boldsymbol{x}) \geq 0$ $i = 1, \ldots, m$

and $h_j(\boldsymbol{x}) = 0$ $j = 1, \ldots, l$

and $\boldsymbol{x}^L \leq \boldsymbol{x} \leq \boldsymbol{x}^U$

Where:

$\boldsymbol{x}$ = a set of design variables
f = an objective cost function
g = inequality constraints (certain $\boldsymbol{x}$ must be more than, or less than if appropriate, a particular value)
h = equality constraints (certain $\boldsymbol{x}$ must equal a particular value)
$\boldsymbol{x}^L$ and $\boldsymbol{x}^U$ = upper and lower bounds on the design variables (and are also inequality constraints)

The alternative, simpler form is:

minimise $f(\boldsymbol{x})$

subject to $\boldsymbol{x} \in \Omega$

Where $\in$ means an element of Ω which is the range of possible values (i.e. all the previously referred-to constraints).

For example, let us consider the optimisation of a simple steel beam. We know the requirements, which are that the beam must span the gap while carrying the loads. The constraints will be that it must be strong enough (strength $\geq$ load), not deflect too much (deflection $\leq$ limit) and, given that we have already said that it will be a steel beam, be an I beam from the manufacturer's catalogue. One equality constraint is that the resulting structure must be robust.

The strength checks for the beam are many and complex (stress in web and flange, lateral torsional buckling resistance, and so on) but can all be reduced to a single number: the utilisation factor — divide the applied loads by the load capacity and the constraint is satisfied if the result is less than one.

For the deflection constraint we might either test each section to determine whether the deflection is less than the target, or work out in advance what the required I value is by using the standard formulas and filter the beams that way.

Finally, we have a choice of what aspect we optimise the beam to. If headroom is a problem, we might minimise the beam depth, if it is a project in our own house we might chose the cheapest section, or without other requirements we might just choose the lightest beam.

[†]   I am not convinced how helpful it is, but some seem to like it. You might find it useful.
[††]   If you want instead to maximise $f(x)$ remember it is the same as minimising $-f(x)$.

What can we glean from this? Firstly, that if a computer algorithm is to control the optimisation, we need to assign numbers to measure how optimum it is: how well the members carry the load as well as how heavy/deep/cheap they are. Second: abstraction. Strength checks on beams can be complex and multi-faceted, but it helps to reduce it down to a single value (the utilisation) where possible.

Next: filtering. If you can predetermine certain criteria, such as the minimum second moment of area, you can minimise your effort by ignoring sections that can never work. Finally: satisfying multiple objectives is difficult. The lightest beam will not be the shallowest and *vice versa*. If your target is shallow and light, then the chosen beam section will be a compromise between the two aspects. Secondary criteria can be helpful though: if you have two beam sections of the same depth, choose the lighter. Your optimisation task will be more efficient if you first eliminate the impossible.

Sometimes the optimisation can be simple. If we have a statically determinate structure, such as a single pin-ended steel beam and we have had a good guess at the self-weight, we know the bending moment, shear force and required stiffness of the beam. We simply choose the lightest beam that meets the serviceability and ultimate limit state criteria.

Optimisation becomes more complex when parts of the structure interact with each other. This might be because it is a moment frame, and the moments and forces are distributed throughout the frame in accordance with the stiffness of the individual elements, or self-weight might be a dominant force. In either cases increasing a member size to meet the requirements increases its stiffness or weight, and in turn increases the forces and moments on that member, requiring it to be even bigger. 'Snowballing' is something to be avoided as you can rapidly get to a solution that is sub-optimal.

Other complexities are introduced when we want to carry out cost optimisation[100]. Steel sections are more expensive when ordered in tiny amounts, meaning that using more of a heavier beam might be cheaper in some circumstances. Similarly, heavier reinforcing bars are cheaper by weight because they require less handling. This is one reason for least-weight designs rarely being the cheapest[†]. On the other hand, if we are optimising to environmental impact this tends to be driven by material usage and thus weight.

Optimising can take place at any stage of the design process, but the crucial time is at scheme stage[101]. This is when you set the topology of the structure, the materials and construction approach. Optimisation at the detailed design stage then allows you to fine-tune the material cost and environmental impact.

Design optimisation is complex, as there are a plethora of choices to be made[102], each improvement possibly having a detrimental effect on other aspects of the construction[103]. For example, fixing a portal column base will lighten the steelwork at the expense of a larger foundation. Optimisations such as these can be tricky to solve and often need nonlinear methods to find the eventual answer. Computers make the exploration of the many options feasible within a project timeframe[104].

For the ultimate in optimisation you cannot do better than to look to nature for inspiration, as all good engineers should. Nature has been creating optimal designs for millions of years. Consider a songbird, which must grow and maintain a body that is strong and yet light enough to fly, with feathers that provide aerodynamic lift, insulation, camouflage and armour protection against predators. The bones must be stiff, strong and light, but that does leave them brittle and unlikely to heal after a break, which nature has deemed an acceptable trade-off. Evolution does not worry if individuals survive, only that the species does: the black and yellow stripes on a wasp do not stop it from being eaten or stepped on, but its death throe sting means that predators will be more wary next time, saving the lives of other wasps.

Biology naturally optimises because that which is inefficient does not survive. Thus, all living species are trying to find an ideal compromise between offence, defence, mobility, material and energy efficiency. For example, limpets famously sacrifice mobility for defence[††], while cheetahs maximise mobility to catch antelope while maintaining enough offensive capabilities to overcome their prey once caught.

[†]   Another is that connections can sometimes govern the required section size.
[††]   Limpets' teeth are the strongest naturally occuring material on Earth.

As a structural engineer it is useful to look at natural structures such as bones and shells. In both cases the cost of manufacturing is not an issue as they just grow themselves. Material usage is key here as more material to build means more food to find. More material also means more weight to carry thus more energy expended and again more food to find. But too little material or material in the wrong place means that the creature is vulnerable.

Nature can also be a good inspiration when it comes to design methods. Evolutionary methods (like genetic algorithms) can be powerful. We can also copy the search methods used by some birds and bees, or the group interactions that ants use to solve optimisation problems. There are many different optimisation methods, linear and nonlinear, and each have problems they solve well and others they struggle with. You need to 'optimise your optimisation' — so you will find it useful to know about a wide range of options, including their strengths and their weaknesses.

8.3 What are we trying to optimise?

The aspects of the design that we might want to optimise could include:

- Constructability
- Cost
- Design life
- Durability
- Environmental impact in construction and service
- Strength
- Weight
- Wellbeing (how happy are the people in your building?)

These are, to some degree, within our control but may need coordination with other design team members, such as the architect or contractor. You must be mindful that some improvements to the structure can have a detrimental impact on other's designs. All designs are a compromise.

The problem with good design is that it costs, but costs manifest themselves in a variety of ways, whether in the fee that we charge the client, in the cost of construction, or in the subsequent maintenance costs. Unfortunately, usually only the fee cost is initially visible to the client, and they need encouraging to pay a little bit more now so that they can (probably) pay a lot less later. It is not unlike Blaise Pascal who finished a correspondence saying: *"I only made this letter longer because I had not the leisure to make it shorter"*. It is the same with structural design: we can knock out heavy and expensive designs very quickly, but light and cheap take longer. It is difficult to make something simple!

> *"It seems that perfection is attained not when there is nothing more to add,*
> *but when there is nothing more to remove."*
> Antoine de Saint Exupéry[58]

Optimisation takes time and effort, and like all design, is a compromise. Do we make it as light as possible and make it expensive to manufacture? Or make it as cheap as possible to build but make it heavy and inelegant? Do we use the minimum material while using the maximum energy and therefore create the most environmental impact? Whatever you do, automating the optimisation is sensible because you can then quickly explore the options.

3D printed steel structures are now reality, making highly efficient forms possible — but they can be very energy intensive. I have seen highly optimised steel footbridges that are very elegant and organic but are made entirely from arc welds[105]. As a research project, I applaud it as it is so material efficient, but it is not a general design option at present because it is also energy inefficient. It must be only a step towards the goal not the end result.

8.4 Optimisation types

Optimisation covers such a wide range of design aspects, and it has come to mean many different things to different people. This is especially true when we look at sub-categories of optimisation and try to say what type of optimisation we are carrying out. The types of structural optimisation may be classified into three categories: topology, shape and size:

- **Topology optimisation:** for discrete structures, such as trusses and frames, searches for the optimal numbers of elements, and how they are connected. Topology optimisation for the entire structure will include location and number of columns, span directions of beams, structural types and materials
- **Shape optimisation:** looks at the overall dimensions and form of a structure, including the locations of its nodes. Shape optimisation is particularly important for tension-only structures, but can also improve the performance of trusses and similar frames
- **Size optimisation:** the optimal design is found by changing aspects such as the sectional sizes of members in trusses and frames, or the thicknesses of plates in a girder

The boundaries between the types of optimisation can be quite fuzzy, so great fun can be had arguing what type of optimisation it is. For example, if we do an evolutionary topology optimisation (Section 8.5.3), we also derive shapes. When Pier Luigi Nervi designed the Gatti Wool Factory concrete floor, where he places ribs along the lines of principal stresses, is he working out the shape of a flat slab or the topology of a ribbed slab?

Just as there are many viable solutions to a design challenge, there are also numerous techniques to help us find the optimum design. There is no single best method, with each having advantages and disadvantages for particular applications. Determining which is which is part of the challenge and pleasure of these techniques — but then you must expect some complexity in the solution when addressing complex problems.

8.4.1 The difference between civil and mechanical engineering optimisation

With all optimisation, upstream costs (i.e. in the design) save downstream costs (that is, in the finished article). Like all designs, there is a trade-off, a compromise between the differing requirements of the project team members. How you balance that compromise depends on the client's requirements.

Optimisation usually comes down to the cost to the project, but that cost manifests itself in myriad ways. As a comparison, it can be interesting to look at other engineering professions to see why they optimise.

For aeronautical engineers, especially those involved in space flight, this comes down to weight. With rocketry, every gram needs fuel to lift it to orbit, and that fuel needs more fuel to lift it plus the tank to hold it in, and so on. The cost of manufacturing is important to aircraft designers but is less of a priority than weight. Economies of scale mean that manufacturing hundreds or thousands of copies of the part drives down the cost of the individual items.

For cars, the weight is important: it saves the driver fuel and improves acceleration, but the cost of manufacturing thousands of the items is key: cheap cars sell better. In addition, if a part fails (and with some notable exceptions) the results are not particularly severe, and the manufacturer gets to sell a replacement at a profit.

On the other hand, civil engineering structures are usually one-offs — the prototype **is** the finished article — which must last for 50 years if a building, or 120 years if a bridge. Major maintenance, such as replacement of beams or columns, is difficult. Finally, the failure of a part is likely to be catastrophic, though redundancy and robustness of the rest of the structure should mitigate this.

When we optimise our structures, should they be:

- As light as possible?
- As shallow or slim as possible?
- As cheap as possible to build?
- As cheap as possible to maintain over their lifetime?
- Producing the least amount of energy/greenhouse gases as possible?

These aims are often mutually exclusive, or at least it is difficult to satisfy more than one.

8.5 Topology optimisation

"The most general forms of structural optimisation are topology optimisation and,
for frameworks, layout optimisation. Both of these involve starting with a
blank space... and generating a geometry for the structure based on mathematical rules."
Fairclough *et al.*[106]

With topology optimisation we do not know what shape the structure will be. We have a blank canvas and know only the applied loading and dimensional limits: it must cover X space and carry Y load.

Topology optimisation is related to shape optimisation, but while the latter keeps the number of elements and their connections constant, topology optimisation allows even this to change. In topology maths, the item is not defined by its shape but by constants such as the number of holes. To a topologist a mug and a doughnut are equal as they both have just one hole[†]: one can be morphed into the other without breaking or tearing.

In the case of these examples (Figure 8.1), you can judge the relative merits of each, using Maxwell's Load Paths.

Figure 8.1: Shape and topology

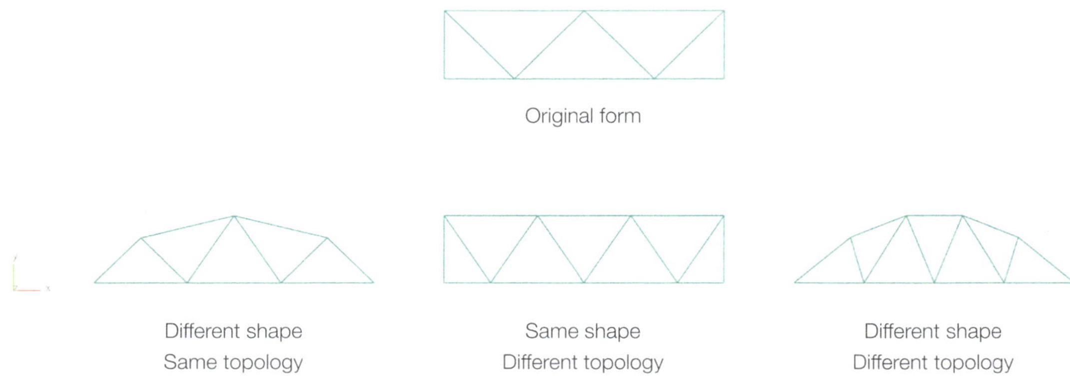

8.5.1 Maxwell's Load Paths and the structural volume

Engineers often say that the shorter the load paths on a structure the more efficient it is. We can quantify how long they are, using Maxwell's Load Paths[107]. Named after the polymath James Clarke Maxwell[††], they are a quick way to assess the optimality of a truss or other axially-loaded structure. Load paths are based on the idea that the structural section will be in proportion to the axial load and thus we can predict the changes in material simply by multiplying the axial load by the element length. The optimum frame will then minimise this formula:

$$\text{Structural Volume} = \Sigma\, l_t f_t - \Sigma\, l_c f_c$$

Where:

$l_t l_c$ = element lengths in tension and compression
$f_t f_c$ = element tension and compression forces

This assumes that both tension and compression forces are positive. If your FEA output uses signs to differentiate between compression and tension, you can instead add the two summations together.

Maxwell's theorem also says that for a given topology:

$$\Sigma\, l_t f_t + \Sigma\, l_c f_c = \text{constant}$$

[†]   This may also explain why coffee and doughnuts go so well together...
[††] When he was not creating structural analysis theories, Maxwell was also busy inventing electromagnetism, colour photography and thermodynamics.

This means you can optimise the frame truss by looking at just the compression or just the tension elements, as the two load paths will stay proportional to each other. This will save pencil and paper, but of course, if you are using a computer to crunch through the options then it can be just as easy to address both compression and tension at the same time, as it saves filtering the results.

Anthony George Maldon Michell used the Maxwell Load Path theorem to derive structural forms that can be proved to be optimal (Figure 8.2)[108]. While in many optimisations you do not know how close you are to the ideal, you can use these as a benchmark for practical trusses. One thing that you might note in these Michell trusses: the elements tend to join at right angles. This is a good rule to follow where possible.

Figure 8.2: Michell ideal trusses

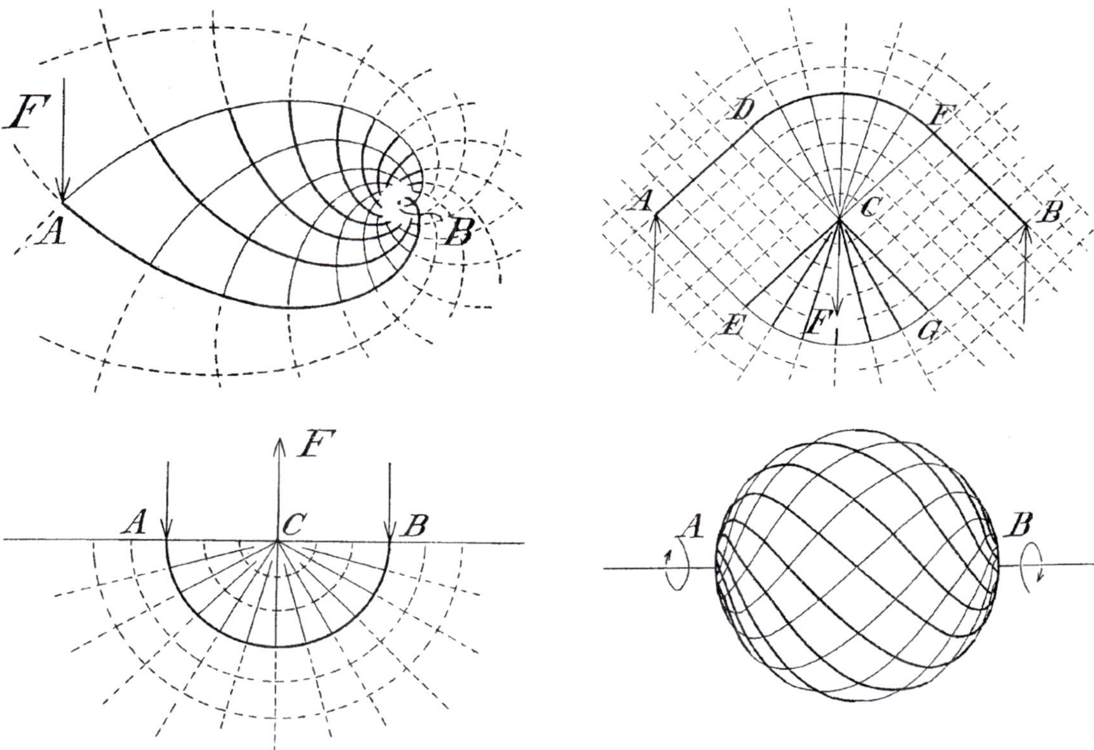

Significant further studies were conducted on these ideal structures by William Prager[109–111], and discrete Michell trusses are sometimes referred to as Prager trusses (Figure 8.3).

Figure 8.3: Prager truss

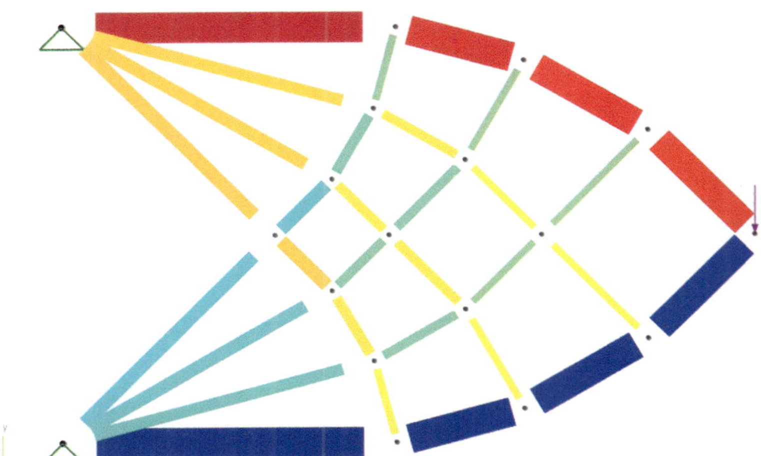

We can use load path theory to answer questions such as: Which bracing arrangement (Figure 8.4) is the most efficient?

Figure 8.4: Bracing alternatives

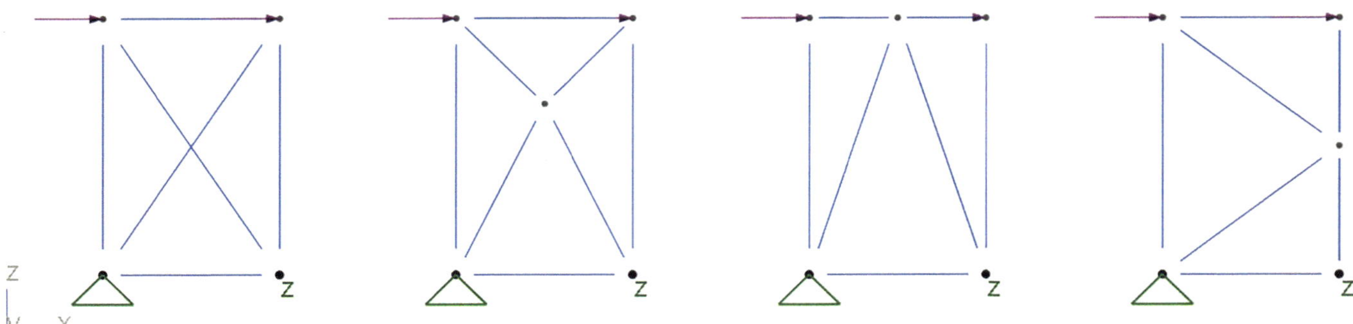

Note that the asymmetric bracing arrangement has two of the elements meeting at a right angle, the X-bracing volume is the same if you use bars or tension-only ties, and the difference between the compression and tension structural volumes is constant between the models (Table 8.1).

Table 8.1: Structural volume/topology constant

Bracing	X	Asymmetric X	Vertical K	K
Structural volume	24.0	22.1	24.0	23.5
Topology constant	2.0	2.0	2.0	2.0

The asymmetrical brace has the lowest structural volume, saving nearly 10%, and therefore, ignoring connection costs, is the most efficient. On a general steel frame, the savings from the asymmetry might be outweighed by the connections, but when designing mega-bracing on tall buildings the savings in material from making the bracing asymmetric might win out.

Similarly, we can use the load path to check the efficiency of different truss topologies, each having the same depth and span, and carrying the same six point loads (Table 8.2).

You may note that the Kingpost, Bowstring and Warren trusses all share the same topology, though are different shapes. Likewise the two Michell trusses and the more efficient K truss are topologically equal. While all the trusses are doing the same job and might share the same topology, the range of structural material required is significant: while the Kingpost truss is a common one for roofs, it needs 50% more structure than Bowstring, 60% more than the Warren, and 75% more than the best Michell.

There are a number of simplifications in the Load Path theorem. The first is that while a tension member is generally in proportion to the load, compression members also need to take buckling into account and so are likely to be heavier than the volume predicted. This does not invalidate the theorem as we can optimise only the tension load path; the compression load path will be heavier, but still optimum. On the other hand, if you want to compare tension and compression versions of the same form (say a catenary vs. an arch) then you may also need to design the members to get a more accurate picture.

Next is that connections are ignored, meaning that Maxwell's Load Paths have problems where connection costs are significant. Timber structures' member sizes are often driven by the connection requirements, and steelwork sections may need increasing to enable efficient connections[112]. On the other hand, 3D printed structures should find this method useful.

Table 8.2: Truss topologies

Truss	Topology/axial forces	Topology constant	Structural volume
Kingpost	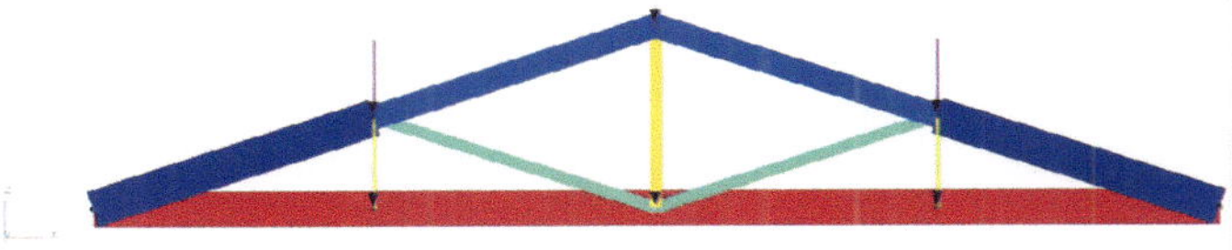	4	236
K truss (1)	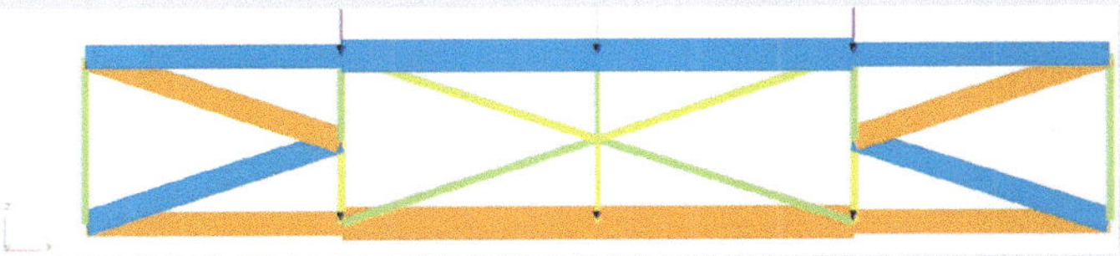	6	220
Pratt	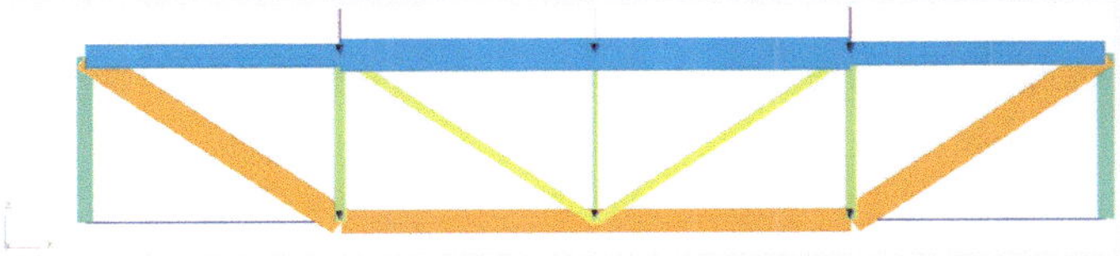	6	164
Bowstring	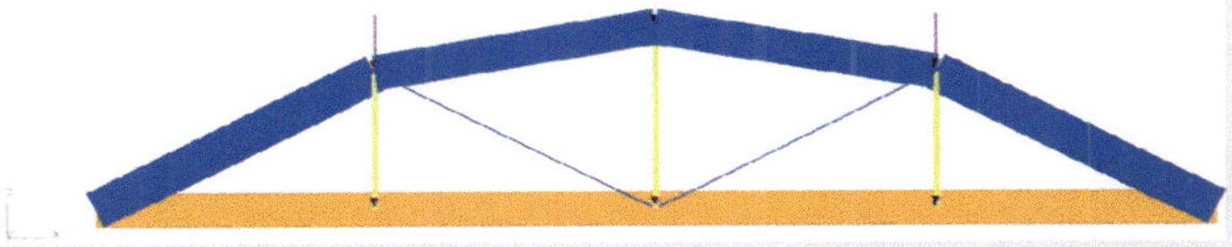	5	159
Howe	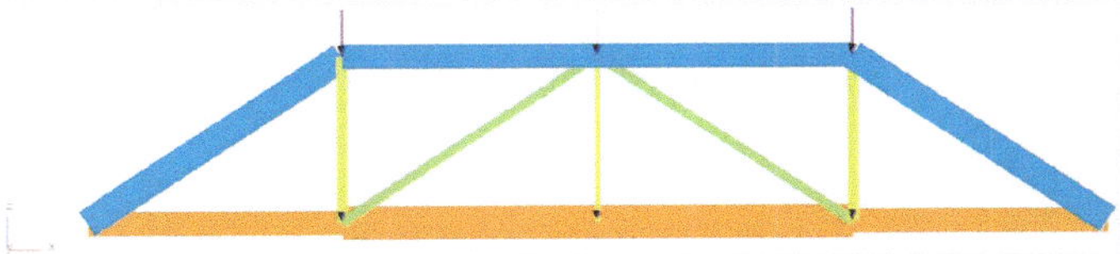	6	152
Warren	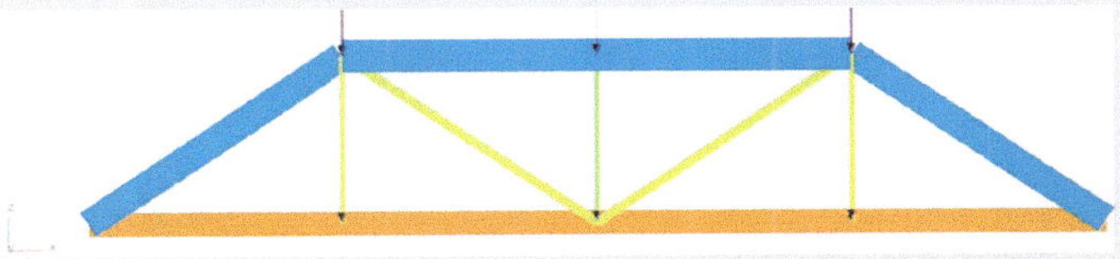	6	148
K truss (2)	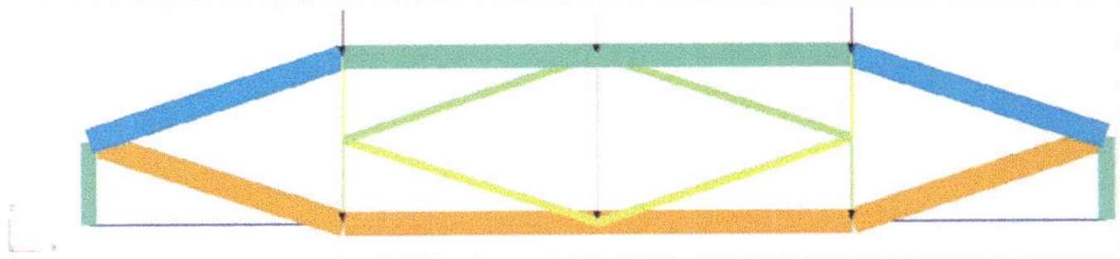	6	142
Michell (1)	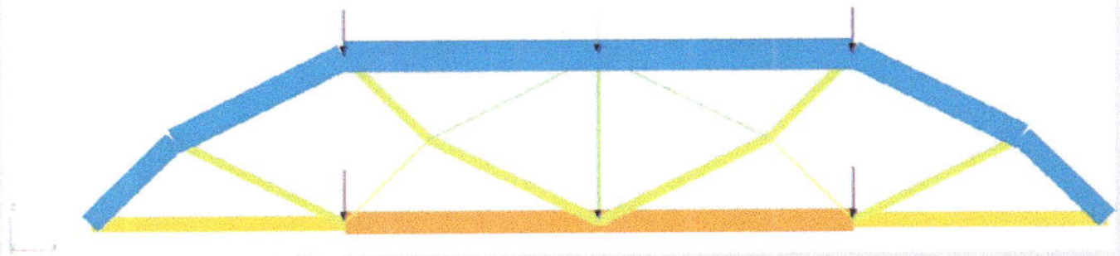	6	137
Michell (2)	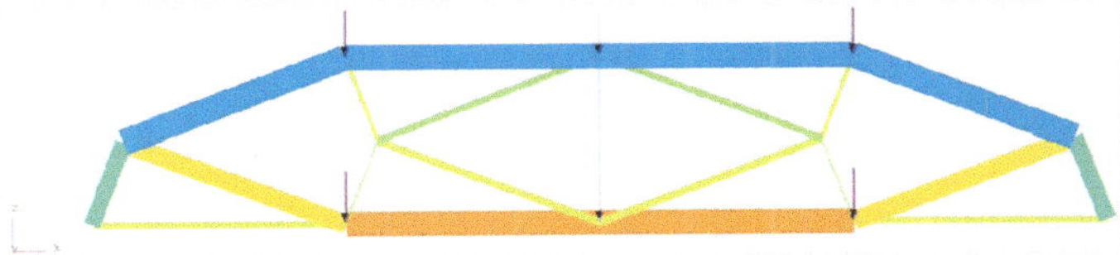	6	135

Finally, the Michell frames are the optimum form for a single load case, but most structures have to address multiple conditions. Tall building frames need to resist both lateral loads and gravity; trusses and arches can experience asymmetrical loading; bridge loads move. These ideal structures are the start of the design, not the final word.

8.5.1.1 CITIC Financial Center, Shenzhen, China

Not yet built at the time of writing, the 311m (65 storey) and 211m (44 storey) tall buildings designed by Skidmore, Owings & Merrill, used load path optimisation aimed to derive external mega-bracing that was stiff, used minimum materials and was aesthetically pleasing. The stability system also used a reinforced concrete core. While the external bracing geometry was initially optimised for the lateral loads, it was also designed to carry gravity loads.

SOM optimisation studies identified a number of efficient bracing layouts, of which two were chosen, one for each tower. These bracing arrangements were then stretched vertically to give the final configuration (Figure 8.5).

Figure 8.5: CITIC Financial Center: bracing arrangements

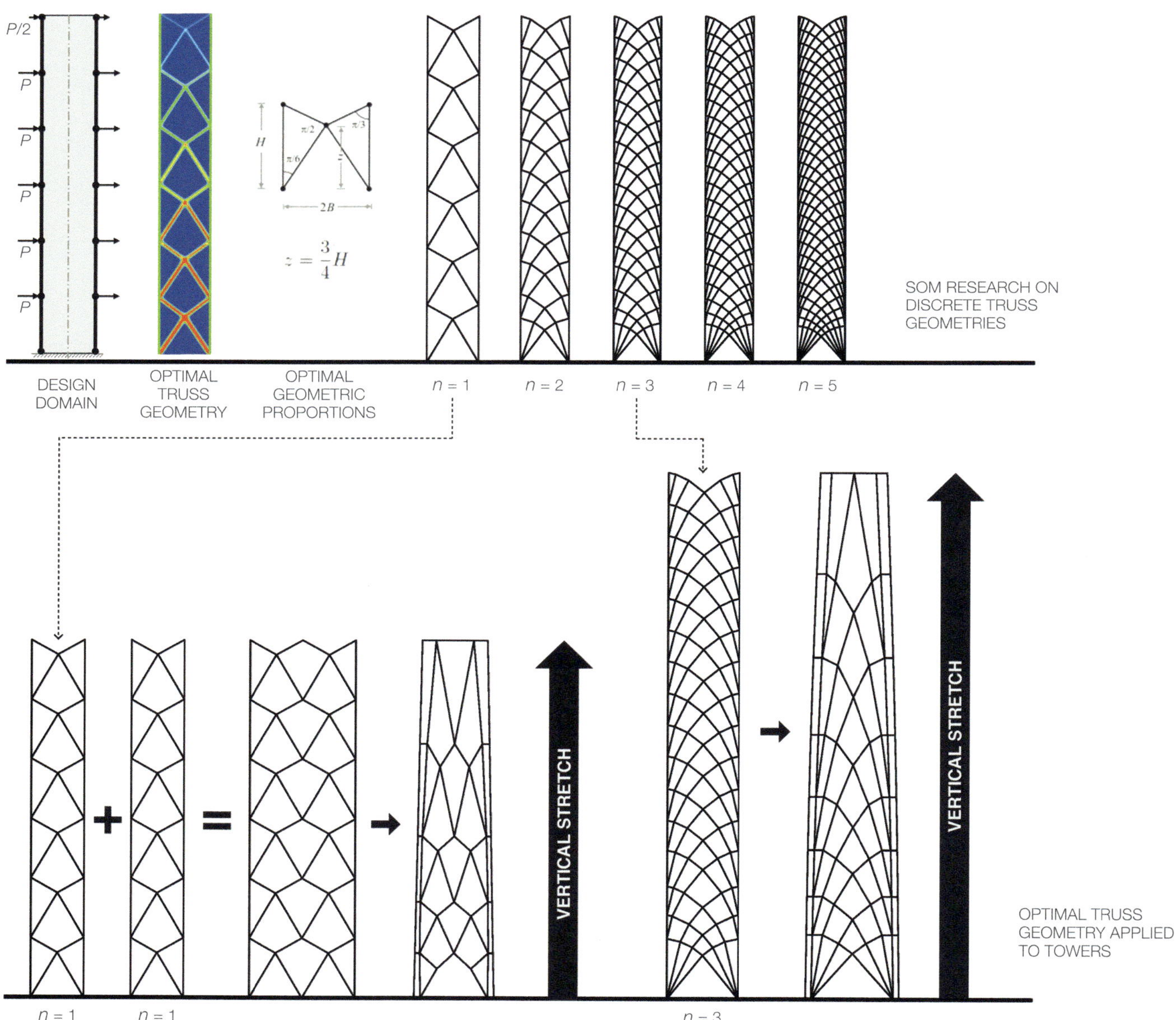

8.5.2 Layout optimisation

Layout optimisation is a form of topology optimisation and is concerned with the best overall form of the structure. Should the bridge be a beam, an arch, cable-stayed or suspension? Should the building be steel or concrete? Flat slab or with beams? What is the best truss spacing? Should we use a deep slab with a few large beams or a shallow slab with lots of small ones? How may bays should there be and in what direction should the beams span?

8.5.2.1 Gatti Wool Factory

Consider the Gatti Wool Factory (Figure 8.6), designed by Pier Luigi Nervi. Today we might consider a flat slab, a beam and slab, or even a waffle slab topology. Nervi instead designed and built a ribbed slab, with the ribs following the isostatic lines between the columns.

Figure 8.6: Gatti Wool Factory, Rome, Italy, 1951

Nervi's factory floor demonstrates how the parameters for cost optimisation change overtime, as it would not be built today due to the labour costs, but at the time it was the cheapest submission in the design competition. It was built in Italy not long after the end of WWII, when labour would have been cheap and materials expensive. Today it is the other way around, so simple but heavy flat slabs are the order of the day. I speculate this may change again soon with the advent of on-site 3D printing (whether of the concrete itself or the moulds) robotic construction to reduce labour costs, and a greater concern with the environmental cost of materials.

8.5.3 Evolutionary topology optimisation

Evolutionary topology optimisation (ETO) is a method akin to sculpting, where you start with a large block of material, gradually removing the bits that are contributing least[113]. These are typically determined by looking at the ratio of maximum to minimum stress or strain energy in a 2D or 3D mesh and removing any elements that fall below a threshold. This threshold may start as low as 1% or even 0.1% of the maximum and is increased incrementally each time the routine fails to remove anything. This process then stops at a predetermined ratio or when an instability develops in the remaining structure due to too many elements being removed. The result is a highly efficient form that is often organic in appearance (Figure 8.7) as well as being similar to the Michell structures (Fig. 8.2).

Figure 8.7: Evolutionary topology optimisation

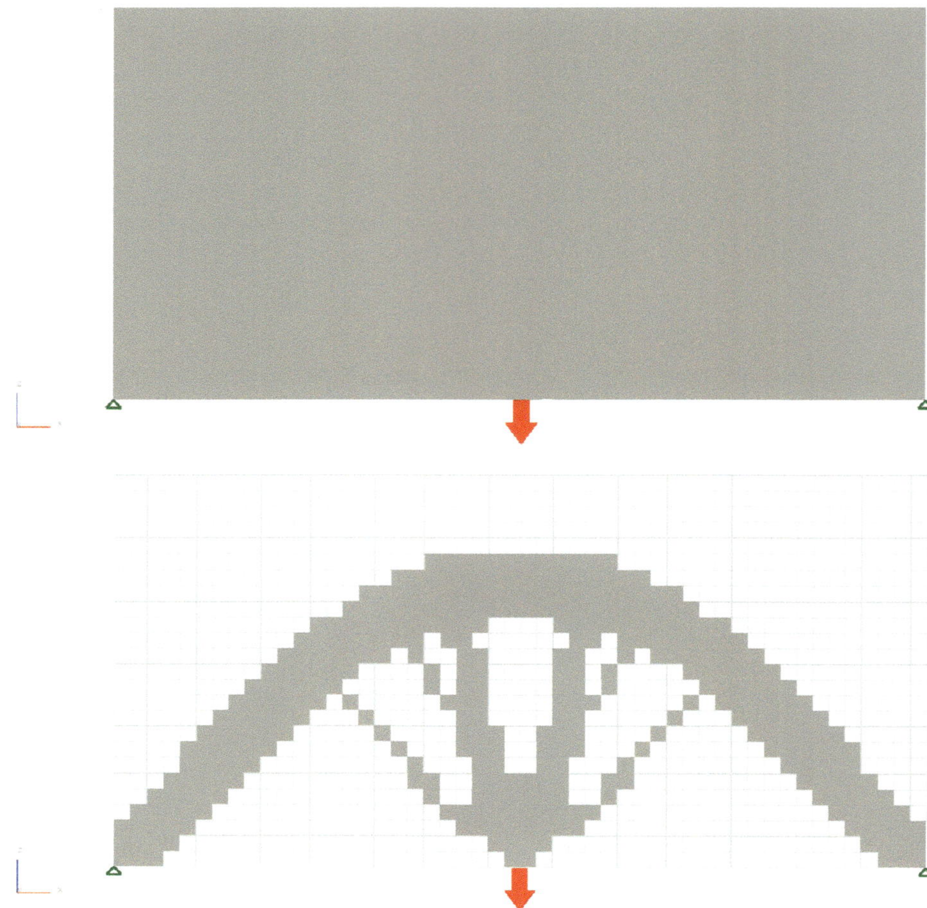

An alternative approach to evolutionary topology optimisation is to use the optimality criteria method (Section 8.7.1), to adjust the thicknesses of each of the elements in the mesh rather than switching them off.

ETO is an excellent method for developing structural forms for additive manufacturing (3D printing). Arup has used it to develop lightweight exposed steelwork connections (Figure 8.8). In this example, the team needed lightweight connections for a tensegrity roof, but the traditional ones made from tubular and flat steel were too heavy (as seen in the left of the image). Using the ETO technique they produced the middle connector, which still connects to the cables with bolts in shear. The second iteration changed the design approach to threaded end connectors, achieving an even more minimal result, having only 50% of the material of the original design.

ETO is not just useful for small-scale problems but also large ones such as tall building bracing. In another study, Arup engineers developed several mega-bracing arrangements with different degrees of stiffness (Figure 8.9). At this scale, the elements are not 3D printed but instead built using conventional steel or concrete members in the building facade. You may note that the bracing naturally forms into Michell trusses, with the connections typically at right angles.

There are some disadvantages to the ETO method: not all the forms generated are buildable by conventional means, and those that are might be vulnerable to problems like buckling. The forms will subsequently need all the usual design checks as it is difficult to include those in the ETO. You may also note that the surface of the resulting form is quite rough; these ETO forms usually need either smoothing, or the parts replacing with conventional members, prior to detailed design.

Figure 8.8: ETO steelwork connections

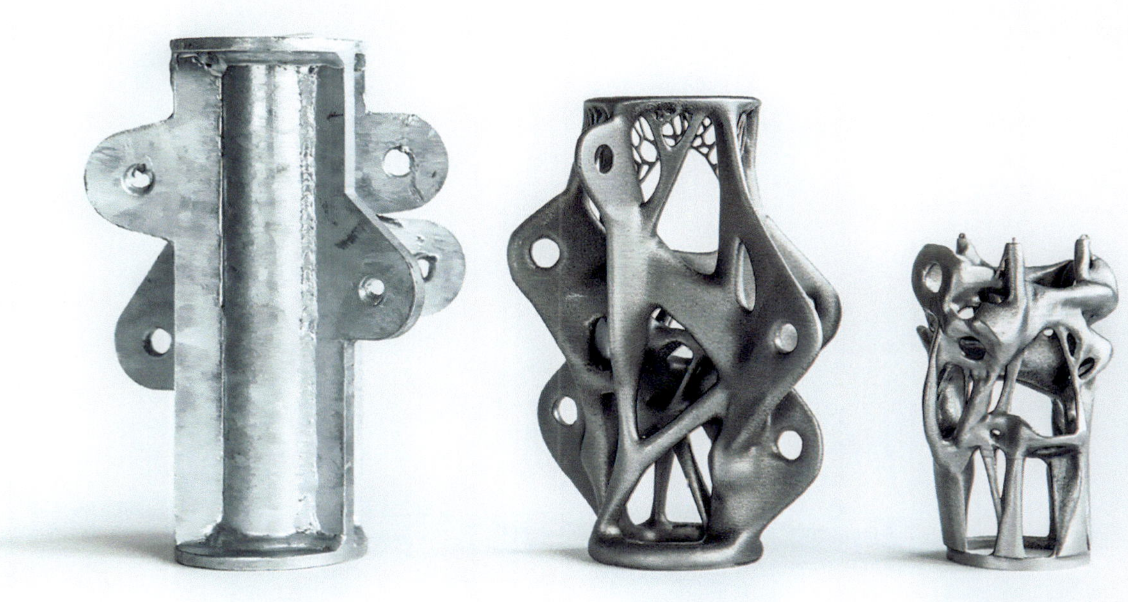

Figure 8.9: Optimised mega-bracing for tall building

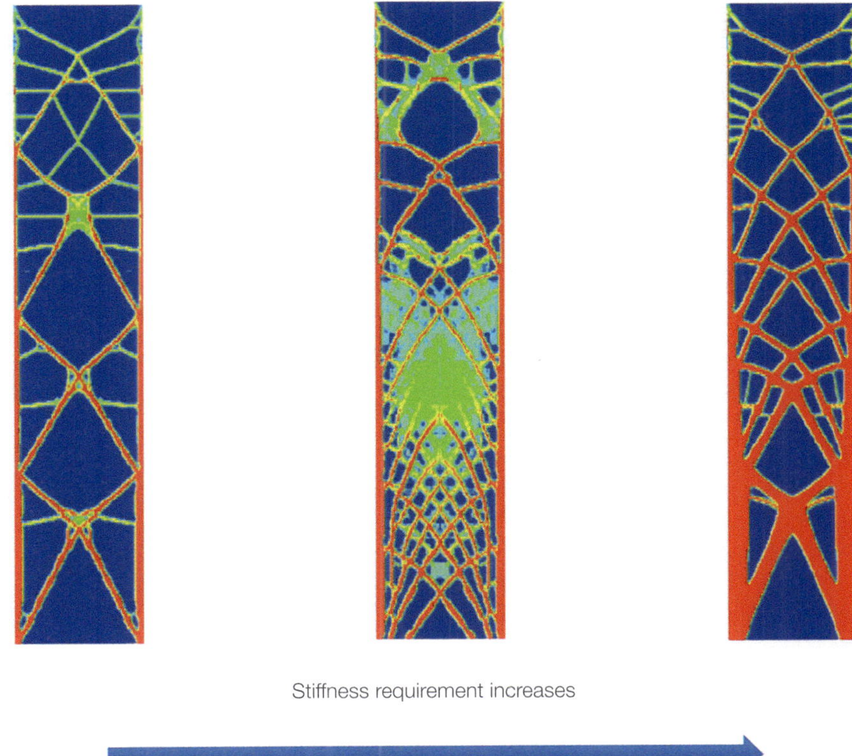

8.5.4 Ground structure

While the starting point for an evolutionary topology optimisation is a mesh of 2D or 3D elements, when creating a truss an alternative is a ground structure (Figure 8.10)[114]. This is an array of nodes that fills the space available, which are then all joined to all other nodes using bar elements. As with ETOs, the elements are removed to leave the optimal structure behind.

Figure 8.10: Ground structure examples

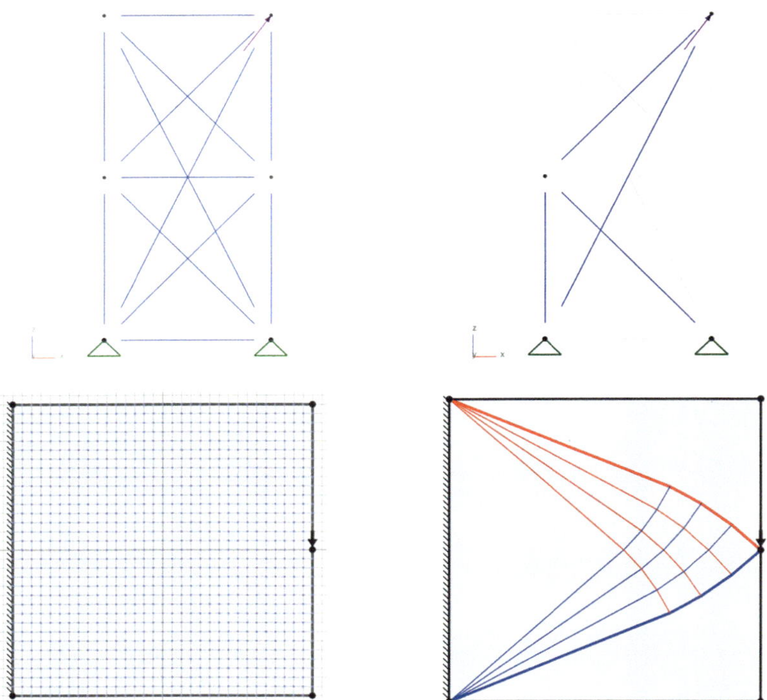

Reprinted with permission from Emerald Publishing Limited, reproduced from 'Layout Optimization of Large-Scale Pin-Jointed Frames' by Matthew Gilbert and Andrew Tyas from *Engineering Computations*, vol. 20 no. 8, DOI: https://doi.org/10.1108/02644400310503017 @ Emerald Publishing Limited, Publication date: 01/12/2003.

If you do design trusses using this method, you will need to do some subsequent rationalisation on them. This may mean removing elements, especially the exceptionally thin ones, to minimise the number of connections, and fine-tuning some of the nodal coordinates[106].

8.6 Shape optimisation

"Achieving strength through geometry requires the structural engineer to regain control of the geometry during the design process."[115]

Shape optimisation, as the name suggests, is working out what the ideal shape of the structure or component should be. Steel I beams are already optimised for bending as they concentrate most of the steel in the flanges, maximising the beam's strength and stiffness. These flanges are then held apart by the web, which also looks after the shear forces. While size optimisation will optimise the individual plate thicknesses of a plate girder, shape optimising will fine-tune the overall dimensions.

If we consider the shape optimisation of a truss, we would look at its dimensions, such as its depth, but can also move the nodes around to find the most efficient shape.

8.6.1 Nodal adjustment

As we have seen in Section 8.5.1, small adjustments to the nodal positions can produce significant gains in efficiency, which shows that there are opportunities to refine the shape of your frames and trusses. We can explore this by automatically adjusting the nodal positions to see if the structural volume reduces.

You can do this by setting up a macro to adjust the location of some or all of the nodes in your structure[†], one at a time, to see if the structural volume, or another suitable measure, reduces. If there is an improvement then keep the node in that new location. If not put it back, before moving on to the next node in the list.

Tip: if your structure is symmetrical you only need to model half of it (Figure 8.11). Add suitable restraints on the cut-line to mimic continuity. Note that if you need to cut any bars on the line of symmetry, you will need to ensure that any changes to the nodes reflect that implicit continuity. For example, in Fig. 8.11 the bisected bottom chord must remain horizontal.

Figure 8.11: Symetrical truss

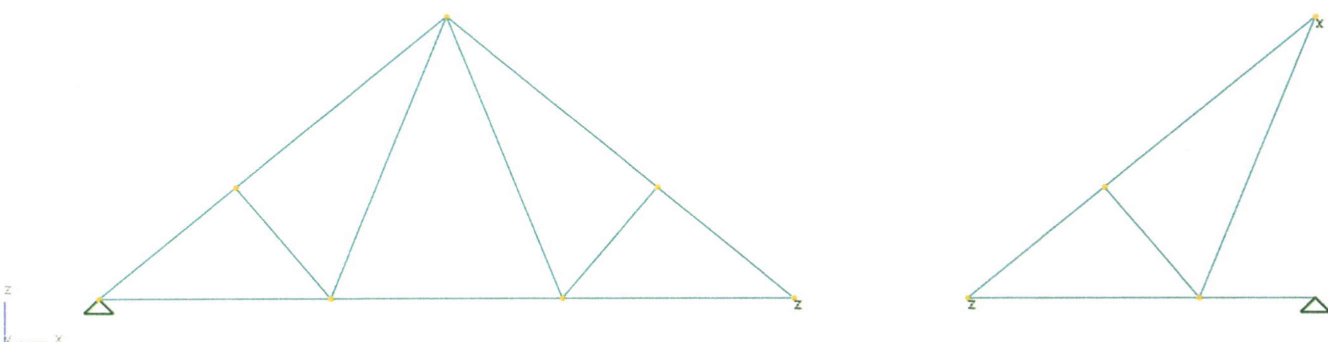

8.6.2 Graphic statics and reciprocal force diagrams

Graphic statics is a method of analysing a structure by drawing it and its forces, and was also invented by James Clerk Maxwell[107]. He realised that for every structure, especially those that work purely by axial loads, you can derive a force diagram that is the reciprocal of the structural diagram. His ideas were then refined by structural experts such as Anthony Michell and Robert Bow (the latter giving us Bow's notation).

Graphic statics was the main structural analysis method of the late 19th and early 20th century, until advances in numerical methods caused it to fall out of favour[††]. While computer analysis is faster and more accurate, it is a method that is still extremely useful to us here as we are talking about optimisation. If we can derive the forces from a structure, then the reverse is true — we can also derive the structure from the forces. This makes it an incredibly powerful method for shape optimisation.

I have spoken to a number of engineers over recent years about reciprocal force diagrams and the majority had never heard of them. I was taught the basics at university, but with the caveat that "no one uses this method anymore" so I did not pay much attention at the time[†††]. As such, I will start from the very basics and expand it from there with a series of examples. The best way to understand this method is to try these yourself, so get yourself a pencil, paper, ruler, set square and compass. Once you understand the principles, have a go at computerising it: three-dimensional reciprocal force diagrams are much easier to draw using a computer than on paper. Here are the basic rules:

Every force can be represented by a vector giving the magnitude and direction (Figure 8.12):

Figure 8.12: Force vector

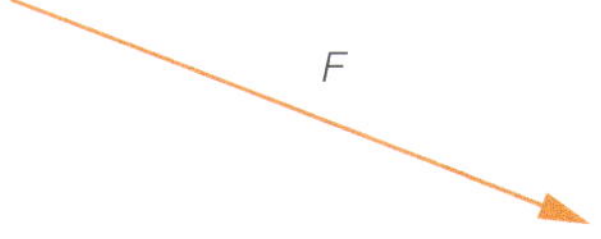

[†] Some positions must remain fixed or remain on a surface. Floors, for example, need to be kept horizontal.
[††] Learning how to create these diagrams freehand will help you quickly and approximately analyse trusses in advance of any numerical checks.
[†††] I was optimising my education!

If you add two or more vectors (Figure 8.13) you can replace them with a single vector (note that the forces form a closed polygon):

Figure 8.13: The sum of two forces is another force

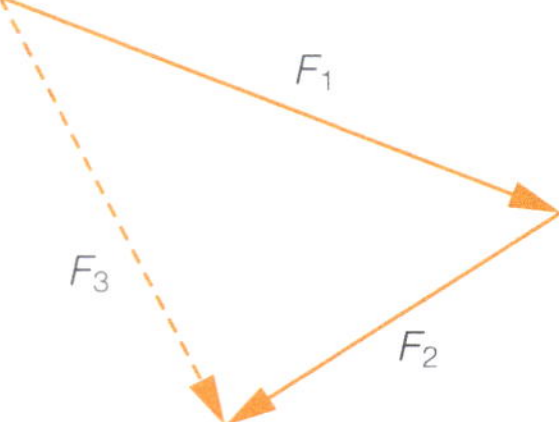

Similarly, a single force of the correct magnitude and direction can resist two or more forces (Figure 8.14).

Figure 8.14: Forces in a closed polygon sum to zero

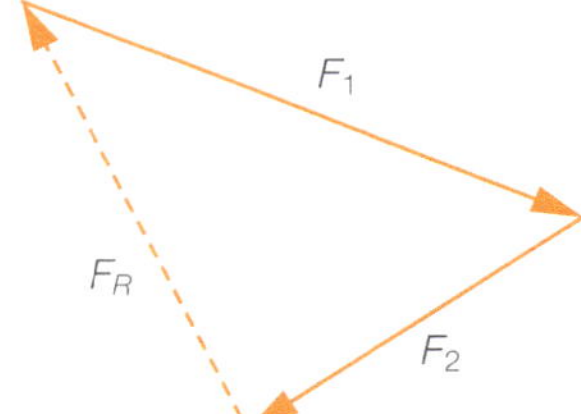

Note that again the forces form a closed loop or polygon; in other words, forces in a closed loop sum to zero. While the forces are clockwise in this instance, they can be anticlockwise or even cross over.

Forces are in static equilibrium if they sum to zero and can be shown to be acting on a single point (Figure 8.15):

Figure 8.15: Forces acting on a single point and in equilibrium

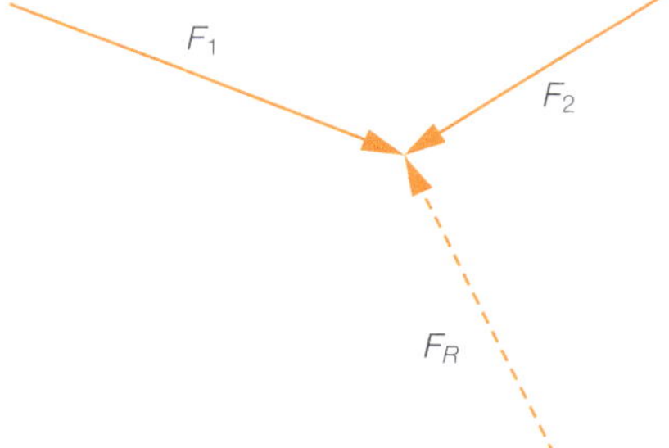

The force in an axially loaded element is parallel to that element.

For every frame there is a reciprocal force diagram. You can create one from the other and *vice versa*.

In the frame diagram the forces all meet at single points (the nodes). In the force diagram the forces on nodes form closed polygons.

As an example, let us draw the reciprocal force diagram for the cables supporting a bridge (Figure 8.16)[116].
Note that the bridge deck is pinned at all supports and at the centre for simplicity, which means the structure is nonlinear and statically determinate.

Figure 8.16: Example tensile structure: suspended bridge

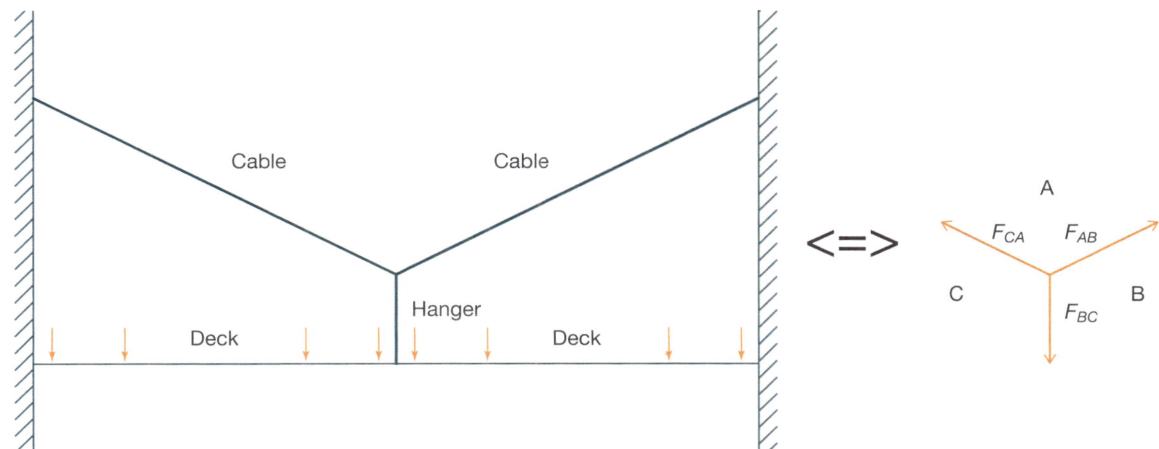

In the frame diagram use Bow's notation to mark up the parts of the structure. Begin by lettering the zones between external loads in a clockwise direction starting at the top.

In the force diagram, the zones become nodes and the structural nodes become zones encircled by the force polygons. Draw the known force vector of the load from the bridge (F_{BC}), with the length scaled to match the load (the reaction from the two bridge halves). Next draw construction lines parallel to the diagonal cables[†]. The point where they cross is point A (Figure 8.17). Join this point to B and C to give us the force vectors F_{AB} and F_{AC}. Measure the line lengths to find the forces.

Figure 8.17: Reciprocal force diagram: suspended bridge

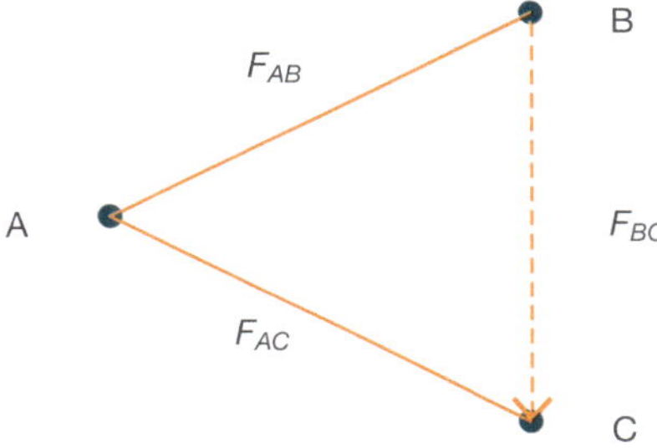

If we repeat that exercise with the support cables at a shallower angle, point A moves further out, and the forces increase accordingly (Figure 8.18).

Figure 8.18: Shallower cables result in greater forces

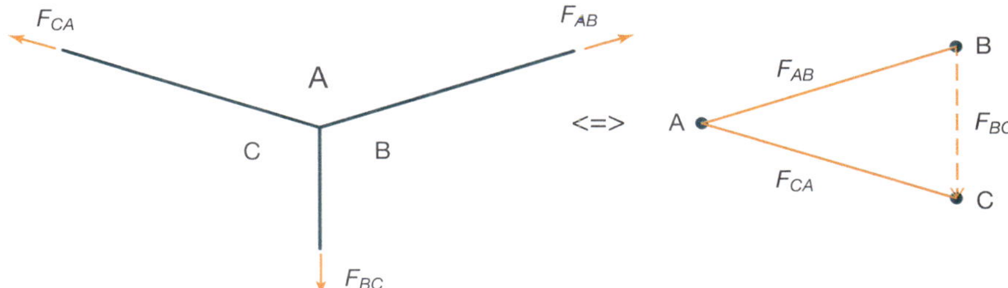

[†] This is where the combination of ruler and set square comes in useful. Position one edge of the set square against the cable line and hold the ruler against another edge. Slide the set square along the ruler until you reach the force diagram.

Similarly, it we use a steeper angle for the support cables, point A moves closer to the applied force vector and the cable forces reduce (Figure 8.19).

Figure 8.19: Steeper cables result in lower forces

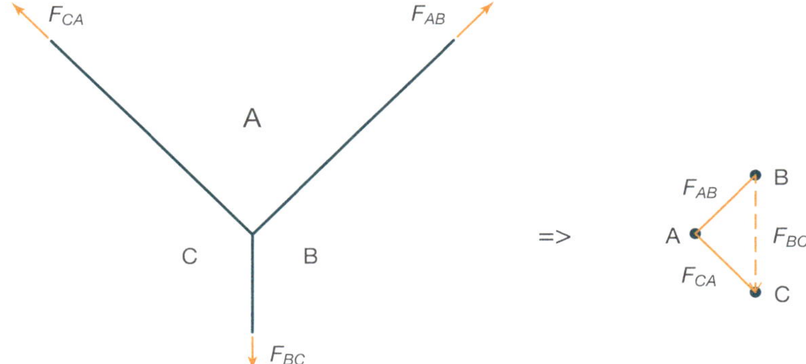

It also works with an asymmetrical cable arrangement. The differences in the cable angles result in different forces, but it is still in equilibrium (Figure 8.20).

Figure 8.20: Asymmetrical cable arrangement

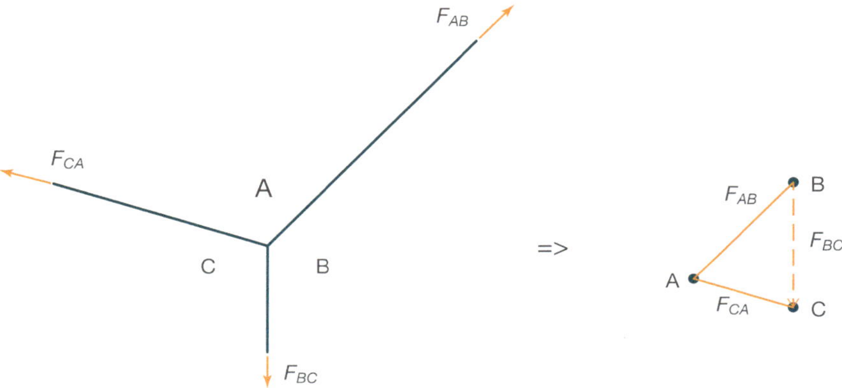

Whichever analysis method we use, when we have the element lengths and forces, we can use Maxwell's Load Path theorem (Section 8.5.1) to calculate which is the best cable angle that requires the least material.

Now let us turn things around and find the geometry where all the cables have the same tension and thus require the same section size. Having drawn the vector BC, use a compass to draw arcs about both points to ensure that the force vectors are the same length. The two arcs intersect at point A, allowing us to draw the force polygon, which is an equilateral triangle. If we then use the angle of the vectors AB and AC on the structural diagram, we get the geometry that generates those forces (Figure 8.21).

Now we can expand this method to find the cable forces when there are two hangers supporting the bridge deck (Figure 8.22). Or we could have drawn the force diagram first, which would have given us the catenary shape of the main cable. We would have to decide where to place point A though.

Figure 8.21: Set vectors to be the same length to derive geometry where force is the same in all cables

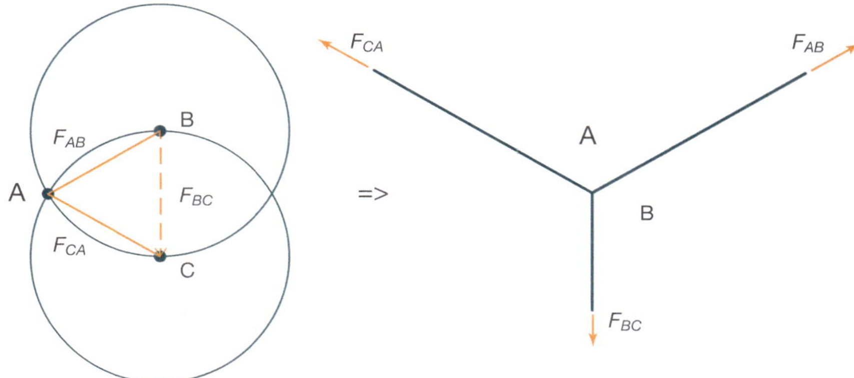

Figure 8.22: Suspended bridge with two cables

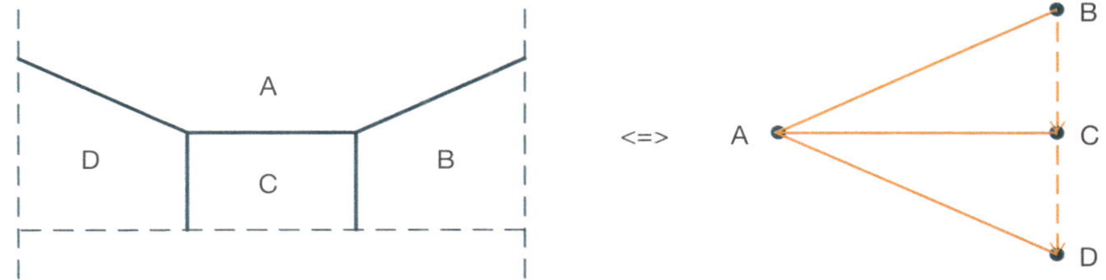

To illustrate this let us find the equilibrium shape and force diagram for the five equally spaced vertical forces. Note that the answer can be either an arch or catenary depending on which side you decide to put point G and whether you assume that the forces are applied from above or below. If the forces are equally spaced, so we know the horizontal location of the forces and reactions, we now need to find their vertical positions (Figure 8.23). Add the Bow's notation, despite not having any structure. A rough sketch can help if you are unsure, but I will assume that you have an approximate idea of the final shape.

Figure 8.23: Initial setup for optimised arch structure

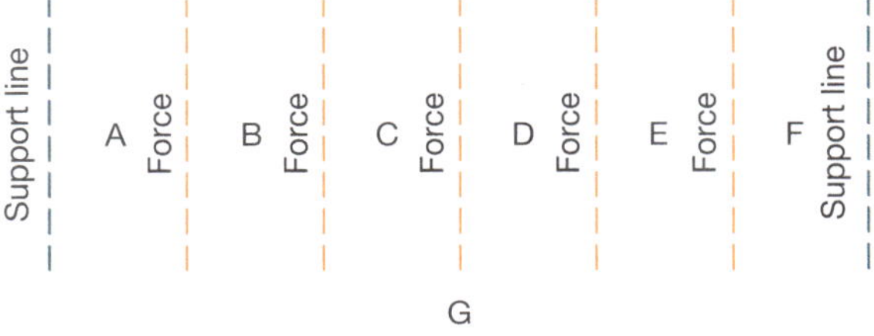

Choose the location of G, which I will put level with the mid-height of the line AF[†]. Join it to all the points on the applied force vectors. Copy the line angles over, starting at one support and continue until you reach the other. In this instance you will get an arch if you place point G to the left of the applied force vectors and a catenary if you place it to the right (Figure 8.24).

Bow's notation and graphic statics also works for structures like trusses; the convention is to use numbers for the internal zones of the structure. To analyse a structure like this Kingpost truss (Figure 8.25), once you have drawn the external force vectors and determined the reactions (by inspection or calculation), start from a corner where you have only two unknown forces (e.g. the left-hand support) and gradually work your way across the truss until you have all the element forces.

[†] See what happens if you move point G, whether horizontally or vertically, on the paper.

Figure 8.24: Derive the force diagram and find the arch

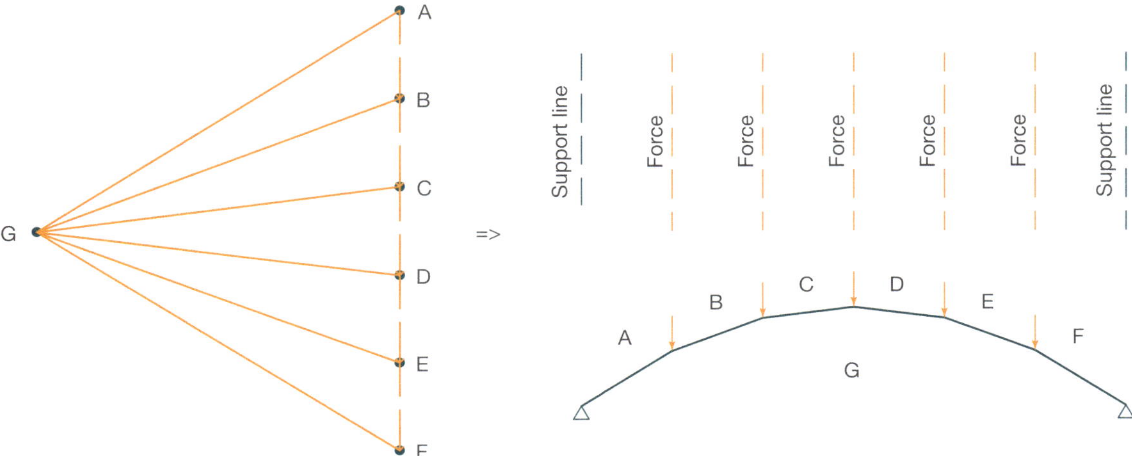

Figure 8.25: Bow's notation for structures with internal panels

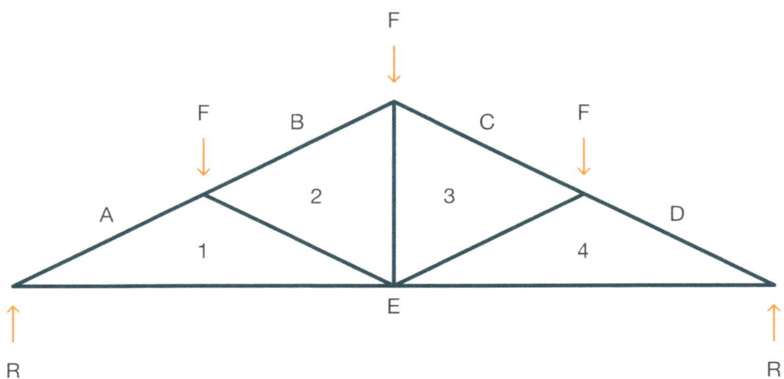

Suppose that we want to find the geometry of a cable beam, ensuring that the cable has a constant tension. We will start with an approximate sketch of what we are expecting, including the Bow's notation, assuming that the floor load is applied only at the strut locations (Figure 8.26).

Figure 8.26: Initial sketch for cable beam

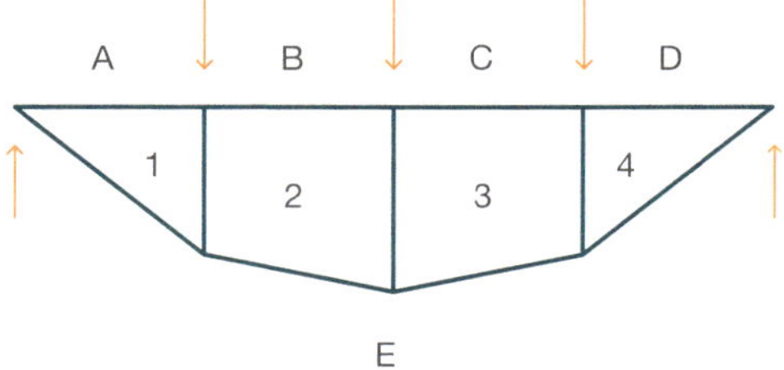

Then draw again the part of the structure that we know: the beam and load locations (Figure 8.27).

Figure 8.27: Fixed aspects of cable beam

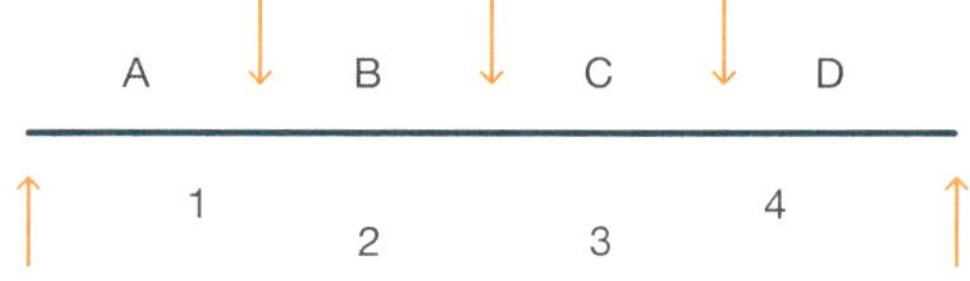

We know the beam is horizontal and thus so are the forces A1, B2, C3 and D4. Having drawn the external force vectors, draw horizontal construction lines from points A–D (Figure 8.28). Point E is our key one for the cable forces; with your compass centred on E, strike an arc across the horizonal lines, giving us the points 1–4 and thus also the cable and deck forces. Join up points 1–2, 2–3 and 3–4, which give us the forces in the struts and their angles. Note that only the central one is vertical, and the others are angled to balance the forces.

Figure 8.28: Force diagram: cable beam

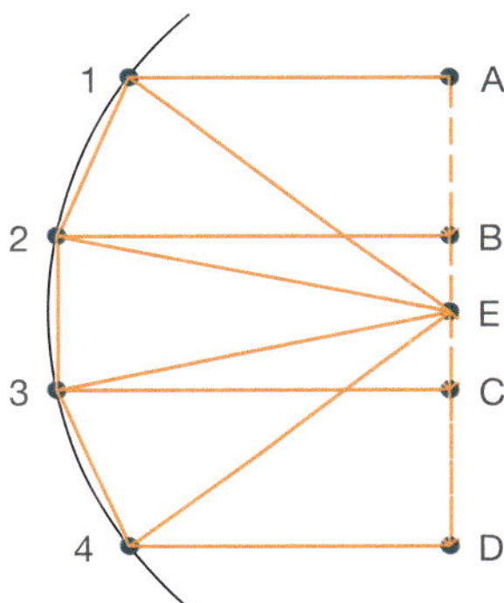

Now draw the elements using the angles from the force diagram (Figure 8.29).

Figure 8.29: Resulting cable beam geometry

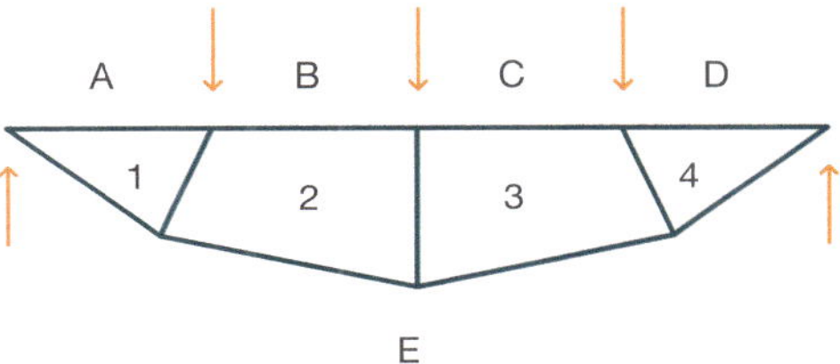

This form of shape optimisation has been used by engineers for a number of years. An elegant example is by the Swiss engineer Robert Maillart, as demonstrated on the Lagerhaus der Magazzini Generali roof (Figure 8.30).

Figure 8.30: Lagerhaus der Magazzi Generali, Chiasso, Switzerland, 1924

We can recreate the basic form of the Largerhaus roof using a reciprocal force diagram (Figure 8.31). The principal variable is the location of the line 1–8 in the force diagram as this drives the slopes of the bottom chord. Picking the location where the force lines C3 and E5 intersect, and likewise D4–F6, gives us a geometry that matches the original. Note that the choice of vertical internal elements ensures that the force in the top chord is constant. There is also a geometry where, starting from the given top chord geometry and load points, the force in the bottom chord is instead constant. I will leave that to you to derive.

Figure 8.31: Force diagram: Lagerhaus structure

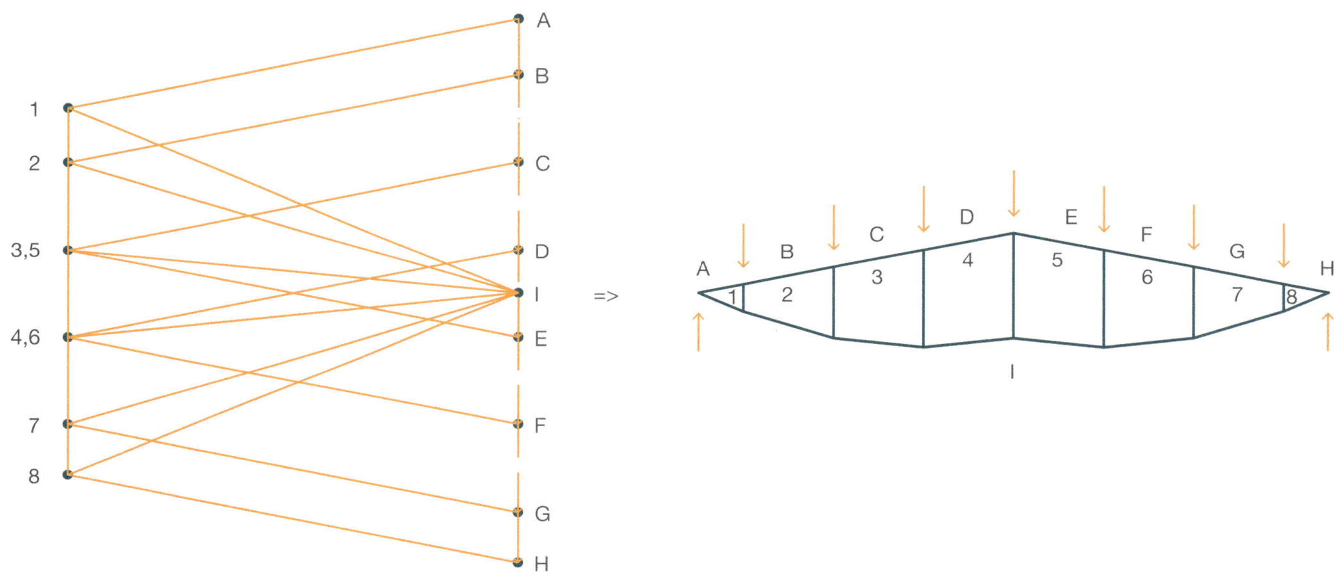

More recent examples of structures where the form was found using reciprocal force diagrams, include the work of the Block Research Group, also in Switzerland (Figure 8.32)[115,117–121] and SOM[122–126].

Figure 8.32: Armadillo Vault, Venice, Italy, 2016

8.6.3 Form-finding

While some of the methods discussed previously can be used to determine the best geometry for a structure, there are methods dedicated to finding the optimum shape for tension structures, along with the necessary pre-stresses. Tension structures are extremely sensitive to the balance of pre-stress, geometry and support stiffness, and anything out of balance will result in slack cables or wrinkles in a fabric surface.

There are two principal methods for form-finding: force density and soap film. Both methods work well for tension-only structures; if the structure is only in compression the trick is to reverse all the loads, as an arch is the mirror image of a hanging chain. The tricky part comes when you want to form-find a structure with both compression and tension elements[127]. The form-finding then either must iterate over the various structural parts (tension structure, compression structure) or form-find just a small part of the structure before gradually adding in more with each successive iteration. An alternative is to try one of the other methods described in Chapter 9.

8.6.3.1 Force density form-finding

The first and oldest method is force density, which is particularly suitable for cable net structures, though it can also be used on fabrics. In this method, the length of 1D elements and the area of 2D ones are set proportional to the applied force. For example, this three-piece tie (Figure 8.33) has an axial load and each piece has a target force density set to be 1, 2 and 3, respectively. The result is that the one with a force density of 1 is twice the length of that set to 2 and so on.

Figure 8.33: Force density form-finding

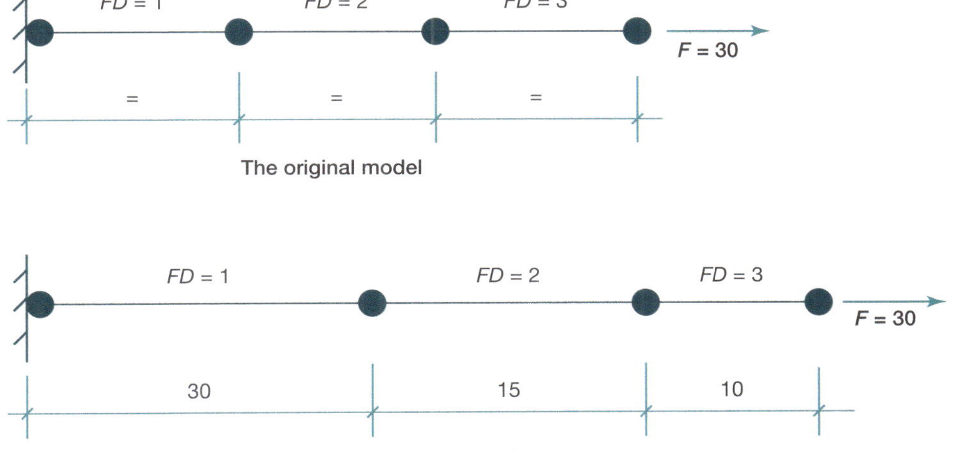

The advantage of force density is that it is fast (and hence popular in form-finding programs aimed at architects) and you can easily fine-tune the lengths of each element. A disadvantage is that the pre-tensions might not be achievable and are not guaranteed to be in equilibrium. Another is that the elements with the highest loads will be longer and need thicker sections. This sets it, in some respects, in opposition to Maxwell's Load Path theorem.

8.6.3.2 Soap film form-finding

Soap film is a more advanced form-finding method, where you specify the structure target pre-stresses (and possibly loads) and then remove all stiffness from the elements to find the equilibrium geometry. The effect is to model something like a soap film (hence the name) bounded by elastic bands, which results in minimum surfaces and cables lengths. It is thus a more suitable form-finding method for engineers than force density.

Soap film form-finding needs a nonlinear static analysis method, such as dynamic relaxation or another explicit solver. These analyses then move any structural element that is out of equilibrium until all forces are in equilibrium and the ideal structural layout is achieved[128].

8.7 Size optimisation

Size optimisation is all about choosing the best section sizes to use in the various parts of a structure. It is possibly the most common form of optimisation as most engineers will use it whenever they design a steel column or the reinforcement for a concrete beam. While shape and topology optimisation make the largest changes to the structure, size optimisation fine-tunes the design. It is also perhaps the type of optimisation that will have the least impact on the final cost or efficiency of the structure.

The process might be straightforward, such as when choosing a single pin-ended steel beam from a catalogue, or it can be a more challenging task, such as designing a moment frame — where changing the member sizes redistributes the moments between the members. Size optimisation also applies to choosing the steel thicknesses for plate girders (changing the overall dimensions is arguably shape optimisation) and the rebar for reinforced concrete sections. The traditional approach is to select some dimensions, such as the depth and width, using rules-of-thumb, such as span-to-depth ratios, prior to optimising the parts. This has the advantage of reducing the number of variables and thus makes the final choice a reasonable one for engineers using pencil and paper.

The size of concrete beams is of particular interest, as typically, more than half the load on a concrete frame is the concrete's own self-weight, while the design of long-span bridges can be driven by the self-weight. This means that a more keenly-sized member high up the building can also reduce the member sizes further down the load path. Considering that concrete production accounts for 5–8% of the world's CO_2 emissions[129–131], any improvement here will be beneficial to the structure and the environment.

When sizing sections it is important to remember the restrictions on sizes. Steel catalogue sections are obviously going to limit your choices, but it is same for steel plates as they also only come in a few discrete thicknesses. Concrete beams and columns can, in theory, be of any size, but the workers on site can typically only build the sections to the nearest 10mm. It can sometimes make the optimisation easier to temporarily relax this constraint and allow any dimension, but remember to round up the dimensions for the eventual answer.

Size optimisation allows you to rationalise the section choices (larger quantities of particular section sizes can attract discounts, while rarely rolled or particularly heavy sections are comparatively expensive[†]), avoid connection problems (welding stiffeners can be more expensive than simply increasing the column size), and deal with serviceability requirements (lighter structures can have vibration problems). Ideally you would carry out the size optimisation as part of the shape or topology optimisation, as it can sometimes influence the final form, but that also adds complication. You may also find that a fully stressed structure is not necessarily the least-weight structure[132].

8.7.1 Optimality criteria

While optimising sections to carry their load is a straightforward concept, a more subtle problem is that of optimising sections to minimise both the quantity of material and the deflection of the whole structure. The optimality criteria (OC) method uses the 'principle of work', which is defined as the displacement of a loaded point multiplied by the load itself ($W = F\delta$) and is measured in joules. Each individual element in a structure carries out work when it has a load and an elongation; the sum of the work done by all the parts of the structure is equal to the work done by the applied load, which means that if we know the forces, we can control the displacement by adjusting the work done by each element in the structure. If each element is optimal the workload is shared equally, so the resulting structure will use the minimum material for the given topology and shape[133].

Consider the truss in Figure 8.34. There is a single point load at node 9 and we want to find the sections sizes such that the deflection at that node is within a certain limit.

Figure 8.34: Cantilever truss

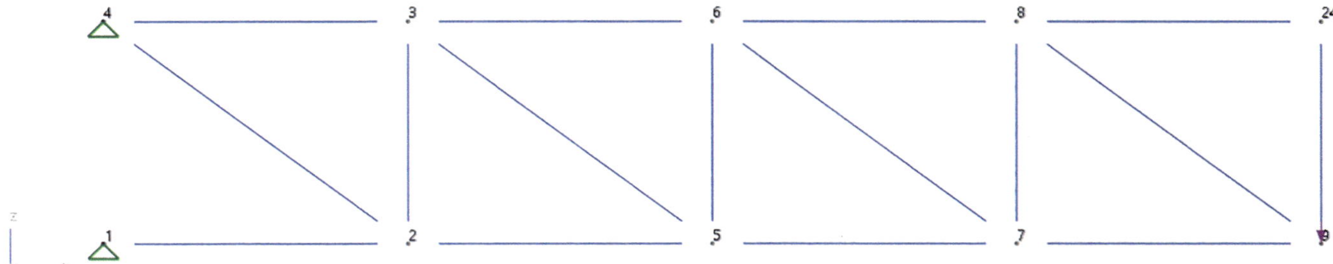

On analysing we can see the axial forces distributed as expected (Figure 8.35). If the bars have the same cross-section, the stresses ($\sigma = P/A$) will also follow the same distribution.

Figure 8.35: Truss axial loads

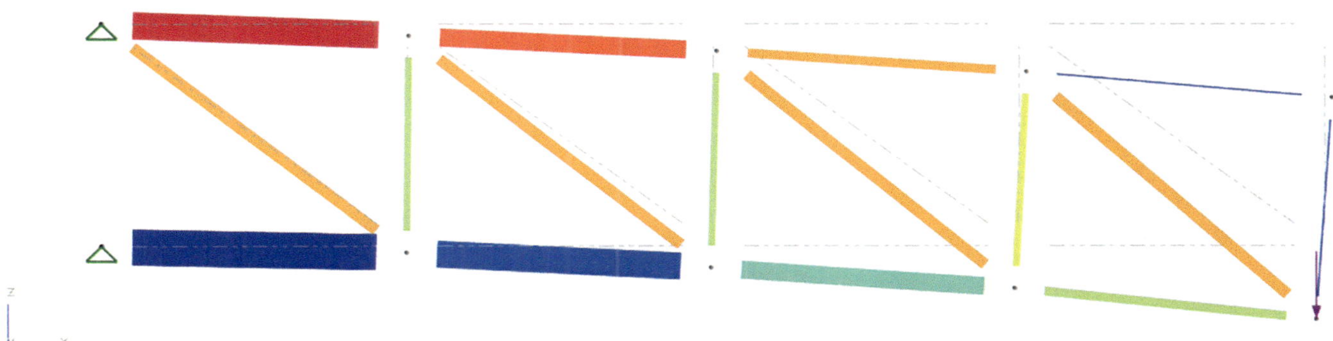

But the contribution to the deflection from each element is not so clear. To find this we can look at the strain energy in each element, which shows how much work is being undertaken by each element (Appendix B).

$$U = \frac{P^2 L}{2EA}$$

Where:

U = Element strain energy
P = Element axial load
L = Length
E = Young's modulus
A = Cross-sectional area

Again, as you might expect, the elements closest to the support have the most influence on the truss's deflection, while those in the top right contribute nothing — but as the axial load is squared, the higher-loaded elements are more significant to the end result (Figure 8.36).

Figure 8.36: Truss strain energy

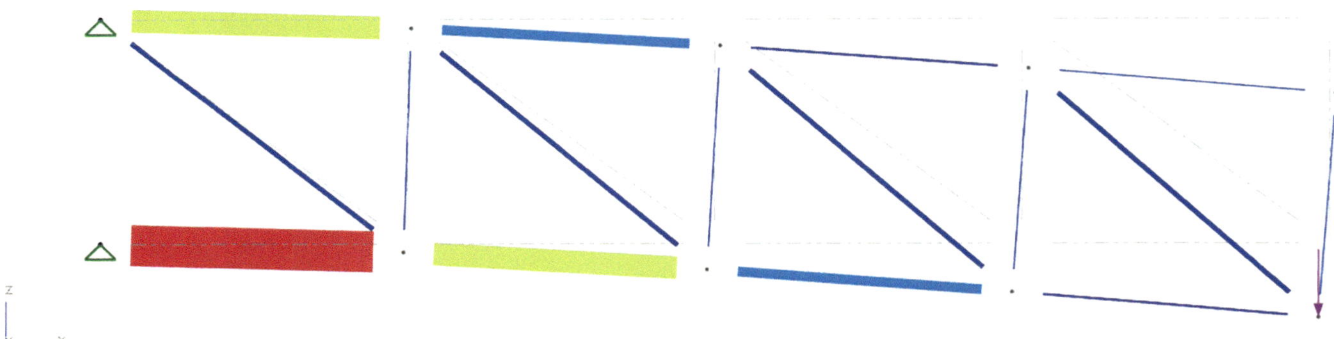

The ideal distribution for the material to control the deflection of the truss is for the strain energy in each element to be the same, which would mean that all the elements were working equally hard. When the structure is statically determinate, the force distribution is not affected by the cross-sections, so it is a reasonably quick calculation to determine the required ratio of the cross-sectional areas ($A = P^2L/2EU$). If you want a minimum weight structure that satisfies the deflection limit δ_T then:

1. Analyse the structure with nominal section sizes.
2. Find the tip deflection δ and the average strain energy of the elements $\bar{U}$
3. To achieve the correct deflection, set the target average strain energy $\bar{U}_T$ in proportion to the target deflection divided by the measured deflection:

$$\bar{U}_T = \bar{U} \times \frac{\delta_T}{\delta}$$

4. Set the area of each element so that the strain energy is equal to the target average strain energy:

$$A = \frac{P^2L}{2E\bar{U}_T}$$

On reanalysing, the deflection should now equal the target deflection (or be close if there are any rounding errors) and the weight of the structure will be minimised (Figure 8.37).

Figure 8.37: Truss with strain energy-optimised sections

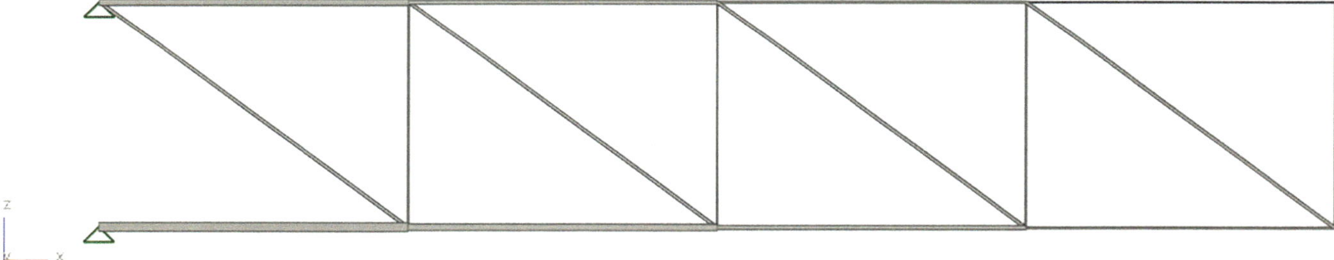

In this example we looked at the OC method with a 1D element mode, but it works just as well with 2D elements, where you adjust the element thickness to suit the strain energy. The method will need several iterations if the structure is statically indeterminate, such as a mesh or a moment frame, as the element stiffnesses will affect the stress distribution. Note that optimality criteria, as shown here, makes no allowance for maximum stress limits, buckling or other design checks. Including these essential steps may increase the number of steps in the optimisation exercise.

In this chapter we explored some dedicated structural methods for optimisation, in the next we delve into methods that can be applied to a wide range of engineering problems — and not just structures.

9 Optimisation methods

"How will you go about finding that thing the nature of which is totally unknown to you?"
Rebecca Solnit[134]

9.1 Introduction

In the previous chapter we looked at optimisation methods that were dedicated to structures; now we will discuss more generic methods, which can be applied to topology, shape and size optimisations, or indeed any problem. These methods, or 'metaheuristics' to use the technical term, require more work on your part to interpret how best to apply them to your engineering problems, as they are general purpose. We will also discuss optimisation concepts themselves, in particular exploring the design space and exploiting what you find there.

Inspired by nature, metallurgy and gambling, many of these methods use randomness to overcome the problems experienced by many optimisation algorithms, such as getting stuck on sub-optimum solutions. Though it seems counterintuitive, chance can help with solving engineering design problems.

9.2 The design space

Engineering optimisation boils down to finding the best design from the options available, preferably in the fastest time. This range of design options is known as the **design space** or **search space**, while physicists would call it the **phase space** and biologists the **morphological space**. The idea is the same though: it is the scope of possibilities or the range of possible solutions to a problem; in our case the range of design options that we want to explore. We do not want to look at every solution, as there are, if not infinite, at least an excessive number of them; an exhaustive search is rarely a viable option.

The design space is essentially a very useful way to think about the design options, and can consider things like steel or concrete, flat slab or beam and slab, column spacing or span directions, down to the size of the members and the material grades used. Establishing exactly which design aspects will be checked and which can be ignored is a key step in any optimisation. Design spaces have one dimension per variable, so that, for example, if there are 100 sections on offer for a steel beam, the design space has one dimension with 100 locations along it. Similarly, if there were two different beams then the design space becomes two-dimensional with 100×100 options, and so on. While reality is three-dimensional, the design space is usually a multi-dimensional hypercube. The trick is to find where our economical, yet functional design sits in this space. The smaller that we can make the design space the better chance we have of finding the optimum. Be warned though: while making the design space too big will slow the process and possibly stop you from finding the answer, making it too small will not provide a useful result.

Some design aspects are continuous, but to solve them we may have to make them discrete; and sometimes we have to treat discrete options as continuous to optimise them, depending on the method used. If we were looking for the best place to connect something onto a beam, we might decide on ten equally spaced locations that we would then check, or check at a given spacing. Real design problems often have dozens of variables and so the size of the design space can grow exponentially. It is thus an important part of the optimisation to first optimise the design space, making it as small as possible while still being sure that you will get a good result. One thing to look at is that of dependant vs. independent variables. The section sizes of the beams are dependent on the bending moment, which comes from the loading positions and can be independent of the section stiffness. This means that we do not always need to include the section as a design space dimension. Alternatively, if we had a moment frame, say a portal, where the distribution of the bending moments is dependent on the relative stiffness of the beam and column, then we do need to include the section sizes in the optimisation, and it becomes a lot more complicated.

We will start with a simple example: the design space for a single steel beam. We know the constraints (it must not fail in bending, shear, etc. and must not deflect more than a certain amount) and parameters (span, support conditions, etc.). We then use our experience and project requirements to reduce the size of design space until it is manageable. For example, we know that if the span is moderate then an I section is more appropriate than a channel. Likewise, the span might not be large enough to consider using a plate girder, so will only look at the steel beam catalogue. And we know which catalogue to use because of the project location. If this is a UK project and we know that we will only choose a universal beam (UB) then we have already reduced the design space to a reasonable size and might even reduce it further. At one company I worked for, we were involved in a number of refurbishment projects; the lead engineer wanted us to only use the lightest weight of the widest flange of any given UB depth. The idea here was that the contractor could acquire beams quickly and cheaply; it also meant that we had only about six beam sizes to choose from.

9.2.1 Exploring the design space

How do we work with the design space to optimise our structure? First, we need to decide what is meant by 'optimise' in this particular case. It might mean the least weight or cost structure that carries the load, or something more project specific. You then need to assign a qualitative value to any solution so that you can compare them. This also means that you can think of the design space as a terrain, where every point in the space has an elevation related to how optimum it is — and thus the optimum location is the highest point (or lowest point if you prefer, as optimisations are usually about minimising quantities).

Imagine that you need to find the highest mountain peak in your country of residence. The problem is that you do not know where to start and know nothing about the landscape. What do you do? This is exactly the problem facing anyone who wants to carry out a design optimisation, except that the geography of a design space usually has far more than just three dimensions. Another problem is that there can be many peaks, so we must be careful not find a foot hill and ignore the mountains. How we explore the design space for the highest peak is the essence of optimisation.

No optimisation is an island, by which I mean that optimising for one aspect may have detrimental effects elsewhere in the project. For example, an optimised truss geometry may result in a curved top chord. The problem with this is that not all cladding can cope with minimal drainage angles, and a curved roof, by definition, will have some parts with zero slope. This will necessitate using a more expensive roof cladding, which will offset the savings on the structure. Similarly, while timber frame sizes are usually driven by the connection requirements, the cost of a steel frame is also influenced by the cost of the connections (and connection requirements can sometimes drive up section sizes), so a cost optimisation of a frame should include an aspect for the joints and not simply focus on the section sizes.

9.2.2 Design space example

For this example, I wanted a structure that had a design space with just two dimensions, so that we can draw it, yet also exhibited interesting or complex behaviour. I chose two beams of equal length joined to each other mid span, with a central point load (Figure 9.1). Obviously one design result is that the two beams are the same size and they carry the load equally; the other is that one beam does all the work while its partner just gives restraint (Figure 9.2).

Figure 9.1: Two-beam problem

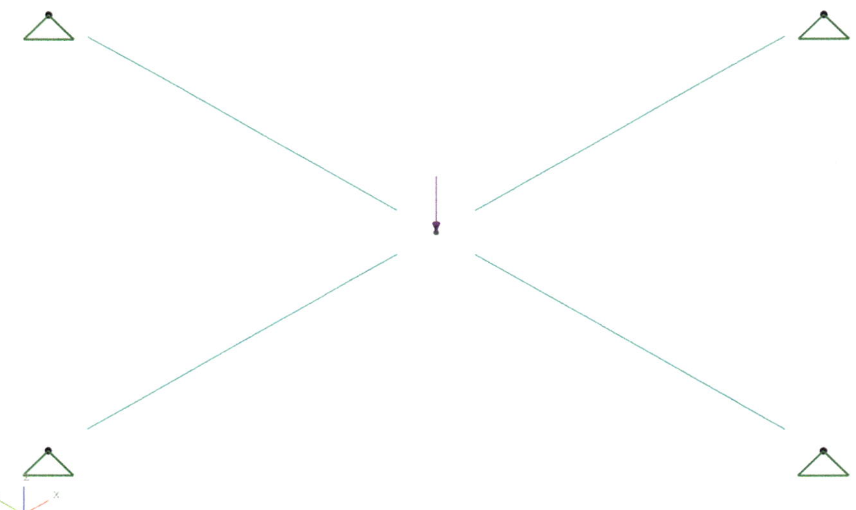

Figure 9.2: Possible solutions for two-beam problem

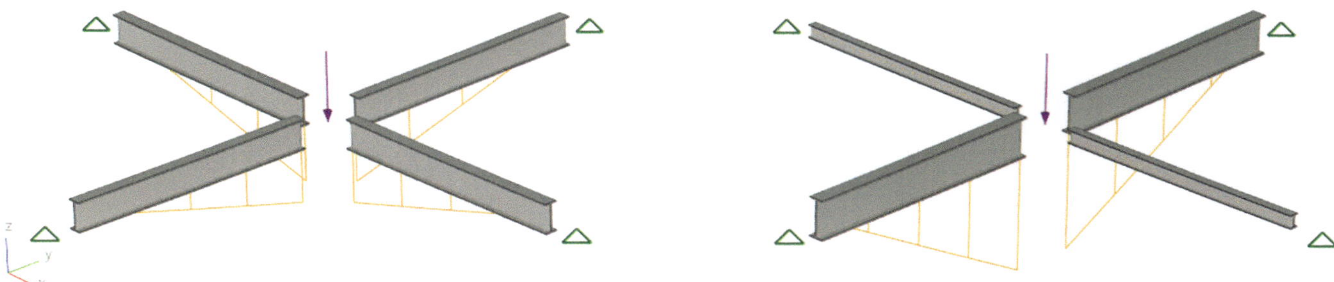

The first step is reducing the design space size as much as possible. To do this we can take the latest prices from your favourite steel section supplier[135] or price guide[136], then sort the sections by cost per metre from the cheapest to the most expensive. Next, we compare the bending capacity[137], and remove any sections where the capacity reduces as the cost increases, as these will automatically be suboptimum and thus never selected.

Then, we investigate the question of fitness[†]: how much better one design solution is compared to another and therefore which one is best. This needs to be an objective measure, such as weight or cost, but must be more than that as the lightest or cheapest beams are useless if they cannot carry the load. While every part of the final chosen solution must satisfy the usual design requirements of a structure, we may have to relax (though not ignore) the constraints to allow us to reach the optimum. It is likely that the optimum is on or close to the pass/fail border, and the fitness calculation we choose needs to be comfortable with exploring in that zone.

Not only do we keep solutions that have elements that fail, but recognise that a structure with one element that has failed is closer to the optimum than one where many elements are failing. Similarly, an element that is only a little overstressed is better than one that is highly overloaded. This means that we also need to take capacity or utilisation into account, heavily penalising understrength members and ensuring they are not selected at the end of the optimisation. The output from the exercise must be a structure that passes all the design requirements.

In this example I chose to measure the fitness as twice the unity factor if it is equal to or below 1.0, and one divided by the unity factor if it is above 1.0[††]. Once we have the fitness of every beam pair, we can use them to contour the design space (Figure 9.3).

On the other hand, if we take all the 96 UB sections, the picture is not so clear. We still get a ridge, but the terrain is considerably rougher (Figure 9.4). Some optimisation methods will struggle to get all the way to the highest peak[†††].

[†] The concept of fitness is drawn from evolutionary theory: survival of the fittest.
[††] Is utilisation the best measure of optimality or can you think of a better one?
[†††] Hill Climb methods, for example, need a smooth slope.

Figure 9.3: Proximity to unity for two-beam problem: refined beam selection

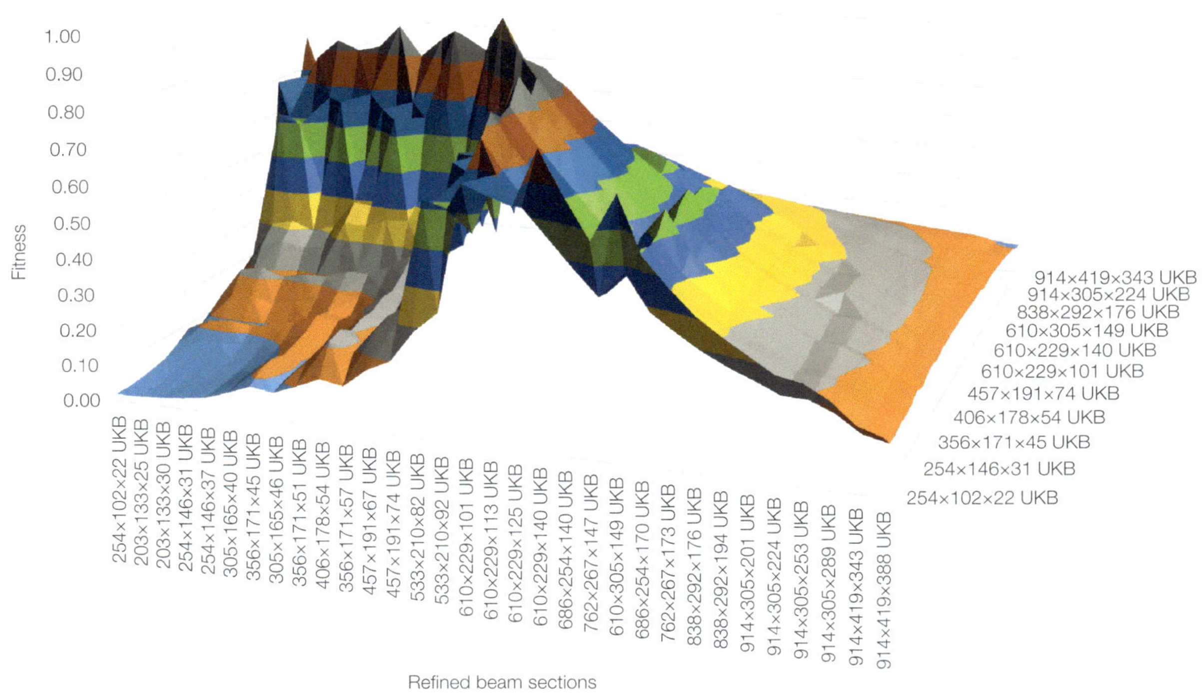

Figure 9.4: Proximity to unity for two-beam problem: all universal beams

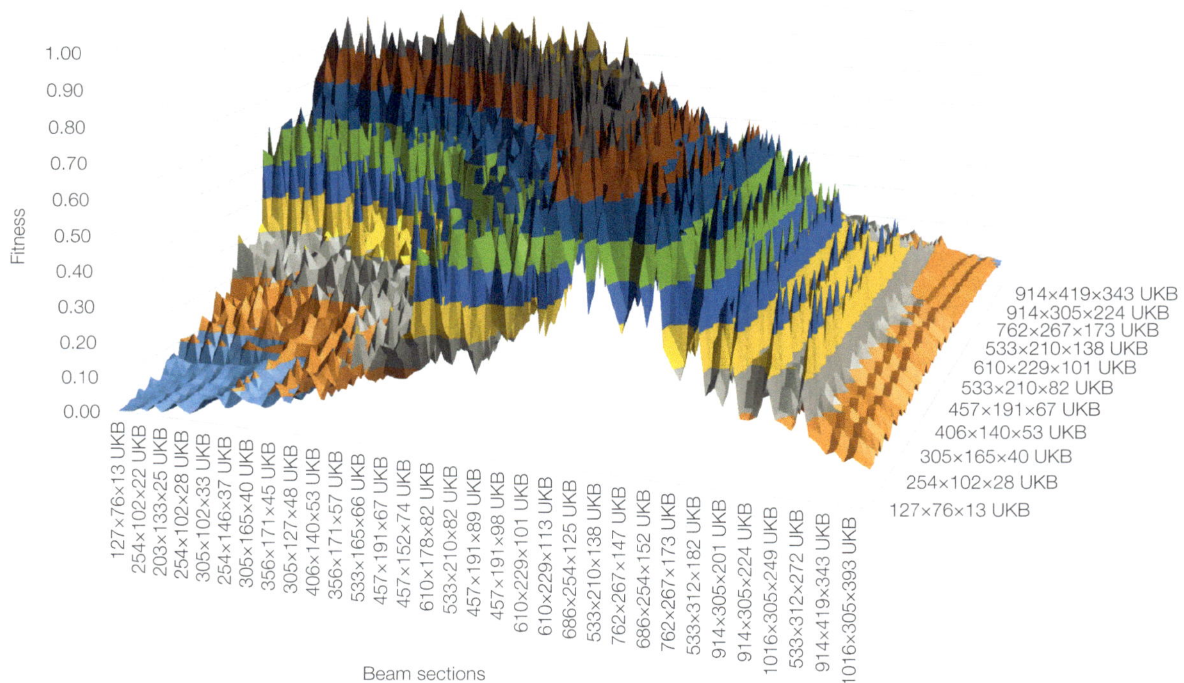

9.3 Optimisation methods

As with any analysis, the first stage in any optimisation process should always be a manual one: What is the problem and how can I simplify it? As Einstein allegedly said: *"Everything should be made as simple as possible, but not simpler"*. You need to optimise the optimisation to get the most efficient results.

What does this mean in practice? It may mean reducing the load cases down to just the significant ones, fixing in position the parts of the structure that are very stiff, and excluding the aspects that are very flexible. If the structure has lines of symmetry then use them to group members that have the same section, or even remove those duplicated parts of the structure and replace with suitable restraints.

Next, what heuristics might we use to address these problems?

9.3.1 Deterministic methods

Deterministic methods will produce the same answer every time. They are usually fast and reliable but can sometimes struggle to find a suitable result. In the previous chapter we looked at some deterministic methods that are dedicated to particular forms of optimisation; here are some more general ones.

9.3.1.1 Calculus

If you can derive a formula to define the problem, you might be able to differentiate it to find the maximum or minimum. Let us take a trivial example; the equation $y = x^2$ (Figure 9.5).

Figure 9.5: $y = x^2$

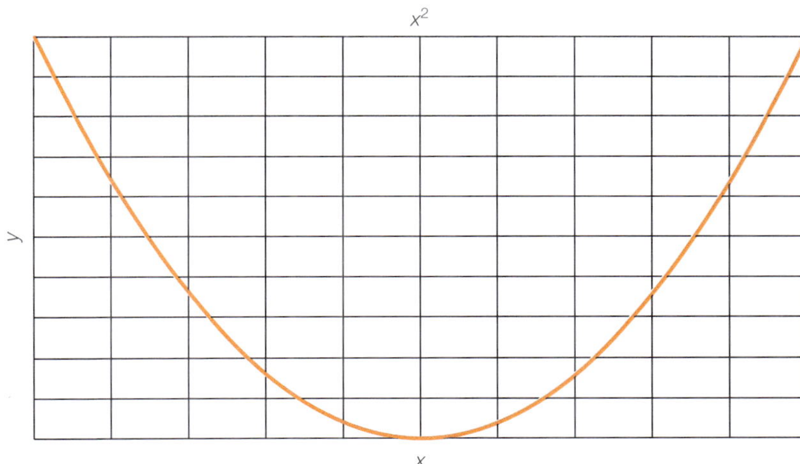

Line	$y = x^2$
Slope	$\dot{y} = \dfrac{\delta y}{\delta x} = 2x$
Zero slope	$2x = 0 \Rightarrow x = 0$
Change at $x = 0$	$\ddot{y} = \dfrac{\delta^2 y}{\delta x^2} = 2 > 0 \therefore$ minimum

Differentiating once finds the slope of the graph: the max. or min. point must have zero slope. Differentiating again tells us which direction the slope is changing in — if it is positive then the slope must be curving up and we are at a minimum. Likewise, a maximum gives us a negative slope change.

Or to take another, similar example $y = x^3$ (Figure 9.6).

Figure 9.6: $y = x^3$

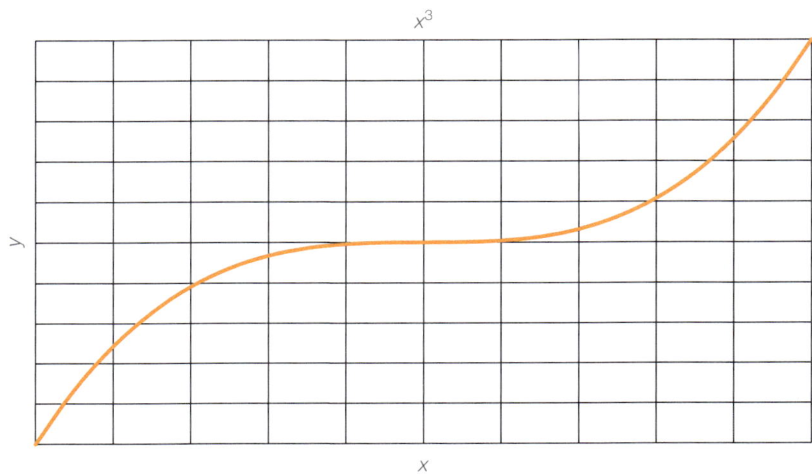

While we do have a point of zero slope, this time the double differential is 0 indicating that we have neither a maximum nor minimum.

Line $\qquad\qquad y = x^3$

Slope $\qquad\qquad \dot{y} = \dfrac{\delta y}{\delta x} = 3x^2$

Zero slope $\qquad 3x^2 = 0 \Rightarrow x = 0$

Change at $x = 0$ $\quad \ddot{y} = \dfrac{\delta^2 y}{\delta x^2} = 6x = 0$ ∴ not max. or min.

Suppose that we want to build a reservoir to hold 10,000m^3 of water that needs a minimum amount of concrete. Ignoring complications such as the wall and base thickness, we will simply optimise the reservoir to have the minimum area of sides and base.

Figure 9.7: Water tank

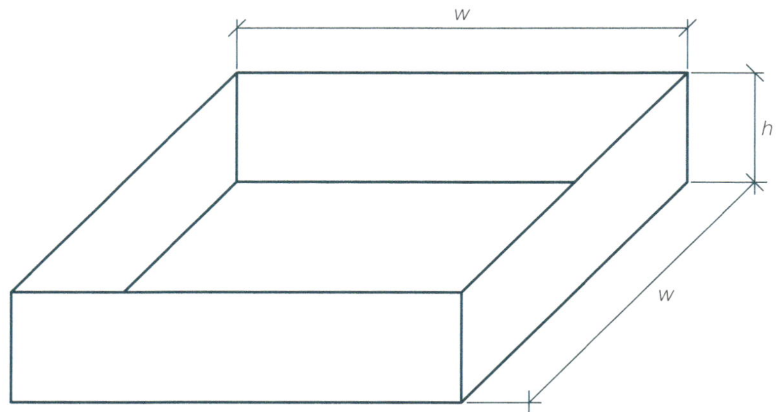

Dimensions:

height $= h$

length $=$ width $= w$

Volume:

$$v = w^2 h = 10{,}000 \Rightarrow h = \frac{10{,}000}{w^2}$$

Area:

$$a = w^2 + 4wh \Rightarrow w^2 + \frac{40{,}000}{w}$$

Differentiate area:

$$\dot{a} = 2w - \frac{40{,}000}{w^2}$$

Max. or min. when:

$$\dot{a} = 0 \Rightarrow w = 13.6\text{m}$$

Check if max. or min.:

$$\ddot{a} = 2 + \frac{80{,}000}{w^3} > 0 \therefore \text{minimum}$$

Calculus works best if the functions are continuous. The reservoir might be any dimension we choose, but rolled steel beams come in discrete sizes, meaning that we need to use a different method for them. Likewise, plate girders can be any dimension, but their plates have standard thicknesses.

9.3.1.2 Gradient methods — climb that hill

Gradient methods can be so simple that you can use them without realising. If you are designing a concrete beam, you work out the area of rebar required, then choose the numbers of particular bar sizes that get you closest to the required area (constrained by maximum and minimum bar spacings) from your handy crib sheet[†]. Similarly, if you are choosing a steel beam, you calculate the required values, second moment of area for displacement and effective length for moment capacity, then choose the lightest option from the capacity tables.

What you are doing is putting all your options in to order then working up the list until you find the first that meets your criteria. A simple computer Gradient method works in the same way. If it was choosing a steel beam it would take all the available sections, put them into ascending order (depth, weight, cost, environmental impact, etc.), then start at the smallest section and work up until it finds the first one that satisfies the criteria. In this way you know that it has found the lightest etc. section for the job.

Gradient methods can also work when you have multiple variables to choose from, though in a slightly different way as there is no obvious 'up' direction and the program has to 'feel' its way. To explain, imagine that you are standing on the side of a hill and you want to reach the top. The problem is that you are blindfolded so do not know the layout of the hill. What you can do is to feel around you to work out which is the 'up' direction and move (carefully) that way. Eventually you should reach the top. The Gradient methods are doing the same thing, though they differ in how they 'feel around'.

[†] The reinforcement manufactures used to publish these; now you can quickly create them for yourself with a spreadsheet.

One of the big challenges in engineering optimisation is that for most problems there is more than one peak; another is that the way up is not always obvious. If you look at the capacity of universal beams in weight order against, say, I value or bending capacity, you will see that the surface is far from smooth (Figure 9.8). When checking a single beam, you can just check the weaker sections and move on as you know the direction to move in. For Hill Climbs with multiple members this is not an option and it can be worth excluding the sections that get weaker while increasing in weight.

Figure 9.8: UB moment capacity/weight

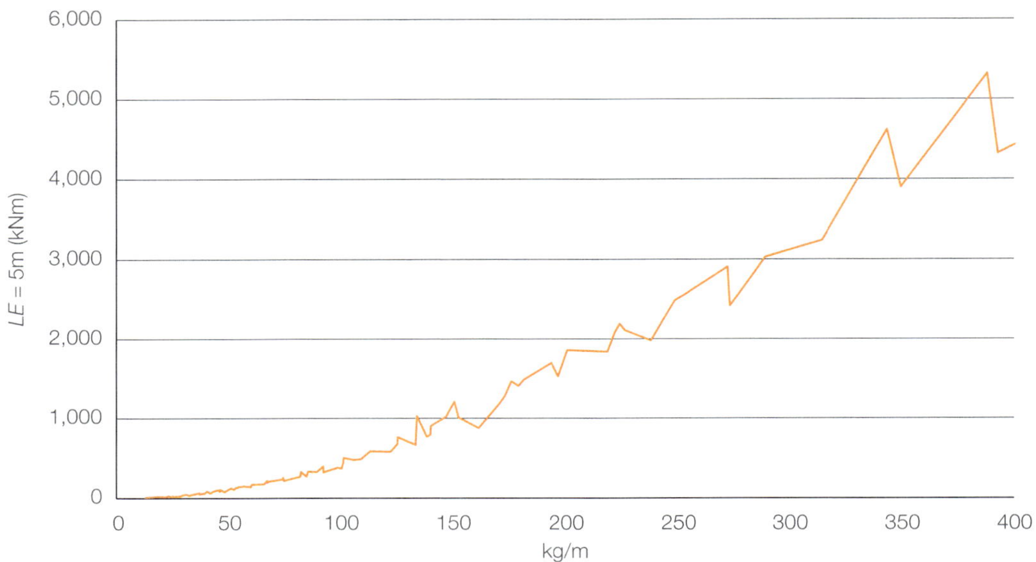

If we look at the section price[135,138] vs. capacity (Figure 9.9) then we see a similar, though different irregular relationship between capacity and cost. These differences are driven by premiums being applied to sections that are either rarely rolled or are particularly heavy for their size.

Figure 9.9: UB moment capacity/cost

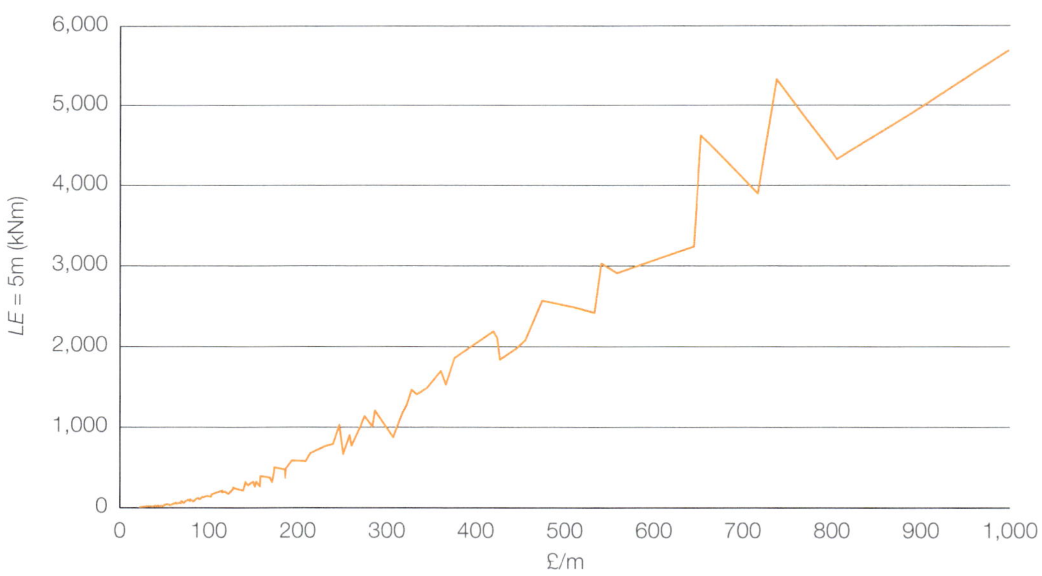

When you have a design space of two dimensions or more, there is no obvious route up the slope, so how do you determine the gradient? One option would be to choose a direction at random and see if that generates a better solution; another is to check all adjacent points to see which the local best is and move to there. When you have checked all adjacent points and found that none are better, then you must be at the top and you can stop.

The Gradient method is fine if the design space has a single, smooth hill. But what if the design space were more varied? If you start somewhere flat, it is difficult to determine the direction of the upward slope. Even the simple two-beam problem discussed previously has multiple peaks. Thus, you might run a few Gradient methods, each starting in a different location, and choose the best of the best.

Another problem is that the method can struggle when the slope is not smooth. Consider the beam weight vs. stiffness. If you used a classic Gradient method it would stop as soon as it reached a downward slope. It would not be smart enough to try bumping over the intermediate valley, thinking that it had reached the peak without finding the solution and thus report a failure. How can we overcome this? As mentioned, we can sometimes reduce the number of sections under consideration to avoid these inefficient beams. Another option is to use a more sophisticated method, such as Simulated Annealing (Section 9.3.2.1).

Gradient methods are fast but do not like rough design spaces. To use them to their best you must either smooth the design space by carefully excluding options or use them in optimisations where the roughness can be ignored, such as choosing the lightest section for a statically determinate beam.

One way to smooth a design space is to use a Neural Network (Chapter 10), trained on select points in the design space, to predict the overall surface. Neural Networks might not tell you exactly which is the best design option, but they can provide a good indication of their approximate location.

9.3.1.2.1 Quasi-Newton

The Quasi-Newton method is an iterative process where each repetition gets you closer to the desired result[139]. Consider a steel moment frame: analysing the structure and then designing each element for the resulting forces and moments will give a new arrangement of section sizes and thus stiffnesses. The now stiffer elements will attract more load, possibly requiring even heavier sections. Eventually the sections will not change from one iteration to the next, and we will have converged on a solution.

Considering our previous two-beam problem, in this case it would not take a computer long to check every single option in the design space for one or two beams[†], but these exhaustive methods would quickly fall foul of what is known as a combinatorial explosion: the numbers of possibilities grows exponentially with the addition of every new member. This is not a problem with all structures, because if you prevent the members from interacting (keeping it statically determinate), you can design them individually, but you might get a more efficient design by allowing interaction. For example, compare a pinned steel frame to a moment frame. Moment frames will usually result in a lighter design, but at the expense of bigger connections and a more complex design.

Rather than testing every possibility, we can analyse the beams, determine the moments, and choose the lightest or cheapest beam that can carry the load, then repeat until the chosen beam matches the one just analysed. In design space terms, we start from one location (two section sizes that we will analyse) and jump to another (two section sizes given by the design). Each iteration will take us to a new point, which then leads to the next, until it converges (the new point is the same as the old one).

Issues can arise however, when there is moment interaction: the moments are proportional to the beams' relative stiffnesses. When we design the beams, the stiffer beam might need to increase in size, attracting more moment, and so might require an even larger beam. Self-weight can also play a part, as more permanent load means higher forces. The result is that it can take several steps to converge to a point where the design stage does not change the beam size.

[†]   My Python script took about five minutes to run.

To illustrate this, let us look for the best solution in the previous two-beam example. In this instance I first reduced the section list by removing any section that has a lower moment capacity (effective length 5m) than a lighter section. This reduced the section list down to a third, and thus the design space by a factor of nine, to just over 1,000. Apart from speeding up the investigation, it should also remove any sections that would not be chosen anyway: a double advantage. It is usually worth the effort to remove the options that cannot be chosen, and the advantage grows with the dimensions of the design space.

I placed all the section sizes I wanted to check in the FEA model, specifying them in weight or price order. By doing this, then always working upwards through the list during the design check, we will either get the lightest or cheapest section that can carry the load. Normally we might choose a starting point, say all beams of the same size, but for the sake of this exercise my script worked through all the possible pairs of beams — analysing and designing them to determine the new section pair.

Looking at the results we can see that if the initial sections are the same, then as we would expect, the resulting beams are also the same (Figure 9.10). Likewise, if the starting sections are quite different, the result is one beam

Figure 9.10: Quasi-Newton tributary areas for optimum designs (darker squares)

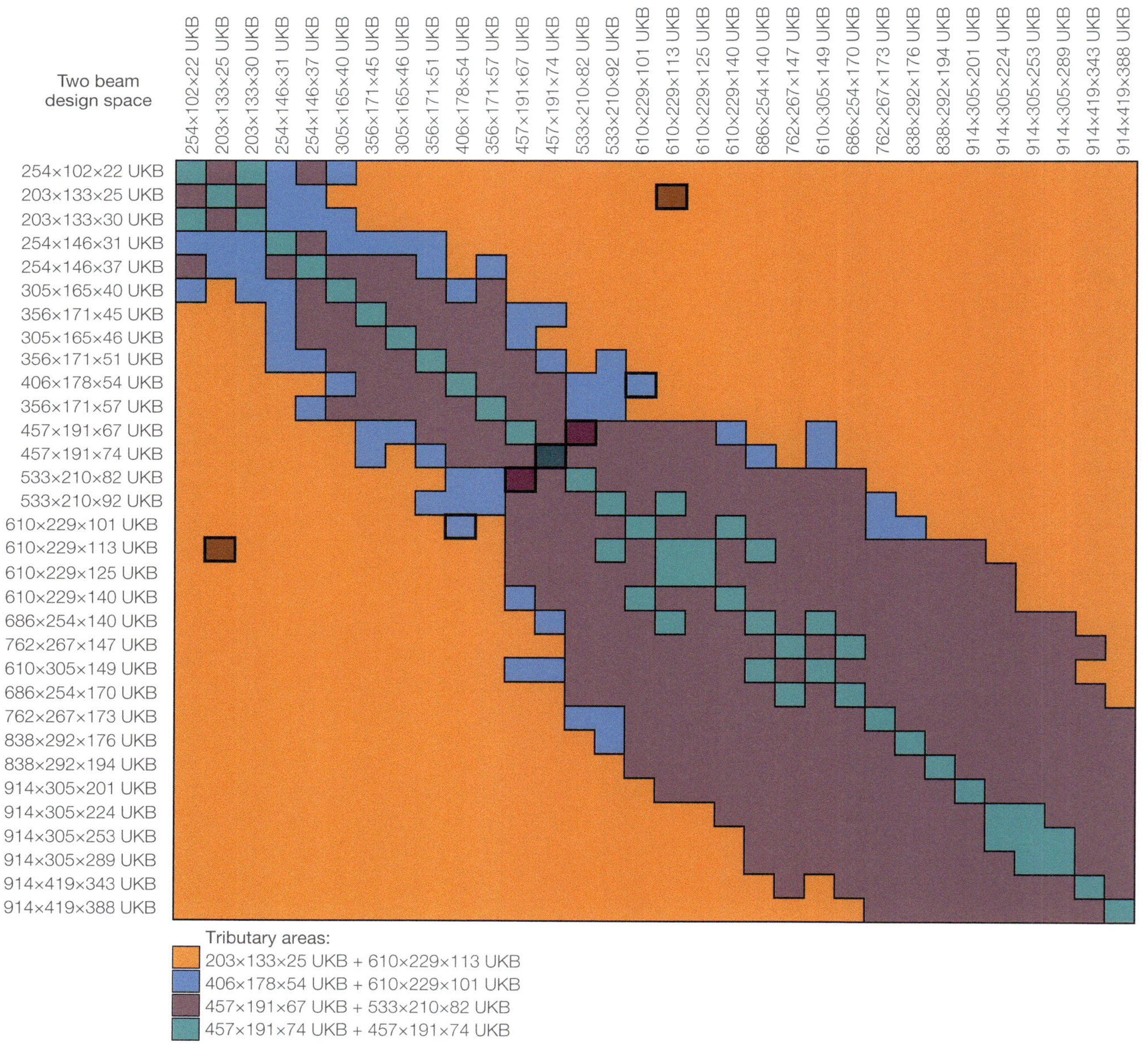

carrying all the load. The surprising result was when the sections deviate only a little from symmetry as the result was also nearly symmetrical.

An especially interesting aspect here, is that there are a couple of section pairs where the Quasi-Newton method would not actually converge: instead, the model oscillated between two sections for the restraining beam. A robust optimisation routine will have to watch out for solutions repeating and stop on the fitter option that can carry the load.

Mapping all the optimum solutions and the tributary areas show that there are five answers: the final solution depends on the starting point in the design space.

The Beijing National Aquatics Center (also known as the Water Cube) project team used a variation on this method when designing the superstructure steelwork. There they started all sections off as the lightest size, analysed the model, design-checked each member, then increased them one section size if it failed the check[104]. This avoided oscillations between section sizes and ensured a balanced design.

9.3.2 Stochastic methods

Deterministic methods should usually be your first option, but as we have seen there are times when they struggle. When that happens, you can benefit from adding some randomness into your method. This may seem a strange idea if you have been brought up entirely with the algorithmic approach to engineering: here is the formula, there are your values, put them together and you always get the same answer. Stochastic methods add a random element to help stop the method getting stuck at sub-optimal answers.

9.3.2.1 Simulated Annealing

Simulated Annealing is a refined Gradient method, where the inspiration is one of annealing cooling metal. With the classic Gradient method, you would check adjacent points in the design space and chose the next point to move to from those. Simulated Annealing takes this idea but instead randomly chooses a point for comparison some distance away. This distance is controlled by a 'temperature' variable that gradually reduces over the time of the optimisation. This has the effect of making the jumps round the design space large to begin with, then as the temperature cools the jumps gradually get smaller[†]. You can also control this temperature change by how successful the heuristic is at finding better values: increase the temperature when it finds a better point and reduce it when it does not[133].

If you were using a Simulated Annealing method to find the highest point in Britain, you might start with jumps of perhaps 200 miles or more to test for better locations. Each subsequent jump would gradually get smaller, until hopefully you find Ben Nevis rather than Snowden, Scafell Pike, or an even smaller hill. You would also improve your confidence of finding the highest peak if you started at multiple points, or ran the optimisation multiple times, before taking the best result from them all.

An example of its use is in the search for the most efficient geometry when adjusting the nodal coordinates, as discussed in Section 8.6.1. To begin with, nothing much is known about how ideal the current geometry is, so the temperature variable is high, giving large dimension changes. As we get closer to the ideal, the jumps get smaller, until we reach a sensible limit. Thus, the routine might start by varying the coordinates by as much as 1m or more at a time and then gradually reduce to a minimum of perhaps 1mm for fine-tuning the frame.

9.3.2.2 Monte Carlo method

The Monte Carlo method was invented by the Manhattan Project scientists, particularly Stanislaw Ulam and John von Neumann, who were faced with predicting the expected yield of a nuclear warhead despite the huge number of variables[45]. Their solution was to pick many points at random in the design space, or 'phase space' as they called it, and then analyse the mass of results as per regular experimental data to determine typical behaviours.

[†]   Not unlike Brownian motion.

With structures, the Monte Carlo method is perhaps best used to conduct a broad survey of the design space to give an understanding of the influence of the various parameters. Other options could be to consider a range of material stiffnesses, e.g. specified vs. supplied, for a seismic analysis to check predicted behaviours, or to determine 'typical' loads for a serviceability vibration analysis. Adding the effect of construction tolerances to an FEA model is another possibility.

9.3.2.3 Genetic Algorithm

*"It is not the strongest of the species that survives, nor the most intelligent that survives.
It is the one that is most adaptable to change."*
Leon C. Megginson (attributed)

The Monte Carlo method provides a snapshot of a variety of positions in the design space, but what if we could build on what it reveals? Genetic Algorithms (GA)[140–144] do just that, taking the solutions from a Monte Carlo distribution and literally breeding a new generation of improved solutions from the best of the previous generation.

Generic Algorithms are inspired by Charles Darwin's Theory of Evolution as well as modern genetics. In a GA you must first describe the variable aspect of the structure using a binary string as the 'DNA', or 'genotype'. The exact form of this genotype will vary between problems: for example, in a study on possible external bracing locations in a tall building facade, each digit might represent a diagonal in one direction in a bay, showing the brace either there (1) or absent (0). An alternative may have the binary string representing a binary number, allowing the selection from an ordered list. Whatever the form of the encoding, each genotype will give rise to a solution, or phenotype, that you can analyse and check.

You will need to decide on a population size for each generation of genotypes. For the first generation, generate each genotype randomly. Once you have a complete generation, you need to assess their fitness, which will be a numerical value of how close they are to your requirements, while penalising but not rejecting solutions that include failing members. For example, if U_i is the design utilisation of a member:

$$U_i = \frac{\text{load}}{\text{capacity}}$$

the fitness f of a phenotype could be:

$$f = \sum_i (\text{if } U_i \leq 1 \text{ then } [2U_i] \text{ else } [1/U_i])$$

This example formula is only looking at utilisation and has no consideration of cost, weight, or anything else that you might be interested in. Your challenge is to find the best cost function for your project.

The parents of the next generation are chosen based on their fitness. A typical approach is to use a weighted roulette wheel process, where the chance of selection is proportional to their fitness. Having chosen two parents, you now need to breed them. We do this using a crossover, which means that you randomly choose a point along the genetic string, then take the code before that point from one parent and the code after from the other. There is also a small random chance that any bit in the genetic string is mutated, or flipped, from 1 to 0 or *vice versa*.

Example:

Parent #1 = 11111111

Parent #2 = 00000000

Crossover at point 5 gives

Children = 11111000 and 00000111

Repeat with two new parents until you have a population for a new generation, then test again. Continue to repeat the process until either the total fitness exceeds an amount, or you complete a certain number of generations. As you might expect, the initial results will be extremely poor, but working with those few that show promise will show rapid improvement, at least at first, but tend to be slow when refining the answers.

There are some refinements you can make to the process:

- Make sure that no genotype is ever repeated in a generation, as you do not want to end up with a lot of clones
- Keep the fittest individual and pass it into the next generation unchanged. This is known as Elitism
- Include mutations: on a random basis, flip one of the bits in a genome string (0 becomes 1 and *vice versa*). The need for this is debatable as crossovers slicing through the genes in the genome will cause effective mutations anyway
- Use Lamarckism[145] to refine the individual results, such as designing the sections to see what sizes are suitable for use. Lamarckism is a discredited theory of evolution that believes that changes experienced by individuals are passed on to the offspring. While biology might not work this way, that does not stop us from making use of the idea

Genetic Algorithms are notoriously slow[†] but can eventually produce impressive results for complex problems, though the solutions will probably need additional refinement.

9.3.2.4 Swarms and agent-based models

"Nature uses as little as possible of anything."
Johannes Kepler

9.3.2.4.1 Particle Swarm Optimisation
The Particle Swarm Optimisation (PSO) method is based on the way that social insects and birds might explore a location looking for the best food to share[146,147]. For example, a vulture will circle around a potential meal because it tells the other vultures that it has found something for lunch. In a PSO you have several particles flying through the design space, assessing each location they land on, but also remembering the best location they have found so far. Their next landing point $x_{i(t+1)}$ is derived from their current position x_i and movement vector v_i modified by an attraction to their own best design solution p_i and to the group's best solution p_g plus weighting factors φ (Figure 9.11).

$$v_{i(t+1)} = v_i + \varphi_1(p_i - x_i) + \varphi(p_g - x_i)$$

$$x_{i(t+1)} = x_i + v_{i(t+1)}$$

Figure 9.11: Particle Swarm Optimisation

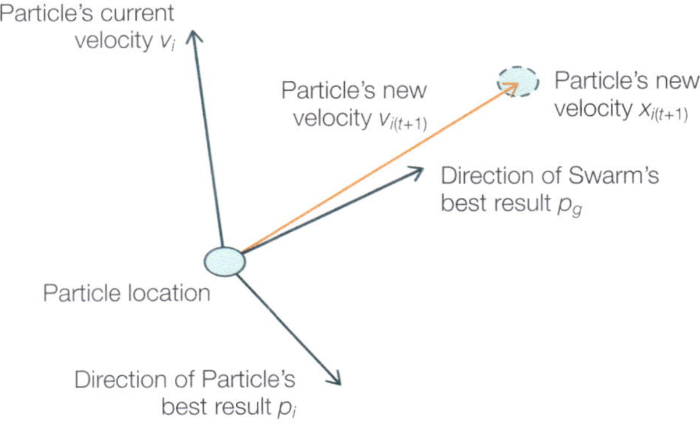

[†]   Look how long it took nature to produce us.

The result is that while some particles spiral into the global best solution found so far, others circle much wider constantly looking for a better answer, orbiting both the global best and their own. If a particle does find something better, the swarm will change its focus to there instead.

Note that while the design space may be defined in terms of discrete values, e.g. beam sections, the PSO works best if you treat the space as continuous, then choose the closest point to be the one assessed.

Some variations and choices that you might make include:

- A drag factor to ensure that the particles' speeds do not get excessive
- Deciding what to do when the particles reach the edge of the design space: do they stop at the edge, bounce off, pass through, or wrap round to the opposite side?
- Some PSOs have the particles in local groups so that they are controlled by local maximums, not the global one

9.3.2.4.2 Ant Colony Optimisation
Ants appear dumb to us because individually they have virtually no brain, but on mass they can demonstrate remarkable intelligence. Ants achieve this by communicating through *stigmergy*: changes they make to the environment. In the ants' case it involves chemical signals, called pheromones, left on their trails (Figure 9.12).

Figure 9.12: The Search

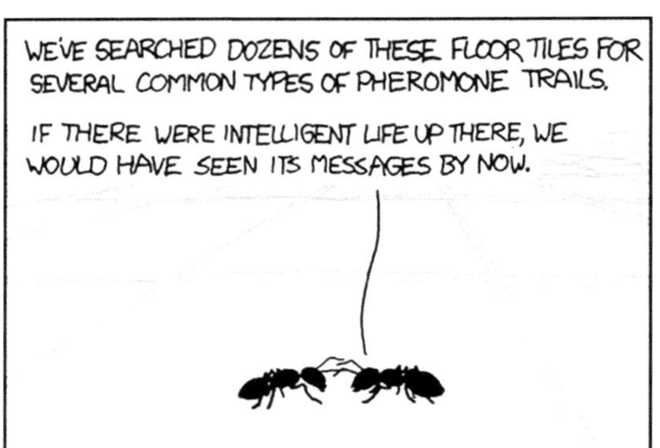

When an ant is deciding which way to go next, it detects the various pheromone trails left by its sisters. Its choice is then random, but the likelihood of one path being chosen is a function of the relative pheromone strengths. These pheromone trails evaporate over time, which means they are stronger when more ants have passed that way recently. This generates a feedback so the shorter routes become more popular, which in turn strengthens the trail.

Figure 9.13 shows how ants might search for food.

1. The first ant returning to the nest carrying food will choose a reasonably random route while following the scent of the nest. It will lay a pheromone trail that says "I have found food".
2. Other ants detect this and follow the trail to the food. When the numbers start to build, there will be less pheromone evaporation on shorter routes, as the transition time is shorter. Also, the busier paths will see more pheromone laid. Together, these effects will encourage the following ants to take the more efficient route.
3. Finally, nearly all the ants travel the shortest route (or the shortest route has 'emerged'), with the occasional ant still going a separate way to search out alternatives and thus avoid the risk of converging on a sub-optimal solution.

Ant Colony Optimisation (ACO)[148,149] has its main application in route finding, such as the classic 'Travelling Salesperson Problem': finding the shortest route passing through a set of points only once. ACOs have applications to civil engineering problems as well: I have experimented with ACOs for finding non-circular slip planes in slope stability problems, where the route is the slip surface with the lowest factor of safety.

Figure 9.13: Ants using stigmergy to find optimum routes

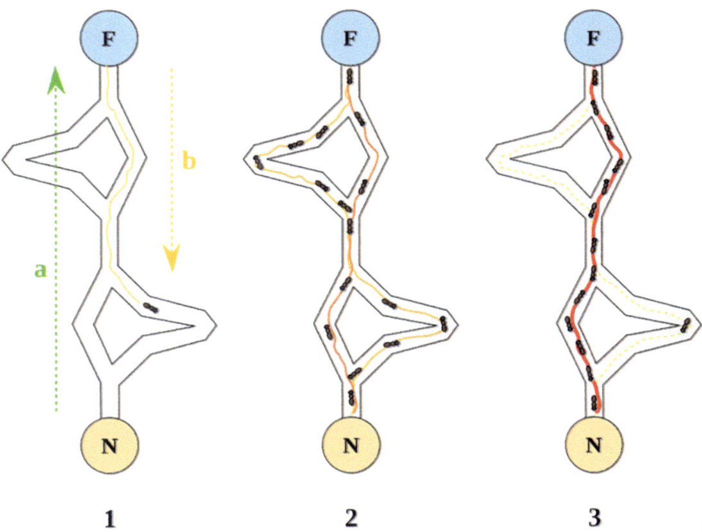

To use an ACO, first create a graph[†,150]: a grid of vertexes that correspond to the design choices or route options. If you are conducting a pure route optimisation, join every vertex to each of its neighbours where appropriate with an edge. Each edge should have a cost associated with it, related to distance or other relevant value.

Graphs have also been used for design space exploration, which views the space not as a multi-dimensional volume but instead as a series of choices known as a 'state space' (Figure 9.14). Make each mutually exclusive option a column of vertexes (for example a particular beam can have only one section size, but what size is it?), then join every item in each column with every item in the next with an edge. In this instance the edges might not have a cost but instead the chosen route will generate a design that will need quantifying.

Figure 9.14: Ant Colony Optimisation for state space search

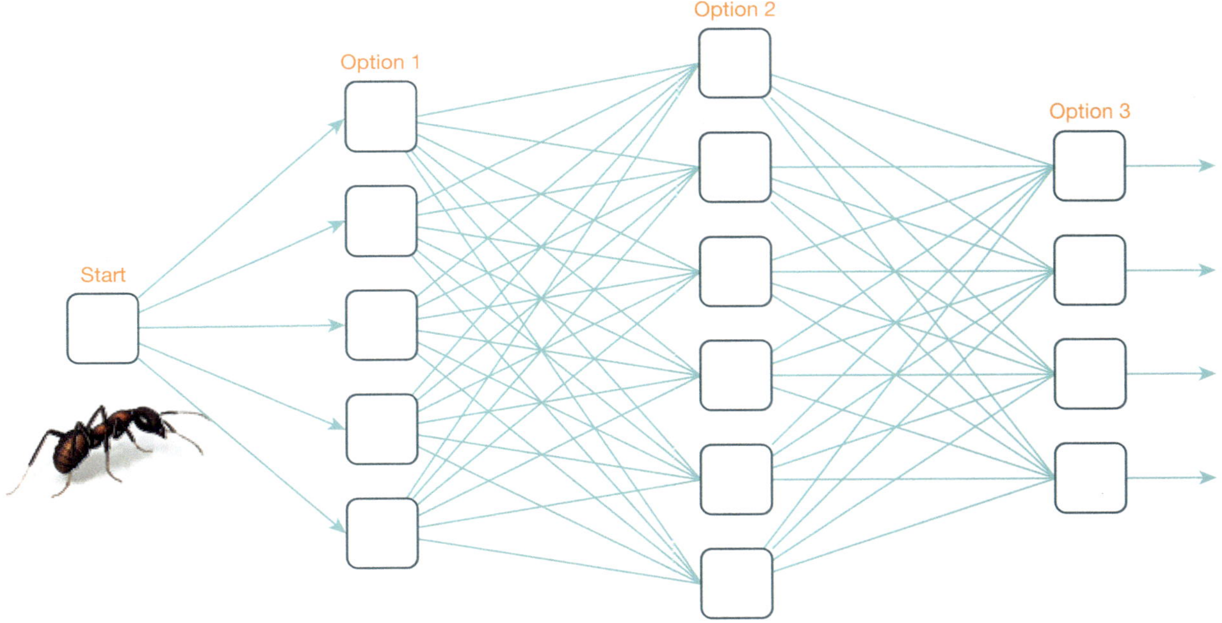

[†]   In this context a Graph is an array of dots (vertexes) connected by lines (edges) and is not to be confused with a graph of an equation.

The assumption is that the relative pheromone strength on any edge is a measure of the fitness of that edge; these edges combined indicate the better routes through the network. When an ant reaches a vertex, their random choice of which edge to follow next will be based on the relative pheromone strength on each[†]. This means that while good routes become popular, there is always a chance that other options are explored.

Before starting the ACO, seed each edge in the network with an initial pheromone strength, making each route equally likely. The population of ants will then individually find a route through the network; each route is then assessed against the desired criteria (journey time, structural cost, etc.). Depending on the ACO variant, either each edge in the route will have a pheromone added in proportion to the relative quality of the route, or only the best route will have the pheromone added.

To prevent pheromone overload, the values on all edges are 'evaporated' by a set percentage every round. The exercise is then repeated until a winner emerges.

9.4 Optimising optimisation

"If you don't know where you are going, how will you know when you get there?"
Lewis Carroll

A key aspect to all computer analyses, including optimisation, is asking the right question. It may be true that with enough time and computing power you might examine and compare all viable solutions and choose the best, but sufficient time is almost impossible to find on an average project. Douglas Adams jokingly suggested that the Earth was a giant computer with all life as part of its matrix, programmed to find the question to the meaning of life, the universe, and everything[56]. The answer was, of course, 42, but after 3.5 billion years we still do not know what the actual question is. Similarly, we cannot ask the computer to give us the 'best' truss. If we ask the question in the right way it might give us the lightest truss, or the cheapest, or the truss with the lowest embodied energy. These are all different answers though: all design is a compromise, and optimisation is no different.

Optimisation involves automated exploration of the design space, but we define what that design space is. No matter how sophisticated the optimisation routine is, it can still only test the possibilities we have set. You cannot have an infinite range of variables and expect an answer in a finite time. The real genetic algorithms, the evolution of life on Earth, has produced some amazingly complex and intricate things, including ourselves, but that took longer than most clients would find acceptable.

"My client is willing to wait."
Antoni Gaudí discussing the design and construction of The Basilica de la Sagrada Família in Barcelona.

Improving the speed of optimisation is something you can address. One of the slowest parts of an optimisation is the analysis and testing of the structural model: Can you find a way to explore the design space while not creating thousands of models? Not only that, but an accurately defined fitness landscape can have a rough surface, making exploration difficult. To speed up the exploration phase you might create an approximate, and smooth, version of the design space surface. One way to do this is to pick a few choice spots in the design space, check them thoroughly, then define a new surface (a meta-surface) of the design space.

You might define this meta-surface as a NURB or spline surface, derive its equation and use calculus to find the peaks and troughs. Another option is to train a neural network to reproduce the set answers. This should then extrapolate between the points, which you can quickly explore. Both options will only find the approximate peak locations, but from there you can use more detailed methods to refine your answer.

[†]   The probability of choosing that edge is the pheromone strength on that edge divided by the total strength of all available edges.

What points should you choose from the design space? I would use hypercube sampling to get a good distribution of points to work with. If a line is one-dimensional, a square is two, and a cube has three dimensions, then a design space with four or more dimensions will be in the shape of a hypercube[†]. The first option is to take your example answers from the corners of the design space and the middle of each edge, each face, each volume, etc. (Figure 9.15, left).

Figure 9.15: Hypercube vertices, edges and faces (left) and Latin hypercube sampling (right)

The other option is to use Latin hypercube sampling (Fig. 9.15, right), which takes one point from every row and column of the space[††]. Of course, you could use both methods to ensure that you have both the extreme values and a good sampling from the interior. If you are using a Genetic Algorithm or Particle Swarm method, you might include the latest best point as a new point on the meta-surface.

Some algorithms can smooth noisy data to find an equation that best fits that data[†††]. You can then use calculus to find the maximum and minimum points.

Do not forget that you can mix methods for the best results. For example, you might use a stochastic method to get you close to the optimum and a Hill Climb to polish the result. It is also worth remembering that sometimes refining an algorithm is not enough and you need to use a completely different approach (Figure 9.16).

Figure 9.16: Travelling Salesperson Problem

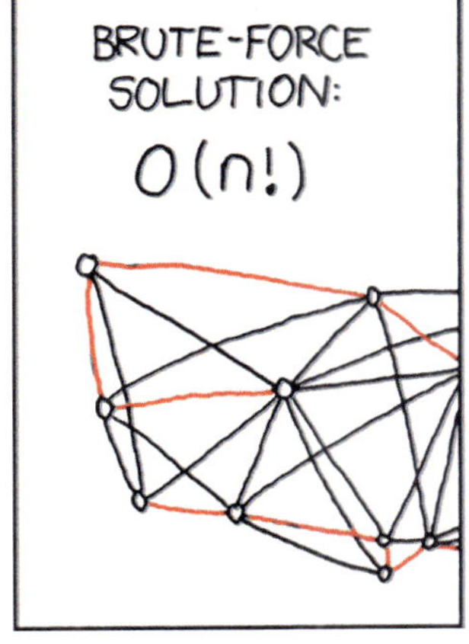

[†]   They are not easy to draw.
[††]   Like a sudoku puzzle.
[†††] The Gauss-Newton algorithm, for example, minimises the sum of the squares of the errors between the data and the model prediction.

9.4.1 Algorithms, heuristics and metaheuristics

As engineers we are used to dealing with **algorithms**, which are a detailed set of steps or instructions that will produce a result, hopefully the correct result, hopefully in a finite amount of time[†]. With optimisation we do not have the luxury of certainty: we might find an answer that is better, but there is no guarantee that it is the best answer, or even that we could recognise it as such if we found it. Here, we often need to use **heuristics**, which are strategies for finding the goal that should produce a result in a finite amount of time. An output of a heuristic is an algorithm.

Heuristics might at first seem like a bad option, until you realise that there are many situations where algorithms fail us. As a general rule you should always use deterministic algorithms as the first option where possible[††]; they are also linear, so work well for linear problems. Nonlinear problems are not directly solvable, which is where heuristics comes in. Heuristics are processes; steps to be followed when we need to search for an answer to a problem. For example, if we consider the optimisation methods that we looked at in the last chapter, a Maxwell Load Path is an algorithm while the form-finding methods are heuristics.

And then we have **metaheuristics**, which are problem-independent methods for producing heuristics: heuristics for heuristics. Metaheuristics often have a story or background narrative to explain their inspiration and implementation.

To put these into context, *"Some foods are better cooked before they are eaten"* is a metaheuristic; *"You can cook potatoes by boiling them in water"* is a heuristic, and *"Wash potatoes and cut them into pieces about 2cm thick, boil them in salted water for 20 minutes, drain and serve"* is an algorithm.

For an engineering example let us consider the nodal adjustment method from Section 8.6.1, which we will use to find the ideal shape of a truss. Adjusting the nodes' position is a heuristic: it is a method but exactly how do we put it into practice? We will need a list of nodes that can move horizontally and a list that can move vertically. Most nodes will be in both sets, but the left support node will be fixed in both directions and the nodes on the line-of-symmetry should only move vertically. If we allow all the nodes to move, the span will be reduced to nothing during the optimisation.

We now have a choice of metaheuristics to use on the problem:

- **Hill Climb**: move each node in turn a set amount in the positive or negative direction. Test if the structural volume reduces; if it has not changed then put the node back. When you have adjusted and tested each node, do it again. Stop when you have tested the complete set of nodes and there has been no change to the geometry in that cycle
- **Simulated Annealing**: implement as per the Hill Climb, but vary the distance moved in accordance with the temperature variable. When the temperature has reached a minimum, use with the Hill Climb to finish
- **Genetic Algorithm**: encode each nodal X coordinate and Y coordinate in binary and combine them all into one string. Note that the binary number must be a power of two, so there is the possibility that some nodes may stray beyond the line of symmetry, but the constraint of the structural volume will probably eliminate such genomes as the structure will be less efficient. Generate an initial population of random genome strings. To test, break each genome into the individual numbers, update the model nodal coordinates and test the volume. Record the fitness, which is the inverse volume (lower volumes will give a larger result), with each genome string (you may find that a dictionary, a key-value store, is useful for this). For the next generation, first include the genome with the lowest volume. Next, randomly choose pairs of genomes from the generation list, each with a probability based on their fitness divided by the total fitness of the population, and cross-breed them to produce a new genome. Add this to the new generation unless it is a copy of one already in there. Repeat until the time or generation limit is reached
- **Particle Swarm**: Each X and Y nodal coordinate is a dimension in the design space hypercube. Populate the space with a number of particles in random locations and velocities. Update the model by setting the nodal coordinates from each particle's position, and test the volume. Record the volume with each particle. Next, move

---

[†] It is easy to create algorithms/programs that never stop. For example:
While x > 1: x = x + 1

[††] This is a rule of thumb, that is, a heuristic (usually a good approach) or a model (wrong but useful). Occasionally non-deterministic algorithms can out-perform deterministic ones. An example is Shor's algorithm, which uses a random number and the probabilistic nature of quantum computers to factor exceptionally large numbers considerably faster than the best linear method.

each particle according to its velocity and the location of its best volume location and the overall best location. Retest each particle, updating them if any find a better volume location in the design space. Repeat until the time or cycle limit is reached
- **Ant Colony Optimisation**: create the state space by making each nodal X and Y coordinate range a column of vertexes, then join each vertex with every vertex in the next column with an edge. Give each edge an initial pheromone strength, then generate a population of ants to traverse the state space. Build and test the model resulting from the state space vertexes for each ant in turn. Increase the pheromone on the route through the state space that gave the lowest volume, then evaporate the pheromone on all edges by a percentage. Repeat for a set number of cycles

You may note that the Hill Climb and Simulated Annealing heuristics will work with (and adjust) only one structural layout, so it will help to estimate the best starting geometry, while the other heuristics generate many random geometries that are then refined. In this particular example I would expect that either the Hill Climb or Simulated Annealing to be the most efficient, but the stochastic heuristics will be better on nonlinear problems.

9.4.2 Exploration vs. exploitation
Any good optimisation routine must be a balance between exploration of the design space and exploitation of the results found so far. This is the same approach as the Double Diamond design framework we looked at in Chapter 2, except that some of these optimisation heuristics will explore and exploit at the same time.

Exploration means searching. The Monte Carlo method is an extreme example of this as it only explores, picking points in the design space at random. Exploitation is the opposite: constantly refining the answer until you cannot refine any more. The Hill Climb/Gradient method is the classic exploitation method, as it will ascend to the local peak, completely unaware of any other option. Good heuristics thus take a bit of both. Genetic Algorithms use the Monte Carlo exploration but exploit the best options by interbreeding them. With Particle Swarms one particle exploits the best option while the others carry on exploring, though in the general area. Simulated Annealing starts by exploring, but then gradually changes the emphasis to exploitation as the optimisation proceeds.

9.4.3 Multi-objective optimisation and the Pareto Front
In a problem where you are only optimising for a single variable, such as weight or a single load case, it is reasonably easy to determine when you have found a good optimum solution. However, it is far more difficult when you are trying to optimise for multiple variables that can be conflicting or independent. Improving one aspect of the design is likely to be detrimental to another.

Multi-objective optimisation can be non-deterministic, which means that each time you run the optimisation you are likely to get a different answer. Plotting these different answers should reveal a region where solutions work and another where they do not. The divide between the two is known as the Pareto Front (Figure 9.17), named after the Italian civil engineer and economist Vilfredo Pareto.

Figure 9.17: Pareto Front for optimisation against two criteria

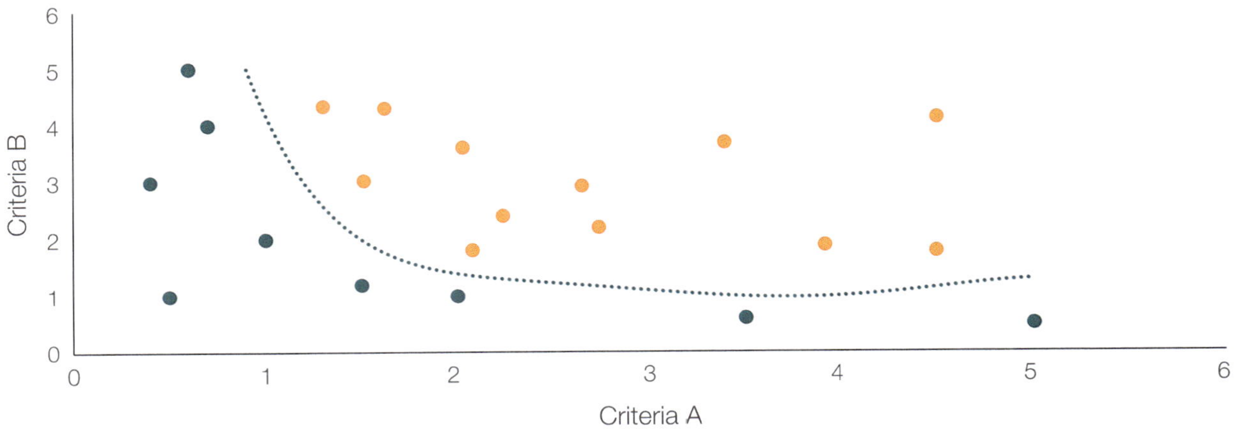

It is not always obvious how to weight the different fitness criteria. Examination of the Pareto Front might help the decision regarding which is the most appropriate compromise, or may even suggest solutions that have yet to be explored[151].

9.4.4 The Stopping Problem

In optimisation you are looking for the global optimum: the best solution. But how do you know when you have reached that point? Some optimisations have a definite stopping point, such as Gradient methods that make no progress, or Simulated Annealing where the temperature drops below a threshold. Other optimisations are open-ended, such as Genetic Algorithms or Particle Swarms.

If you do not know that you have the optimum do you just keep going forever on the off chance that there may be something better? Of course not: an optimisation that never delivers an answer is worse than never starting: you do not have an answer but have burned up a lot of time and energy. With optimisation there can be the uncertainty: have you found **the** optimum or just **an** optimum? Also, computation can be expensive, whether in time or money: it may be quite cheap but slow on your own laptop but on a cloud server or with some commercial programs you will pay while you are using them. You cannot allow optimisations to run forever in the hope that they might find something better, but when should you stop?

What we have here is something called a Stopping Problem, and these have challenged mathematicians for a long time. One option is to decide how long you want the optimisation to run; when that time is up you stop and take what you have found. The problem with this is that you may have found the optimum quite early on and have spent a long time getting nowhere, which is suboptimal.

Another solution is described in a thought experiment conducted by Ted Hill[152]. Adapting Hill's scenario, assume that you are looking for a new place to live and that you need to have settled within a specific timeframe. There are probably hundreds of properties out there, and you could spend lots of time and energy viewing any number of them. The problem is that you never know if there's a better house/apartment waiting for you — and any that you reject are likely to be taken by someone else. How do you decide when you've found 'the one'?

Interestingly the heuristic to give you the best chance is to reject the first 37% (approximately) of the ones that you view, then choose the first one after that, that is better than all the ones you have viewed before. This is not a perfect formula. You might have seen the ideal match in your rejection phase and therefore continue until choosing the very last one, who has a small chance of being your ideal. Or you might unluckily see only poor options in the first phase, leaving you to choose a property which is only marginally better than the first.

But how does this apply to optimisation problems, where we can record the best option found and always go back to it? You might have a limited time or computer budget to run the optimisation, or you are running an optimisation that keeps getting stuck in a different local optimum each time. Thus, you need to keep restarting the optimisation, hoping to get a better result next time, but how many times? One answer is to decide on a maximum number, record the best from each run, and stop after the first 37% when you get an improved result. The worst that can happen is that you run the optimisation for the complete cycle, but the chances are that you finish in a shorter time.

If your optimisation routine is delivering improvements there is little reason to stop, but you do not typically get improvements from an optimisation at a steady rate. Instead you get lots of progress early on, followed by steadily diminishing returns (Figure 9.18).

That curve paints a somewhat optimistic picture of what happens within an optimisation. More usually you get prolonged periods without progress punctuated by sudden jumps in peak fitness (Figure 9.19). But if there has been no improvement recently, have you gotten stuck, reached a limit, or are still waiting for the next jump?

The question is then: How long are you happy to accept no progress until you call it day and stop the optimisation routine? You might set a maximum number of iterations, but if it is too low you will still be in the rapid improvement stage. A maximum time is recommended: set it to run overnight or at the weekend and then stop regardless. Optimising your optimisation is an interesting problem in its own right.

Figure 9.18: Theoretical optimisation results

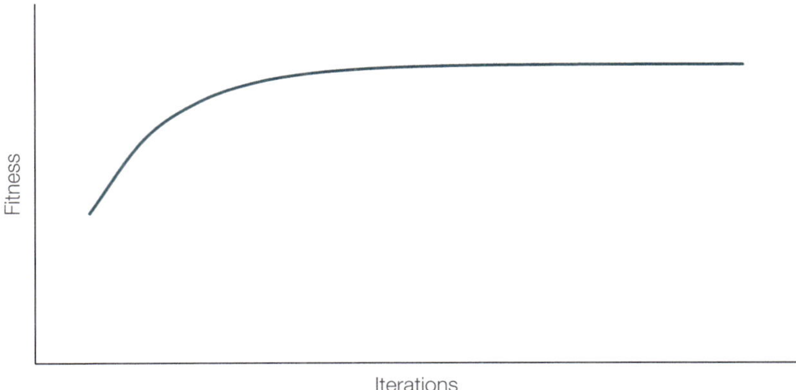

Figure 9.19: Realistic optimisation results

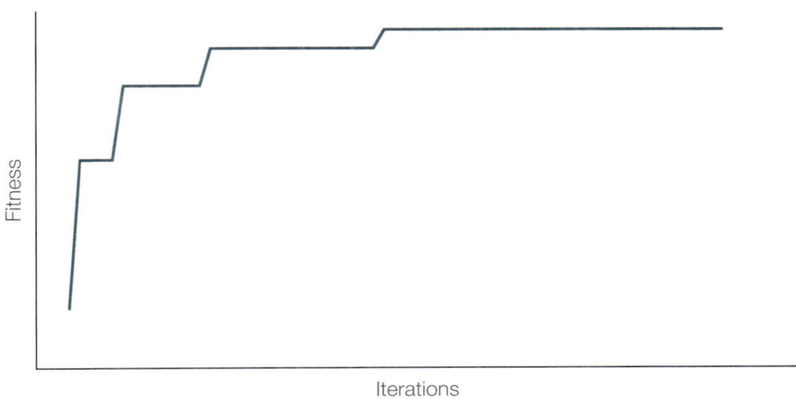

9.4.5 Critical optimisation questions

Before devising any optimisation, consider first these questions:

1. What is the best way to objectively measure the design fitness of the structure?
2. What is the fastest way to sufficiently explore the design space?
3. What is the most robust way to exploit the findings of that search, while avoiding both premature convergence and excessive run times?

9.5 Conclusion

"The best is the enemy of the good."
Voltaire

Which is the best method for design optimisation? Alas the No Free Lunch theorem[153] states that all methods considered over all problems average out the same. Or to put it another way, certain methods are better for some problems, but no method is best for all problems. Deciding which method to use is a matter of judgment and experimentation. You may find it useful to combine methods. A Genetic Algorithm or Particle Swarm might get you close to the answer and then a Gradient method could fine-tune the result. Or you might use a Genetic Algorithm to determine the ideal number of hidden nodes in a neural network, and so on.

There is also no such thing as the 'best' design. There will be a lightest and a cheapest, a lowest embodied energy and a lowest embodied carbon, a most recycled and a most recyclable, and they will all be different. There might also be a strongest and a most robust, but those are quantities we need to satisfy not maximise: safe enough is safe. There will be beautiful and ugly solutions, but these are hard to quantify and even harder to agree on. The cheapest structure may lead to a more expensive building to construct or use; the most efficient structure may lack robustness and alternative load paths.

Does this mean that we do not need to optimise? Not in the slightest! So many designs are woefully and needlessly over-designed. Improving the structural shape, for example, can result in huge savings without impacting robustness, and sometimes a stronger structure can be more economical. Construction accounts for a sizeable proportion of the damage we do to the environment and it is our duty to reduce that. It is our duty to design that which is safe and efficient, robust and elegant, and that the final product does what it needs to do and no more.

Some might worry that they cannot take everything into account with their optimisation, and so do not start. Something is better than nothing, so a partial optimisation is better than none. Do not worry that the contractor will request changes to your design: they will be starting from the improved place that you have found for them.

It is also good to remember the principle of 'satisficing', as you will struggle to wring those last few improvements out of your model. One of the arts of engineering is knowing when close enough is good enough.

10 Artificial Intelligence and Machine Learning

"All right," said Deep Thought. "The Answer to the Great Question…"
"Yes…!"
"Of Life, the Universe, and Everything…" said Deep Thought.
"Yes…!"
"Is…" said Deep Thought, and paused.
"Yes…!"
"Is…"
"Yes…!…?"
"Forty-two," said Deep Thought, with infinite majesty and calm.
"Forty-two!" yelled Loonqual. "Is that all you've got to show for seven and a half million years' work?"
"I checked it very thoroughly," said the computer, "and that quite definitely is the answer. I think the problem, to be quite honest with you, is that you've never actually known what the question is."

Deep Thought/Loonqual[56]

10.1 Introduction

Though it may seem magical, Artificial Intelligence (AI) is like any other programming task: it involves a tremendous amount of coding, algorithmic design and data processing. The difference is that instead of the programmer sitting down and deciding how the code will calculate the answers, they create a program that either learns from existing data or rapidly searches options to predict the best results. The rest remains the same.

But will AI help us as structural engineers, or is it going to take all our jobs? Surely engineering is far too complex for any computer to understand?! It was said that the games of chess and Go could only be comprehended by a human brain, but programs have beaten the world champions of both. Humans have not stopped playing though; on the contrary they are both learning from and working with AI to improve their game. In an online open chess competition in 2005, Grand Masters and the top AI programs were beaten by a team of two amateurs working with three moderate chess programs[7]. It shows that reasonable humans working with reasonable AI can beat the best, whether silicon or carbon-based. Garry Kasparov calls these human-AI teams 'centaurs' (Figure 10.1).

Figure 10.1: The centaur

The question is: How might we as structural engineers benefit by working as centaurs?

Let us first look at what AI is…

10.2 What is Artificial Intelligence?

"Cogito, ergo sum — I think, therefore I am."
René Descartes[154]

Artificial Intelligence is a much-misunderstood term, which is not helped by the meaning morphing over the years. Its original aim was to understand and replicate human thought[155], while today it has the more modest and focused goal of producing smart behaviours. It has also not been helped by science fiction, over-inflated promises and media hyperbole. AI will change the way we work but is not going to usher in the 'Robot Apocalypse' in the near future.

AI has also suffered from the so-called 'AI effect', where something is classed as Artificial Intelligence while it is a research aim, but once it has been achieved it is reclassified as 'programming'. Research into natural language processing gave us grammar checkers, which are now so ubiquitous that we might rarely think about them. Playing chess was held up to be a pinnacle of human thought, but when Big Blue beat Kasparov using state searches to estimate which was the most promising move, it was dismissed as 'brute force'. Working out how to get from A to B needs a route-finding algorithm, but now it is just the satnav telling us where to go. It is no accident that Google is a major leader in current AI research, because internet search is itself an AI tool.

Artificial Intelligence is often conflated with Machine Learning. Machine Learning is a phenomenally successful subset of AI where intelligent-seeming behaviour is produced by training from large datasets, while AI is a much broader subject. To attempt a definition, I would say that Artificial Intelligence is a computer program that deals with ambiguity to predict what the answer is likely to be; solving problems and performing tasks that seem to require intelligence. AI achieves this using a variety of methods, some of which are preprogramed and some that are learned by finding patterns in massive amounts of data. In many respects, the optimisation heuristics that we looked at earlier are examples of Artificial Intelligence.

While neural networks, and in particular 'deep learning', have come to dominate the field because they enable Machine Learning, early AI research focused on symbolic logic and heuristic search methods because that was their understanding of how the brain worked at a high-level. This proved too difficult. More recent research has focused instead on low-level behaviours and working up from there. Using a variety of methods — including search, testing options, and using data to extract empirical rules — AI is not one thing but a myriad of technologies grouped into two main categories: General AI and Narrow AI.

10.2.1 General AI
General AI is an attempt to mimic human intelligence in a computer, consciously learning about and acting intelligently regarding anything that it turns its digits to. It was the original aim of AI research, when it was believed that all human thought was computation. Thus, if we could create a machine that computed in the same way as human thought then it would think like a human and achieve consciousness.

Alan Turing, father of both electronic computers and AI, proposed a test for General AI: The Imitation Game, which is often referred to as the Turing Test[3]. If, in a blind test, the computer fools a human into thinking that it too is human then it was said to be deemed intelligent. There are now programs that can pass the Turing Test[156], but they are still not intelligent.

Since the start of AI research, General AI has been about 30 years away[157,158], and 60 years later we are still no closer. Some leading AI researchers suspect that General AI might be a century away[159] while others argue that it is not even possible[160,161].

10.2.2 Narrow AI

"How do I get to Carnegie Hall?"
"Practice!"

But if you asked Google Maps the same question you will get a d fferent answer; despite its sophistication it can only do one thing, which is giving geographical directions. Successful AI is narrow in application, such as playing games, having conversations, flying drones and recognising objects.

Narrow AI has been compared to Dustin Hoffman's character in *Rain Man*[162]: incredibly clever in a small number of things, but hopeless in everyday tasks most people find simple. The reality is that Narrow AI is far more focused even than that: both Deep Blue[163] and AlphaGo beat their respective champions[8,164], yet neither program knew that their human opponent even existed. Games are beloved to AI researchers for several reasons: they are complex, formal (there are definite rules to follow) and bounded (the extent is the game board). Other successful AI applications are (to date) similarly narrow.

10.3 AI methods

AI covers a very wide range of processes. The following is an overview of some methods relevant to structural engineering.

10.3.1 Predicate Calculus and automated reasoning

"The only way to rectify our reasonings is to make them as tangible as those of the Mathematicians, so that we can find our error at a glance, and when there are disputes among persons, we can simply say: Let us calculate, without further ado, to see who is right."
Gottfried Wilhelm Leibniz[165]

A principal early aim of AI research was abstracting and manipulating *qualitative* information by working with symbols that represent concepts, relationships and objects. The technique is called Predicate Calculus, where statements are expressed in a form like Boolean Logic. For example, the statement *"All humans are mortal"* is rendered as:

$$\forall x(human(x) \Rightarrow mortal(x))^{\dagger}$$

and thus *"If Socrates is human then Socrates is mortal"*:

$$human(Socrates) \Rightarrow mortal(Socrates)$$

We can reverse the statements, such as *"There exist some mortal creatures that are human"*:

$$\exists x(mortal(x) \Rightarrow human(x))^{\dagger\dagger}$$

test truth and falsehood:

$$mortal(Socrates) = T$$

$$alive(Socrates) = F$$

and combine statements with 'and' ($\wedge$), 'or' ($\vee$) and 'not' ($\neg$):

$$mortal(Socrates) \wedge \neg alive(Socrates) = T$$

The result is a set of logical expressions that we can use to either test compliance or state requirements.

[†] Where $\forall$ means 'all' and $\Rightarrow$ means 'implies'.
[††] $\exists$ means 'there exists'.

10.3.1.1 Natural Language Processing (NLP)

Before any interrogation of text, it must first be automatically broken down into its constituent parts (syntax) before working out what it means (semantics). To demonstrate this in action consider this sentence:

Jack ate a banana

We can parse, or break this down, into a subject (noun phrase) and a predicate (verb phrase), and hence into nouns, verbs and so on (Figure 10.2).

Figure 10.2: Sentence syntax

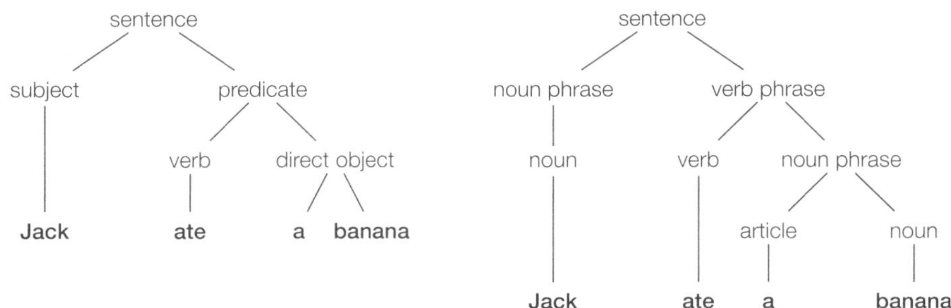

Once we have those parts, then we might extract answers to questions such as: "Who ate the banana?". A leading example of this is IBM's Watson[166], which in 2011 won the American quiz show *Jeopardy!*[167], beating two former champions. To do this, Watson needed to interpret the question and search its database to find the most likely answer. While it did misunderstand some questions, it got enough correct to win the match.

A result of breaking sentences down using the rules of grammar, is that the program can check whether the rules have been followed. For example, these two sentences both use the same words, but only one order is acceptable:

- A dog likes people
- A people likes dog

While AI started with the aim of reproducing human thought, it has been most successful when replicating animal behaviour (grammar and sentence parsing is an exception as this seems to be an exclusively human trait). Most forms of life communicate with each other to some degree — some apes have been taught sign language and some house pets are very good at 'asking' for food! — but we must be careful to not anthropomorphise animals or computers.

AI has had success with parsing sentences to answer questions, but semantics, the understanding of those sentences, is currently an unsolved problem. They can extract answers from the text, but cannot read between the lines to extract the meaning. An example test for this is the Winograd Schema Challenge[168], where questions are asked that we might find easy due to our extra knowledge, but programs struggle[†]. For example, what is 'it' in each of these sentences?:

- The large ball crashed right through the table because it was made of steel
- The large ball crashed right through the table because it was made of foam

Apart from the internet, text search is now a business tool. For example, the legal profession use it for checking case law[169]. Similarly, you can automatically check BIM models against standards[170,171]. As long as the source materials are well written, it should be possible to use NLP to take text from documents, such as client specifications, and express them in this formal way so that we might automatically test our documents to see if they meet the brief[††].

[†]   Some sentences, known as Garden Paths, are difficult for humans to understand on first viewing. Examples include 'The old man the boat' and 'The florist sent the flowers was pleased'.

[††]   It could also be useful to take political speeches, render them down, and see what is left…

10.3.1.2 Expert Systems

With Predicate Calculus you might also take the experience of human experts and create a program known as an Expert System. While promising, these have proved difficult for a number of reasons. One is that an expert's time is expensive, they are needed by their organisations, and Expert Systems need a lot of examples to work well, meaning that gathering enough data is difficult. Expert Systems work by matching up questions with answers, but they are extremely hard to make general-purpose. They are brittle and error-prone when dealing with problems that they are not explicitly programmed to address, while humans find it easy to carry knowledge over from one case to something similar[172]. We use metaphors to understand and recognise when something is similar to what we have seen before.

Experts also struggle to explain things because they are so familiar with the subject: the so-called 'curse of knowledge'. They naturally take short-cuts, leaving out what is background knowledge to them but potentially critical to the problem at hand[173]. This is related to the Dunning-Kruger effect, where a little knowledge leads to overconfidence, while knowledgeable people recognise their own ignorance and over-estimate the knowledge of others[174].

10.3.1.3 State space search

Once we have knowledge encoded, whether in well-formed Predicate Calculus expressions or a series of viable options, we can then search these in an organised way to find the answer to questions or problems. We can think of these options as a state space, where each option leads to further choices to be made. To explore our options, we start from the trunk and progress out, or down, searching until we find complete solutions that we can judge (Figure 10.3). One approach is to implement a depth-first method, where a particular option is explored first with all its subordinate options, before exploring other options. The other way is breadth-first, where we explore the state space level by level. Depth-first works best if there are a finite number of levels in the state space, even if the width is very wide. Breadth-first is best when the number of options is small, but the number of levels is large or unknown.

Figure 10.3: State search

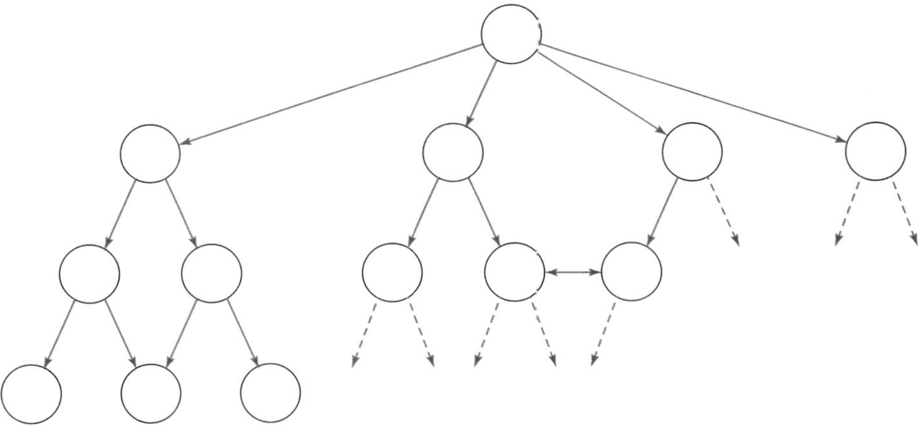

Unlike a real tree you can have additional connections: to reach some nodes you need to satisfy two or more conditions, which is an AND constraint. OR constraints mean that you can reach a node from a number of starting points, and some nodes at the same level might be joined because there is more than one way to reach some states.

An example of breadth-first state space is the Deep Blue chess program. In chess you have a maximum of 16 pieces that you can play, each of which has a limited number of squares that it can move to. This means that at any stage in the game there is a definite finite number of moves that you can make. On the other hand, it is not possible to know how many moves there will be in the complete game, so a depth-first search is not an option. Once Deep Blue established each of its viable moves, it then looked at the next level of the state space and calculated each move that its opponent might make, then each response that it might make to those, and so on. To give an idea of the numbers of states involved, suppose that there were just 16 moves available each turn. The first level in the state space has 16 branches, the second 16^{16}, the third $16^{16^{16}}$ and so on. Even supercomputers can only explore a certain number of levels in a reasonable time. Once the state space reached a suitable depth or a time limit was reached, the move chosen was the one that led to the most promising result: those involving check or checkmate scored highly while losing pieces or the entire game were penalised.

Suppose that we want to design a building. We might start with a decision concerning the material: steel, concrete or timber etc. If we decide to explore the concrete options, we might then consider using a flat slab, or beam and slab, and so on. These options would then give choices regarding *in situ* or precast manufacture. They might also impose restrictions on column spacing or be checked against architectural requirements.

10.3.2 Machine Learning

"When the facts change, I change my mind."
John Maynard Keynes (attributed)

When media articles refer to AI, they are mostly talking about Machine Learning. This is a family of methods where the AI system is not told how to determine the answers but is instead either trained to give the right answers from given examples (supervised learning) or find relationships in a large dataset (unsupervised learning). In both cases Machine Learning is concerned with pattern recognition, whether from known patterns or discovered ones. But just as aeroplanes do not fly like birds, computers do not learn like humans.

There are three ways that we humans know about things: we see them in the world, we remember them, and we make up metaphors to understand what they are like. The first robotic experiments tried to get the machines to have complete knowledge of their environment, but this proved too difficult[175]. On the other hand, expert human typists can use a keyboard without looking yet will struggle to recreate one that has all the letters removed[†].

Robotics researchers realised that, rather than memorising the environment, it was easier to use sensors to check what was around the robot, which is how most insects and human babies function: if I cannot see it then it is not there. This has proved a successful strategy for robotics and novice typists.

The third way of knowing is via metaphor and understanding. In *The Design of Everyday Things*[176] Donald Norman gives the example of an indicator switch on a motorbike handlebar. The usual design is to rock sideways, so riders naturally match up rocking the toggle to the left with the left indicator. The book also discusses a variant, where the indicator switch rocked up and down. There is no direct way to map an up-down movement to left-right, but as the switch was by the left grip, the rider can map the up movement of the switch to a push on the left handlebar and thus a right turn. Computers (at present) cannot make these conceptual links but can only learn or remember explicitly: AI may 'know' things, but it does not 'understand'.

"I wonder whether or when AI will ever crash the barrier of meaning."
Gian-Carlo Rota[177]

Despite the recent successes of Machine Learning, it still requires a lot of human creativity and hard work to program the learning environments and algorithms, as well as ensuring that the training data is both suitable and as clean of bias as possible. Machine Learning has a lot of overlap with data science and statistics, but while the latter's output is a report, Machine Learning enables automation of difficult tasks.

"Once it is accepted that 'learning' is a synonym for statistical optimisation using noisy data, and 'understanding' is synonymous with function approximation of the source of the data, then 'recognition' is simply interpolation on this function."
David Lowe[178]

Speaking of data, in this chapter we will use a 'toy' dataset[179] (Table 10.1) to explore how Machine Learning might be used — in this instance to help understand what led to some walls cracking — and hence make predictions of future behaviour.

[†] Similarly, I once forgot my staff number, but could remember it by typing it into the photocopier.

Table 10.1: Toy data set

Location	Foundation	Soil	Contractor	Cracked?
C1	Trench fill	Clay	J. Wayne Ltd	Yes
C2	Strip footing	Clay	J. Wayne Ltd	Yes
C3	Trench fill	Silt	J. Wayne Ltd	No
C4	Trench fill	Clay	J. Wayne Ltd	No
C5	Strip footing	Silt	B. Kid & Ptnrs	Yes
C6	Strip footing	Silt	J. Wayne Ltd	Yes
C7	Strip footing	Silt	B. Kid & Ptnrs	Yes
C8	Trench fill	Clay	B. Kid & Ptnrs	Yes

10.3.2.1 Decision Trees and Random Forests

While Predicate Calculus seeks to embody existing knowledge, Decision Tree learning extracts knowledge from existing datasets to solve classification problems. Decision Trees can work with continuous or categorical data to find how the data relates to the correlating variables. Here each choice or variable is thought of as a branch in a tree, which we traverse via decision nodes from trunk to leaf (Figure 10.4).

Figure 10.4: Decision Tree

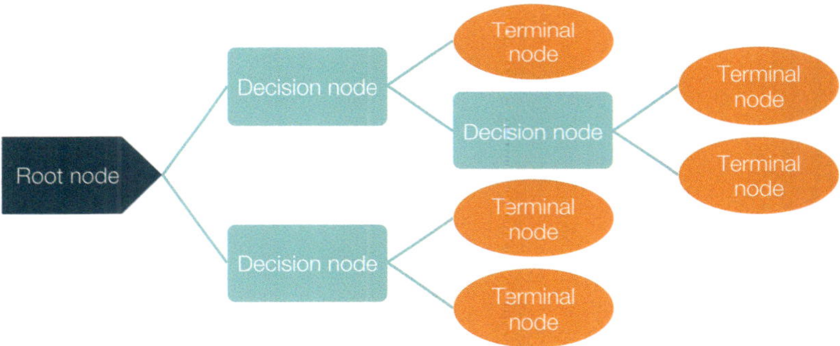

There are some problems with Decision Trees as a form of Machine Learning. There can be many ways that we can break down the results against the variables, they can be sensitive to changes in the training data, and they can be over-fitted to the data meaning that they exactly represent the training data and cannot say anything useful about new cases[180] (Figure 10.5).

Figure 10.5: Data fitting

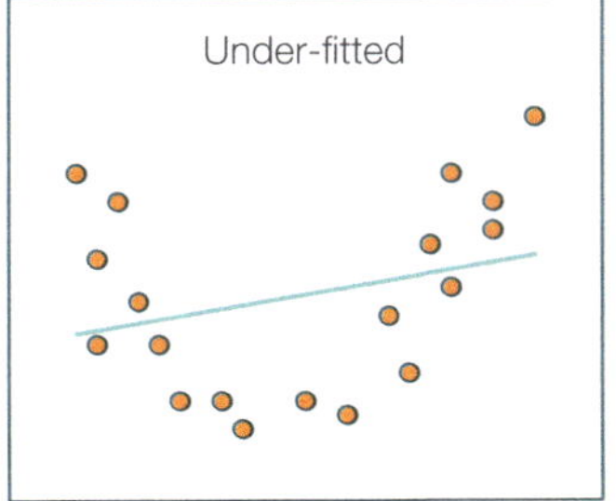

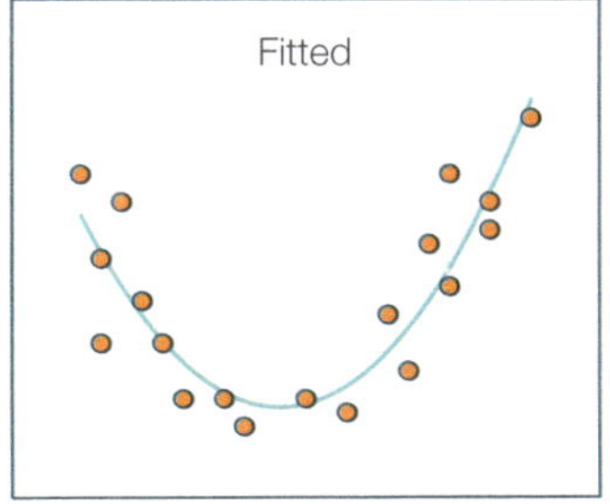

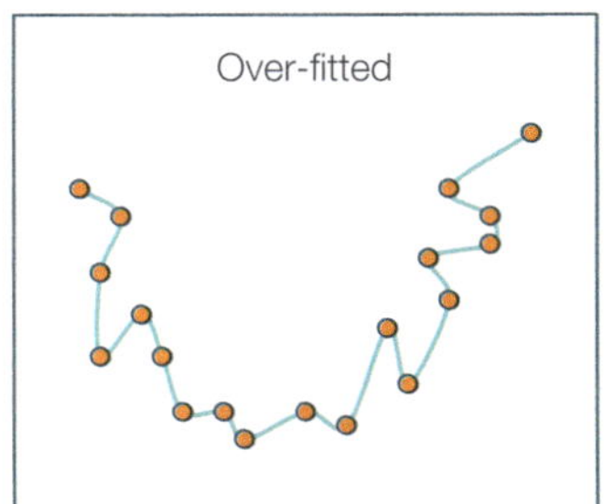

We can make the Decision Trees more general and thus more useful for prediction by pruning, which in this case means replacing a branch with a terminal node set to the removed branch's majority decision. The danger with pruning is that over-pruning can add errors. Luckily, we can overcome these using the wisdom of crowds, or as we have trees, the wisdom of a Random Forest. Here we create many pruned trees from the data. For any given new case we submit the data to the forest and see how the trees "vote". While individual trees might have flaws, the forest overcomes them to give reliable results. This matches what we find in human estimations: the crowd tends to give the right answer even though the individuals are wildly wrong[181]. For example, if there is a 'Guess the weight of the cake' or 'Guess the number of marbles in the jar' competition, the correct answer is usually close to the average of the submitted answers, though you can be influenced if you are not careful[182,†].

For our dataset, choosing one order of options can obscure causes (Figure 10.6) while a better order and some pruning might give insight (Figure 10.7). In both cases a ' ~ ' shows that there was no consistent result. Noting which pruned trees in the Random Forest that contribute most to the answers, can indicate which are the most important parameters.

Figure 10.6: Full Decision Tree

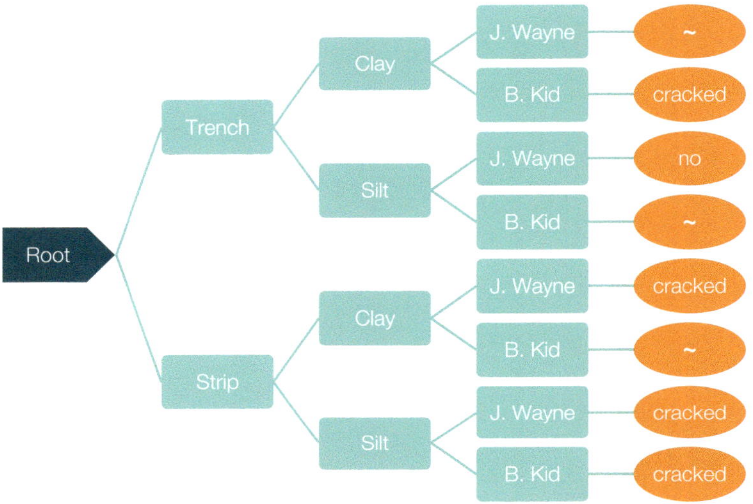

Figure 10.7: Pruned Decision Trees

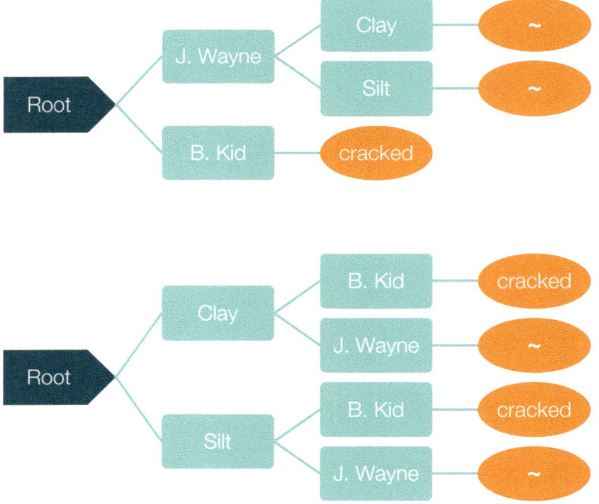

10.3.2.2 Neural Networks and Deep Learning

> *"...if you can take a picture of it and stick a label on it,*
> *you can create an algorithm to find it."*
> Hannah Fry[183]

Neural Networks are a Machine Learning method inspired by how brains work. Their success has been made possible by the vast amounts of data collected on the internet by technology companies such as Google and Amazon, backed up by the parallel processing power of Graphics Processing Units (GPUs) and cloud computing. They have transformed problems that have taxed AI researchers for decades into everyday tools, such as speech-to-text dictation, language translation, and image recognition.

It is fair to say that every generation has interpreted the workings of the mind in terms of their prevailing technology. In more spiritual ages the mind was the ghost (aka guest) riding in the physical body, during the Industrial Revolution it was a machine[†], and since the invention of electronics it has been seen as a computer[††]. The truth is, as always, far more complex. While most of the pioneers of AI resisted the idea of copying how brains work[184], neural computing dates back to 1943 with the McCulloch-Pitts neuron, followed by the Rosenblatt perceptron[185], which sought to simulate real neurons. These connectionist networks were mostly ignored until the invention of multi-layer neural networks in the 1980s, and finally achieved their current dominance of AI in the late 2010s with Deep Learning Neural Networks.

Before we start looking at how artificial Neural Networks work, we should look at their biological inspiration. Brains are an array of specialist cells called neurons and are the central control system for our bodies, regulating bodily functions but also providing us the means for learning and communication. Containing about 100 billion neurons, each of which might be connected to up to 10,000 other neurons via approximately 1,000 trillion synaptic connections, the human brain forms the most complex living thing known in the universe[186,187].

We have a reasonable idea of how neurons work (neuroscience) and how people think (psychology), but the hard part is joining them up (neuropsychology) to say how the neuron activity gives rise to phenomena like consciousness. So how do neurons work?

Neurons are nerve cells, which animals use to transmit signals throughout their bodies, whether messages of touch and pain to the brain, or instructions to the muscles to act. They collect electro-chemical signals from other nerves via their dendrites. When there is sufficient input the nerve 'fires' and sends a signal down its axon to other cells. The connection is via synaptic terminals (synapses), which might then trigger that nerve to fire as well. This process takes about 25 milliseconds, then it rests for about five milliseconds, before it is ready to fire again[187].

In a similar way, Neural Networks take a number of inputs and process them through a number of layers before outputting an answer. The best way to understand how they work is to look at a single 'neuron' (Figure 10.8). There are a number of numerical inputs (X_1 to X_n, typically adjusted to be between 0 and 1) that are multiplied by the weights on their connections (typically between -2 and $+2$), summed and processed; if the result passes a threshold then the neuron is activated and outputs a number.

The original Neural Networks only accepted binary input (0 or 1) and so also gave a binary output (Figure 10.9). This made them difficult to train, so most modern Neural Networks use some sort of activation function that allows a degree of fuzziness. One typical activation formula is the Sigmoid function, which returns negative numbers as zero, positive as one, and a sliding value around zero:

$$y_k = \frac{1}{1 + e^{-netinput_k}}$$

[†] Hence references to people being under stress or strain, 'letting off steam', etc.
[††] Time to unplug and get some down time.

Figure 10.8: Simple artificial neuron

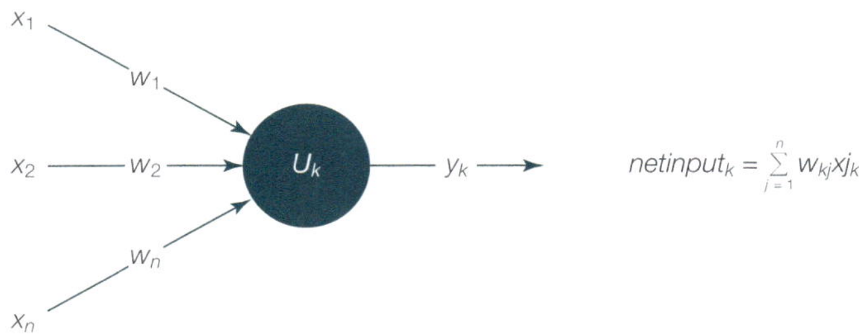

Figure 10.9: Binary and Sigmoid activation thresholds

The Sigmoid function can also be used for squashing the input values into the range 0–1.

As an example, let us consider how we might use a simple Neural Network, with just two inputs and one output, to represent either an AND or an OR gate (Figure 10.10). Working in binary, an OR gate will output 1 if either of the inputs are 1, while an AND will only output 1 if both inputs are 1. Otherwise, they will output 0. It is useful to show these relationships in what is known as a Truth Table (Table 10.2).

Figure 10.10: Simple gate

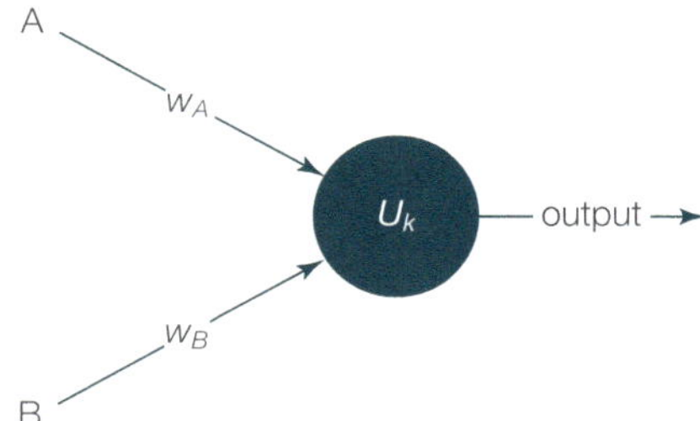

Table 10.2: Truth table

OR			AND		
A	B	Output	A	B	Output
0	0	0	0	0	0
0	1	1	0	1	0
1	0	1	1	0	0
1	1	1	1	1	1

As we will not be dealing with negative numbers in these examples, the unmodified Sigmoid function with a threshold of 0.5 is not useful, as all non-zero positive weights will give 0.5 or more. To fix this we can either change the threshold (e.g. to 0.75), or change the function by doubling and subtracting 1[†]. The result is a graph that is divided into two regions (Figure 10.11).

Figure 10.11: Graphs for truth gates

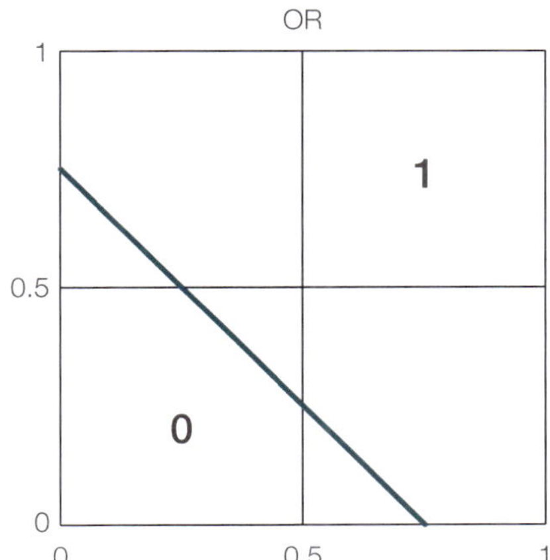

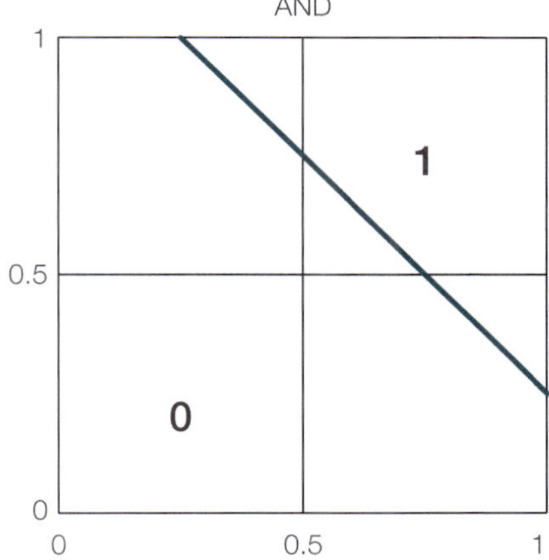

A single-layer Neural Network like this can only give a single division across a range of values, which means that it cannot deal with more complex problems[188]. The solution was to add at least one hidden layer of nodes (so called because we neither input directly to them nor see their output) (Figure 10.12). Neural Networks with a single hidden layer enjoyed moderate success, but it was not until algorithms were developed that could train multiple hidden layers, 'deep learning' (Figure 10.13), that Neural Networks assumed their current dominance of Machine Learning.

Figure 10.12: Neural Network

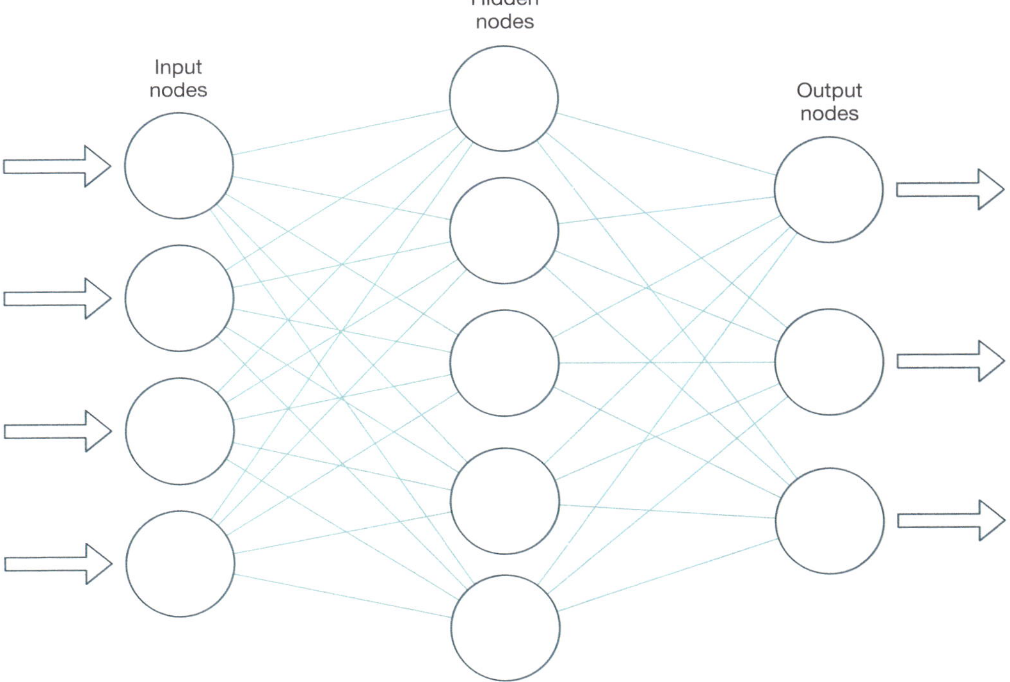

 I will leave it to you to calculate what weightings give the required outputs, though a range of values will work.

Figure 10.13: Deep Learning Neural Network

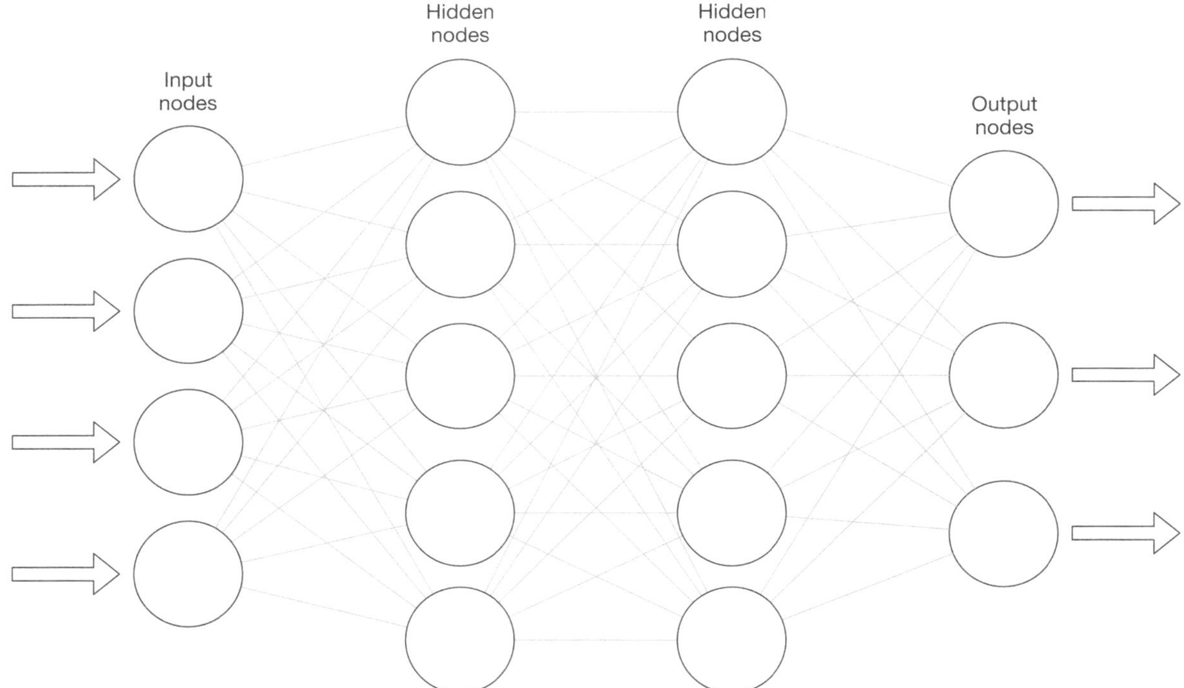

With possibly millions of connections, it is far too difficult to determine the weights on all the connections ourselves. Instead, with supervised learning, the initial randomised network is fed a set of training examples[†]. If it gives the correct answers, the weights that lead to that answer are strengthened. Similarly, weights that lead to bad answers are reduced. With enough training the Neural Network consistently gives accurate results.

The way we do this is to calculate the network error e_k on the kth pattern each output unit by comparing the desired value d against the actual output y:

$$e_k = (d_k - y_k)y_k(1 - y_k) \text{ or } e_k = (d_k - y_k)\tfrac{1}{2}(1 - y_k)^2$$

The change to the weights that lead to that output is then:

$$\Delta w_{jk} = \eta e_k x_j$$

$$w_{jk}(t + 1) = w_{jk}(t) + \Delta w_{jk}$$

Where:

η = a variable to control learning speed
x_j = input value from node j
Δw_{jk} = change to weighting of edge between nodes j and k
$w_{jk}(t)$ = weights at timestep t

[†] A Machine Learning algorithm walks into a bar. "What will you have?" says the bartender. The algorithm replies "I'll have what everyone else is having".

The procedure is the same when training the hidden layer(s), except calculating the error e_j on the hidden unit is a little more involved:

$$e_j = y_j(1 - y_j) \sum_i^o e_i w_{ij}$$

Where:

y_j = activation on that unit
o = number of units in layer above

Neural Networks are mostly used for pattern recognition, such as image or text recognition, to predict answers. One example use is the scanning, reading and cataloguing of old drawings and documents. Another is training a bridge monitoring system to spot corrosion problems combined with drone-based video[189] — the engineer would not have to watch hours of footage but instead check just the clips flagged by the system.

If we were to use a Neural Network to predict future cracking on the toy data set, we might have input fields for each contractor, soil and foundation, then train it using the existing cracking records.

What we have looked at here have been 'feedforward' Neural Networks, but while they are possibly the most common type, they are by no means the only option. Brains can suppress signals that are not interesting by using neurons that feed back to lower levels. This is the reason we can be aware of a conversation that is happening behind us but not pay attention until our name is mentioned. Brains can also respond to how things change over time to the particular order in which they occur. In a similar way, Recurrent Neural Networks feedback some or all outputs as an input to an earlier layer.

Artificial Neural Networks do not work continuously like the brain does. Instead they tend to process a single set of inputs before moving onto the next. When Recurrent Neural Networks feedback results, this is processed in the next cycle of calculations. This makes them useful for processing and predicting values that change over time, e.g. for translating text[172] and music composition[190].

Other types of Neural Networks include grid arrangements, where all nodes are connected to their neighbours. Unlike the supervised learning discussed previously, grid-type networks work well with unsupervised learning, where the answers are not known in advance. Instead, if nodes activate together then the links between them are strengthened; alternativly nodes immediately adjacent to active nodes are boosted while those further away are suppressed. Both options tend to find correlations between various parts of the data.

10.3.2.3 Bayesian Inference

"Probability is expectation founded upon partial knowledge."
George Boole[191]

Bayesian Inference is a statistical method that takes the data we have and makes predictions, using it for new situations or combinations of values.

If the probability of two events is $P(A)$ and $P(B)$, and that the probability of event B given that A has occurred is $P(B|A)$, we can find the probability that A will occur given that B has happened. You can also get to the probability by summing the number of times when both A and B occur ($|A \cap B|$) by the number of times that B occurs ($|B|$).

$$P(A|B) = \frac{P(B|A) \times P(A)}{P(B)} = \frac{|A \cap B|}{|B|}$$

In this example we can use Bayesian Inference on our dataset to determine which factors, or combination of factors, were the most probable cause of the cracking. While all the work by B. Kid cracked ($P(cracked|Kid) = 100\%$) they were not involved with all the work ($P(Kid) = 38\%$). The structure cracked three quarters of the time ($P(cracked) = 75\%$) which means that we can determine the likelihood of B. Kid being involved if we find a crack:

$$P(Kid|cracked) = \frac{P(cracked|Kid) \times P(Kid)}{P(cracked)} = \frac{100\% \times 38\%}{75\%} = 50\%$$

Suggesting that while all B. Kid's worked cracked, it is not conclusive that it is their fault. Conversely the strip footings are a little more suspicious:

$$P(strip|cracked) = \frac{P(cracked|strip) \times P(strip)}{P(cracked)} = \frac{100\% \times 50\%}{75\%} = 67\%$$

Bayesian Inference is a powerful yet simple way to interrogate data and measure the correlations between events. For example I have used it for discerning customer buying patterns, but it also has applications in analysing and predicting material properties and fault diagnosis.

10.3.2.4 Genetic Algorithms

We have already met Genetic Algorithms in Chapter 8. While, in engineering, they have mostly been used for design optimisation, they are also useful for Machine Learning, such as creating weighting for Neural Networks[192] or optimising other Machine Learning parameters. In the context of Machine Learning, Genetic Algorithms are breeding behaviour into a routine, so that it learns how to solve a problem over many generations.

In the natural world we see the results of this process in animal instincts and other precocious behaviours. On the other hand, we see social learning (learning from others) in very few species other than dolphins, chimpanzees, and of course, humans.

10.4 The problems and limitations of AI

> *"So far as the laws of mathematics refer to reality they are not certain.*
> *And so far as they are not certain they do not refer to reality."*
> Albert Einstein[193]

I am met by a variety of views when I speak to people about AI in engineering. Some are enthusiastic, some wary, some have over-optimistic expectations, some dismiss it out of hand. You will also see a similar range of opinions among AI experts concerning when AI might achieve general intelligence and what this might mean to society[194].

I am not surprised that there is such a wide range of reactions as this reflects our media: the press is full of either hyperbole or doom when it comes to AI. This in turn is derived from our history and culture. In the last 700 years we have seen waves of technology bringing changes and improvements: the Renaissance, the Agricultural Revolution, the Industrial Revolution, and the Digital Revolutions. Conversely, our culture has been steeped in fear of intellectual and technological advance since the dawn of history. We have the stories of *The Garden of Eden*, *The Golem*, *Frankenstein*[195], *R.U.R.*[196],[†], *The Terminator*[197], *The Matrix*[73] and more. The constant themes are that the created will rebel against the creator, and that gaining knowledge results in punishment. The truth is that technology does bring changes, some of which are bad, but most are good. History shows that advances in efficiency reduces or removes some jobs[††], but creates many more. Like FEA before it, AI is poised to carry out some of our tasks for us, freeing us up to concentrate on new opportunities.

[†]   *Rossum's Universal Robots* is the origin of the term 'robot', derived from 'Robota', the Czech word for 'forced labour'.
[††]   For example, sewers removed the jobs of Gong Farmers, who had to empty out cess pits by hand. Automation may make some jobs obsolete, but that is not always a terrible thing.

What are the actual problems with AI? The principal objection is that it is empirical. This is a valid point if you assume that AI will replace FEA and detailed design programs, which is an area where we intensively use computers presently. AI, particularly Machine Learning, makes predictions of the answers using existing data. This means that it is best suited to situations where we might otherwise use our own intuition, experience and guesswork: scheme designs, structural inspections, business analysis, archiving old documents, and so on. AI will be there to highlight potential problems, spot unusual inputs, make suggestions, and do the mundane work.

Machine Learning may find correlations but that does not necessarily imply causality[198] (Figure 10.14). There is, for example, a correlation between ice cream sales and shark attacks[199], but one does not cause the other. While we can get the reasoning from some AI systems, it is not currently possible to determine why a Neural Network has given a particular decision[172]. This means that any important decisions by Machine Learning programs must be checked.

Figure 10.14: Correlation

Another problem with Machine Learning is the quantity and quality of its training data — our toy dataset is far too small to make any good conclusions. Also, as we have also seen from "fake news" and propaganda, data that is biased, whether consciously or not, will affect decisions. There are several biases that we need to be aware of when generating data for a Machine Learning project[200].

Data that is ambiguous will produce dubious results (Figure 10.15). 'Garbage In-Garbage Out' (Section 1.4) has long been an expression in the computing industry, highlighting that computers will mindlessly give you answers from whatever you put in. Even worse, Machine Learning that is not sufficiently robust can be deliberately fooled. For example, a research group at Berkeley has shown that adding random pixels to a traffic sign can confuse autonomous vehicles[201]. Such research will help to make these and other Machine Learning systems more resilient to hackers in future.

Figure 10.15: Rabbit and duck: Both Machine Learning and humans can be fooled by ambiguous data

Machine Learning will only recognise the things it has been specifically trained to recognise, with all the biases that entails: voice recognition systems that only understand certain accents, and light-activated soap dispensers only recognising certain skin tones. You might also seek the wrong answer from the data: in WWII, the US Air Force asked mathematician Abraham Wald to optimise the armour on planes that has returned from sorties with damage. Instead of focusing on the areas of damage, Wald told them to armour the undamaged parts, as it was hits to these areas that would likely prevent the plane's return. An AI program would have placed the armour to match the measured damage[202]. The effect is now known as 'Survivor Bias', which is also the reason old buildings are of such high quality. Similarly, Genetic Algorithms deliberately use Survivor Bias to achieve results.

As we have seen with Neural Networks, Machine Learning does not learn the same way that human intelligence does. Humans can recognise the same object in multiple pictures even though they are taken from different angles. For example, a Neural Network might recognise two pictures as both being of a pepper, but not that they are the same pepper, or mistake a picture if a few pixels are changed, though the differences are so small that a human might not even notice. Humans can also recognise a symbol after seeing it only once, while a neural network needs thousands of training cases.

As an example of how humans learn and recognise things differently to computers, consider these pattern quizzes devised by Mikhail Bongard[203], which are designed to be easy for humans (though getting increasingly difficult) and yet extremely difficult for Machine Learning programs (Figure 10.16). The figures on the left of each set have a common aspect, which is in contrast to that on the right. These are large/small (2), white/black (3), vertical/horizontal (7), and triangle/square (10).

Figure 10.16: Bongard pattern recognition puzzles

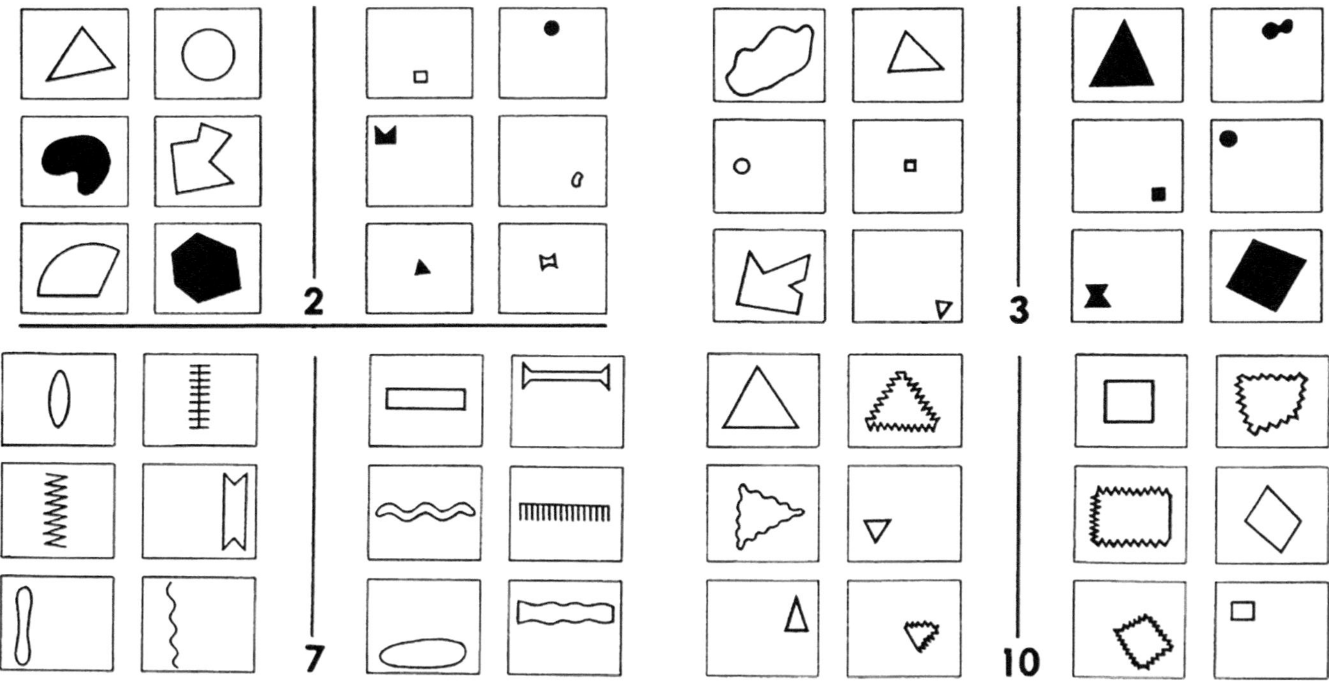

Figure 10.17 shows patterns that Machine Learning programs really struggle with. Puzzle 84 requires the recognition that the circles form a loop and the square is either inside or outside. Puzzle 91 contains the concept of three vs. four. Puzzle 96 needs the program to see not the individual lines but the triangles and other shapes that the lines imply, while for puzzle 98 the program must ignore the background (something that is extremely difficult for a computer) to see the remaining triangles and quadrilaterals. As humans we can recognise the patterns in each puzzle with just 12 examples, but neural networks have been trained on 40,000 examples in each puzzle type yet often only score slightly better than guesswork[204].

Figure 10.17: More complex Bongard pattern recognition puzzles

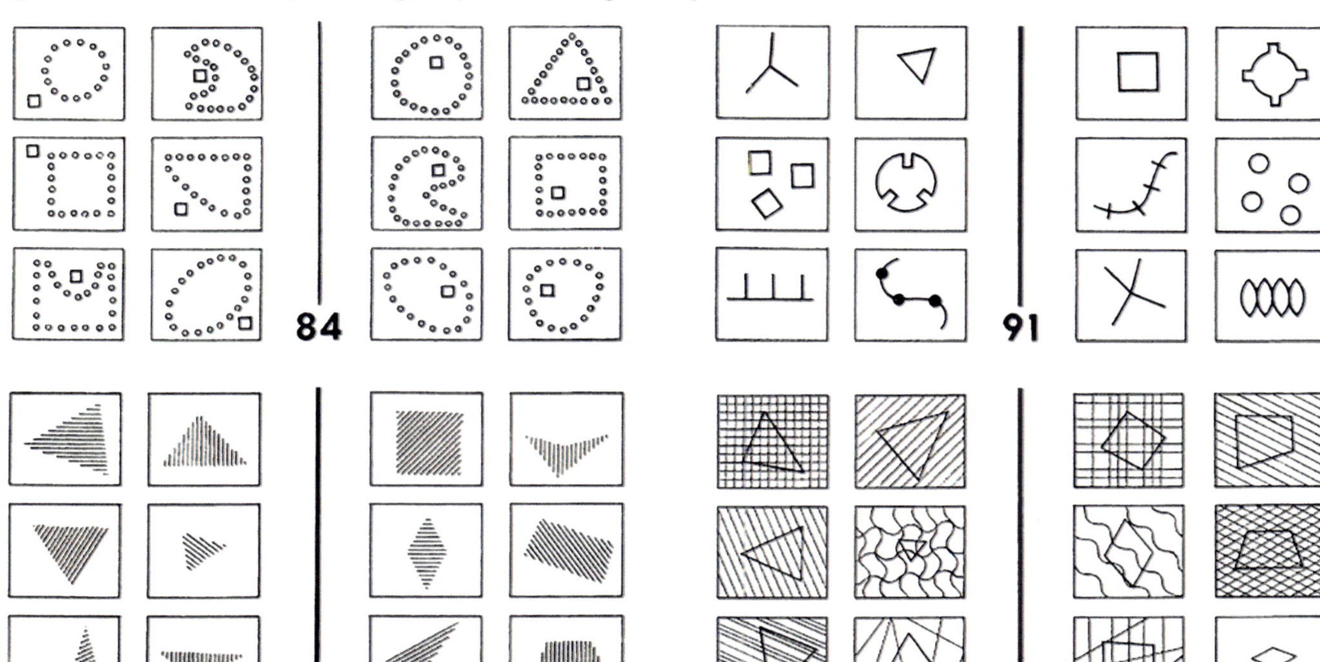

Computers also interpret visual 'illusions' differently to humans. For example, a basic image recognition program looking at a Kanizsa Triangle (Figure 10.18) sees a matrix of dark and light pixels, and might be able to identify the edges. A more sophisticated program might add an association with the text strings 'V' and 'Pie', or even three 'Pac-Men'. It might also see the text bleeding through from the other side of the page. A human brain however, can see what is not there: the subjective contours of lines that are not actually drawn. In this case a white triangle on a white background overlaying an outlined triangle and three circles. It also can dismiss the text behind as irrelevant or focus on it to extract information. These are behaviours that are difficult for computer programs at present.

Figure 10.18: Kanizsa Triangle

While Machine Learning can learn to recognise objects and situations, and some researchers believe that if computers can compute the same way that we do then they will think the same way (Strong or General AI), it is also reasonable to believe that they will never understand things like we do. The most famous example of this is John Searle's thought experiment: The Chinese Room[160].

Searle imagines he is locked inside a room with an instruction book. A slot in the door opens and some symbols are posted in. On opening his instruction book, he find the symbols, along with instructions on what symbols he should send in return through the slot. Those outside the room understand the symbols, which are Chinese writing; they had posted a question, and received an answer, yet Searle himself supplied the answer without ever understanding it. In the same way, an AI computer program may give intelligent-sounding answers to questions, but it does not understand what it is doing. It is acting like the person in the room giving out Chinese characters without understanding what they mean. It appears smart and behaves intelligently, but it is just mindless computation.

> *"For me, AI is a field of outstanding engineering achievements that helps us model living systems but not replace them. It is the person that designs the algorithms and programs the machine who is intelligent, not the machine itself."*
> Noel Sharkey, Emeritus Professor of Artificial Intelligence and Robotics, University of Sheffield

10.5 How might structural engineers use AI?

> *"Never send a human to do a machine's job."*
> Agent Smith, *The Matrix*[73]

There are many areas where we might use AI to automate our business and we should take inspiration from our fellow engineering professions: geotechnical engineers use AI to predict soil behaviour from the parameters[205,206], water engineers process aerial photographs to inform flood risk management[207], road management systems analyse live data to adjust traffic signals or redirect autonomous vehicles[208], seismic engineers predicting aftershocks[209], and so on. Earlier in this chapter I included some suggestions concerning text processing, forensic engineering and structural inspection, but where else might AI be beneficial?

One area where we may see changes in the industry is that of scheme design. It is a relatively straightforward process to automate the principal design of standard buildings or small bridges, as they have few options. Machine Learning routines might read old designs and recommend options and details based on those. Parametric routines will be able to quickly explore options such as bay sizes and span directions to find the best arrangement. Of course, automated designs will need the full rigour of analysis, detailing and checking that we carry out on all designs.

Another area is checking. We might use AI programs to spot errors in our calculations and values. The problem is that automated checkers can only assess what has been a problem in the past and will find it difficult to anticipate unfamiliar problems. The Tacoma Narrows and Millennium Bridges both had novel problems where AI would not have helped us, but then humans missed the impending problems as well. Might an error-checking program have spotted the columns that are in the analysis model yet missing from the BIM file[49]?

The 2018 collapse of the Morandi Bridge in Genoa highlights the importance of corrosion monitoring in major structures. One possibility is for sensors to record how the vibration frequencies change over time[210]. Working with a digital twin FEA model, methods like Genetic Algorithms or Particle Swarms can then be used to determine the stiffnesses of the elements to produce those frequencies and hence what weakening has occurred.

Construction work and site inspections are likely to be more automated in future. Drone systems that can roam the site and record progress are already being tested[211,212]. Fabricators are increasingly automated[213] and there are numerous robotic bricklayers in development[214,215]. We could measure concrete or timber strengths against its parameters including site or surface conditions, then use Random Forests[†] or Neural Networks to predict strengths of new items — and these applications only scratch the surface of where AI can be used.

On the other hand, there could be structural problems caused by the misuse of AI just as the misuse of FEA caused the Hartford Civic Center, Sleipner A and CTV Building collapses. These were not the fault of the computer programs but the lack of engineering rigour and checking. We can and must learn from these mistakes of the past to prevent those of the future.

[†]   There would be a certain satisfaction in getting a Random Forest to predict timber material properties.

Some people fear AI, inspired by popular horror stories, but any danger that AI might pose will not come from its artificial intelligence (because it looks like they may never be as intelligent as us) but from artificial stupidity. As humans we can recognise when an answer is wrong or that we are in an unusual situation where the usual rules do not apply, but a computer will plough-on regardless. Many of the self-driving cars are being developed by technology companies in California, which means that they will become particularly good at driving on Californian roads: wide, straight and bathed in sunlight. To be truly autonomous they must also deal with narrow, winding roads that are subjected to fog, flood or the occasional flock of sheep. The other danger will be from natural stupidity, where humans assume that the AI knows best and surrenders all authority to them. This may not be a problem with non-critical applications, like a camera autofocus or reading text, but for safety-critical applications such as structural engineering design this is not acceptable. Any AI must be treated as an engineering assistant, exceptionally smart in its own way and capable of making good suggestions, but also, in its ignorance, occasionally making bad ones. When it is important, the qualified human must always make the final decision.

10.6 Conclusion

"It's tough to make predictions, especially about the future."
Danish proverb

Computing has always been concerned with the automation of tasks. Initially this was focused on repetitive calculations, but over the years has grown into one that helps with analysis, design, communication and leisure. To date, engineering software has focused on deterministic problems; with AI and Machine Learning we can start to automate tasks that are complex or deal with uncertainty, such as categorisation, recognition or recommendation.

AI is not going to replace our existing methods of analysis: we know how beams work, and direct solutions are always best. AI will instead help us automate in other areas of design and business. Likewise, AI will not replace engineers in the near future: a study by Oxford University[216] found that only 1.9% of civil engineering work is vulnerable to automation, though this may be an underestimation. AI is a tool, like FEA, that can make our lives easier in some design tasks. Also, we need to remember that different methods work better in different situations: one size does not fit all. The engineer must select the right tools to tackle a problem, choosing the right data to get the best results. In many respects we should rename Artificial Intelligence as 'Augmented Intelligence'.

Chess masters did not stop playing when Kasparov lost to Deep Blue but are embracing computers to help them play better[217]. Go experts were surprised by Sedol's defeat by AlphaGo but now rejoice at the new tactics revealed to them[218]. How will AI change structural engineering? What innovations and advanced designs might we create by working with AI? Will the profession accept them?

Like other programming techniques before, working with AI will free us from the burden of some tasks, make other tasks possible, and give us the freedom to concentrate on what humans do better than machines: imagine and create. The best engineers will work with AI and other digital tools: the engineer of the future, as now, is a centaur.

11 Engineering the future

"As for the future, your task is not to foresee it, but to enable it."
Antoine de Saint-Exupéry[219]

11.1 Change[†]

Digital technology has transformed structural engineering over the last 50 years: calculators replaced slide rules, FEA analyses any structure, apps size beams and rebar, CAD removed drawing boards and was supplanted by BIM. These tools have all been automating designs and workflows. Now AI, including Machine Learning, optimisation, digital fabrication, cloud and quantum computing promise to do the same for engineering design and practice, especially for situations that are too complex for the current generation of programs. The profession is entering a new and exciting age where innovation and change are the rule, but what does the future hold?

The 19th Century brought us the Industrial Revolution: steam power and mass transit, cast iron and steel. The 20th Century gave us nuclear energy and jet aviation, reinforced concrete and carbon fibre, and computers. We would be foolish to expect the 21st century to be any less revolutionary. Charles H Duell, commissioner of the US Patent Office, is supposed to have said that the patent office would have to close because *"everything that can be invented has been invented"*. Actually, that was a joke in an 1889 edition of *Punch* magazine, but quotes are so often misattributed[††].

I started my life as a digital engineer working on the university mainframe, yet wrote the first draft of this book on my phone. This seems incredible to me. In the 15 years or so since the first commercial smartphones, global society has changed almost beyond recognition. We have instant access to information, whether true or false. We have democratic politics and revolutions run through social media. People living in some of the most remote places on Earth have online bank accounts. Who could have predicted all this 20 years ago? The one thing we can be certain of is that the pace of change is accelerating.

How will technologies change over the next 20 years, and how will these technologies change the industry? Imagine maintenance robots with magnetic feet crawling over the outside of long span bridges, cleaning off corrosion and pealed paint before spraying on new metal to replace and reinforce weakness. Imagine that paint mixed with stress gauge fibres that wirelessly report on excessive strains and cracking. Imagine that you are not restricted to prismatic beams, designed not only for the midspan peak loads but optimised all along the length. Where structure is not fabricated but printed. Where material usage, not manufacturing, drives the cost. I am sure you can see problems with these ideas and also possibilities that I have not thought of. Of course, that is what engineers do: identify problems and prevent them, identify possibilities and create them.

11.2 Digital fabrication

"Anything that can be automated, will be."
Michael Betancourt[220]

Automated machinery has been with us since the invention of the Jacquard loom in 1804, where punched cards were used to control the weaving of complex patterns into textiles. Today, fabricators use Computer Numerical Control (CNC) machines to cut and drill steelwork, with the dimensions coming directly from the detailed CAD models. This results in a

---

[†]   This chapter is perhaps the hardest to write for longevity. 'Correct at time of writing' is unlikely to be 'correct at time of reading'. I make little mention of blockchain, which is still an answer looking for a question, and skip completely over that wonderful (but yet-to-be-invented) technology that you now cannot live without. On the other hand, in the 1990s I was writing about BIM and playing with virtual reality vision, yet over 20 years later people are still treating them as amazing new innovations that they might get around to using at some point. With the 20/20 vision of hindsight, some of this chapter might seem obvious, some laughable, and some a good idea if only others would agree to go along with it. Some things do not change.

[††]   Winston Churchill, Oscar Wilde and Mark Twain are responsible for only a fraction of the quotes attributed to them.

series of parts that the workers can then weld or bolt together as appropriate. Car manufacturers have been using robotic welding for many decades, but cars on a production line are highly repetitive. Originally these robots were programmed by a human who guided the machine through the moves, but there are an increasing number that are autonomously working out for themselves how to control themselves, even to the extent where they are safe to work alongside human co-workers[221,222]. Structural steelwork involves a lot of variation so has been slower to automate detailed functions such as welding, though more robotics is being used in fabrication workshops.

It is harder to automate construction on site than in factories: in factories the environment is controlled, and you can take the parts to the robots. Robotic site welding requires getting the device tens of metres up in the air attached to a crane that might be swaying in the wind. Robotic bolting will need the prior coordination of beams, bolts, washers and nuts. Despite the slow progress this is an area that needs more robotics as, according to a 2003 Health and Safety Executive report[223]: *"one third of all work fatalities happen in construction and construction workers are six times more likely to be killed at work than employees in other sectors"*. Falling from height accounts for about 50% of deaths on site, though accidents with heavy machinery are also common, so site robotics could reduce risk quite significantly. This is an area that needs more research and development.

We are already seeing drones used for site inspections. The main use to date has been overviewing the site, especially for civil works, but the drones are being used increasingly for detailed inspections on hard-to-access points on building frames and bridges. Drones have also started to be used for actual construction (though to date, only in experiments and demonstrations). Companies like Amazon are already experimenting with drone deliveries[224], so it is likely that drones may start to deliver lightweight construction equipment[†].

Robots have been used, at least in experiments, to lay bricks[214,215]. They have proved successful in building walls both simple and complex as they can measure the location of every brick. While current bricklaying robots can work about six times faster than a human bricklayer, they still need, at present, humans to feed them with bricks and mortar as well as help with tidying the joints.

Additive manufacturing, also known as 3D printing, is likely to make the biggest change to construction industry methodologies. Printed parts are starting to be used in automotive and aeronautical engineering, which means that, on past performance, we should see them extensively used in the construction industry in around 20 years' time. The first steel 3D printed bridge has been built in Amsterdam[105], but the process is currently very energy intensive.

3D printing of concrete has also been used for a small number of buildings. While concrete may seem like an ideal material for 3D printing, it is hindered by its lack of tensile strength. Printed concrete walls thus work like masonry, where the self-weight and layout provide strength and stability.

3D printing of formwork is another application for concrete construction as it allows the creation of complex, optimum beams and slabs, with the concrete just where it is needed. You can also make highly finished 'fair-faced' concrete with this method. The printed formwork options include sand with a binder and plastic shapes. Another alternative is to carve the formwork out of a solid material, such as foam or even ice[225].

3D printing has also been proposed for building on the Moon[226] and Mars[227]. Lunar construction faces many challenges that terrestrial construction does not. Unlike terrestrial concrete, the most expensive component in lunar concrete might be the water, while astronomical transportation costs means that the concrete will need to be made from regolith (lunar soil). The proposed buildings might be domes printed over inflated formwork, providing the internal dwellings with protection from meteorites and extreme temperatures.

11.3 Mobile computing

No one would have believed, in the last years of the twentieth century, that human affairs would be so radically changed by the small, humble telephone. That the phone was used for speech was obvious, that it could send small text messages was innovative, that it would become a ubiquitous handheld computer was revolutionary. An iPad in the 1980s would be one of the world's 10 most powerful computers[228], so what will computing power give us in 25 years' time?

[†]   So, no more sending the site junior down to the builders merchant for a 'long weight' and other traditional jokes.

Mobile phones have been around since 1983 (though the first mobile phone call was in 1973 by Motorola engineer Martin Cooper on an experimental device[229]) but the impact of mobile technology on society was slow until it all came together in the 2000s. We had wireless phone calls, then texts, then mobile computing, but the mobile phones had no computing power and the mobile computers had non-rechargeable batteries. Mobile computing was around for nearly 10 years, with Psion Palmtops and Apple Newtons failing to gain popular acceptance. It was not until they were combined with mobile phones to create smartphones that they really took off.

In 2000 I had an Oregon Scientific Osaris palmtop computer, but was stymied by the poor battery life, difficult backup (you needed a special cable and software to save your work), and lack of existing applications. To add insult to injury the memory was wiped if it lost power, which it did frequently[†]. To be fair, the convention is that RAM chips are wiped as part of a computer restart, otherwise *"turn it off and on again"* would not work. But on the other hand, when everything is held in RAM and backup is difficult or impossible, it is a recipe for disaster. Smartphones however, recharge and do not wipe the RAM chips when power is interrupted; they also back up directly onto the Cloud, which admittedly didn't exist at the turn of the century. The smartphone revolution is only about 15 years old, but modern life now totally depends on them.

I have heard it said, admittedly without any justification, that 65% of those born in 2020 will go on to perform job roles that do not currently exist. Take those that were born in 1982. The first commercial mobile phone was a year away, the invention of the World Wide Web was another six years after that, yet the jobs of many of that generation revolve around the internet and smartphones. There was no Amazon, Google, Facebook or Twitter, nor indeed any hint among the brightest of computer gurus of the revolutionary changes ahead. But now, imagine what mass unemployment, never mind mass boredom, would result if the internet was to be permanently shut down.

11.4 Cloud computing

Cloud applications are changing the way we undertake engineering tasks, making them more collaborative and freeing us from the processing restrictions of our own hardware. The classic approach for data handling on computers is for both the data and the processing to be done by your desktop or laptop computer. All models are stored on your hard disk and no network connection is required.

The next approach is known as a 'thick client'[††]: your computer is the client of the online data server, whether this is on your company network or an internet server. Data files are stored on a network and opening them downloads some or all of the data into your computer's memory. You work on the data, and then save it back onto the network drive. Interrogating and editing the data can be fast because it is all held locally in memory, but very large models are going to be difficult or impossible to run as you are still dependent on local hardware capacities. Another problem with large data files is that you need to wait for them to be downloaded and returned, plus any interruptions to the transfer risk corrupting them.

The third option is the 'thin client', where nearly all the processing is done online and only the display instructions are sent to your program. This is the standard approach for webpage applications, but cloud applications do not have to be run in a web browser. Working with a thin client machine means that local hardware becomes much less important as nearly all the processing is being done online. There can be a problem with slow transfer speeds if you want to interact with the data, but this is less of a problem if you are sending only small packets of data.

The cloud is a shortcut way of saying that the data is being stored on an internet server and you access it via a thin client. It is an enabling technology for program and data sharing among the team regardless of the team members' locations and type of hardware. The cloud application might be a project model or a database containing events, timings, people and materials. It might be a way to access all the project information via your phone while you are on

---

[†] My first attempt to write a book was wiped when the battery failed.
[††] I will let you make your own joke here.

site, which combined with GPS location enables an Augmented Reality view of the site, overlaying model data on to the video streaming from your phone camera. It might be a way to run large analysis models that need far more memory than your computer has access to, or enabling parallel processing of optimisation options, so that you can get to the solution far faster than is possible on a single machine.

Cloud computing has a number of uses, advantages and risks:

- **Data sharing:** data that is stored on a cloud sever can be accessed and edited by any device with an internet connection (and suitable security rights), including mobile devices if the device has an appropriate app. The data might be a BIM or FEA model, information about a client or prospective client, or construction progress reporting. It may be that some devices, such as laptops, are better for creating models, but that does not mean that mobile phones cannot view the results
- **'Unlimited' hardware:** your laptop might have very little memory and a slow processor, but you can hire just about any size computer to run your model for less than buying a new computer. Only using the processing power needed at that moment, and sharing that hardware with other cloud customers, means that high-powered computing has a lower environmental impact than everyone owning their own hardware. Quantum computers are going to be mostly accessed via the cloud, at least in the short-term
- **Data security:** comes with benefits and risks. The cloud providers should have dedicated security with knowledge and measures not available to small/medium-sized engineering companies. But because your data is on the internet it might be accessible by anyone. Controls are needed to ensure only authorised people and devices can access your data
- **Machine Learning applications:** this is going to be a controversial one. If you want your software provider to supply you with engineering software that has Machine Learning built in, then that needs a lot of data. Where will they get the data from? From you, the rest of your company, and all your competitors. The problem is that most project data is private and may even relate to sensitive projects, such as those for government or big businesses. Your company may decide to keep all your data on a private cloud, but that will not help create the future smart applications. A balance is needed to ensure that we have both the benefits of Machine Learning plus appropriate data privacy
- **Location of data:** just because something is stored in the cloud, it still has a physical location. Data protection legislation and good practice means that there may be unacceptable locations for your data to be stored, especially if it involves personal data
- **Data recovery:** accidents can happen to any computer — mobile or desk-bound — so data only stored locally can be lost as a result. Storing your data on your network is more robust, but what happens if the building burns down, or even just the server room catches fire[†]? If your data is duplicated over multiple sites, with protection such as uninterrupted power supplies, then data loss is considerably less likely
- **Automatic updates:** you need to ensure that all local devices are kept up to date with the latest programs and drivers. Running your programs on a cloud means that only it needs the latest software version, and if that cloud is run by the software provider they will ensure that it is updated continuously
- **Low bandwidth:** this can be a problem if you want to transfer large amounts of data, especially if your local connection is poor. Thin client applications are going to make the most of the available bandwidth, as they transfer little data, but the problem comes if you have no internet access at all, making your data inaccessible as well

Cloud applications are already regularly used for administrative tasks and are starting to be used for engineering tasks, especially on large projects.

[†] Server rooms can get very hot, so fire is a risk.

11.5 Quantum computing

"If you think you understand quantum mechanics, you don't understand quantum mechanics."
Richard Feynman

Quantum computers are coming[230],[†]. There are many predictions about how they are going to revolutionise computing and usher in a new golden age, but after 30 years of research they are still at the experimental stage. They are large, slow, extremely difficult to build, and have to be chilled close to absolute zero. In 2020 the largest quantum computer had just 53 qubits, which is large enough to run experiments, but we will need quantum computers to be orders of magnitude larger before they are useful.

11.5.1 Why might quantum computers be useful?
Quantum computers are not expected to replace classical computers as there are many problems that are better suited to current approaches, but where quantum computers are likely to shine is where classical computers struggle: exponentially complex problems.

Suppose that you write down a number, e.g. 9. Assuming that you write very neatly it would take you 1 second to write the number. As such, if you were to write all the numbers from 0 to 9 it would take 10 seconds. Next consider a two-digit number, e.g. 99. Writing it will take two seconds but writing all the numbers to 99 will take almost 200 seconds. As for writing out all the numbers to 999, I will leave that to you to calculate. At these speeds writing 1,000,000 will take you just seven seconds but writing the numbers to a million will take you over two months.

Writing the number itself increases linearly with the number of digits but writing all the previous numbers increases exponentially. We are looking here at work rather than complexity, but the principle is the same. Writing down the number is an example of a problem that grows linearly with size of the data: we would solve this in *linear time* and might formally describe this problem as $O(n)$ in Big O notation, where n is our problem size. This would mean that if we doubled our data size then the algorithm would take twice as long. Similarly writing down all the numbers grows in *exponential time*, which we can notate as $O(k^n)$, where k is a constant. In addition, there are intermediate problems where the solution time grows in *polynomial time*: $O(n^c)$, where c is another constant. Finally[††], we can describe an algorithm that takes a fixed time regardless of the data, as $O(1)$.

What quantum computing promises to do is to take some problems that are $O(k^n)$, that is take exponential time to solve on a classical computer and solve them in $O(n^c)$ polynomial time instead. Similarly, it looks like some polynomial problems might be solved in better polynomial time. Conversely there are many problems where classical computing might always be more efficient. It depends a lot on the problem and the algorithms used.

11.5.2 Classical computers
Before we examine quantum computing in detail and what it might mean for engineering, let us recap on how computers, which I will now refer to as 'classical computers', work.

Classical computers store data and instructions in binary strings, which are collections of 0s and 1s. Each character is known as a 'bit', or to put it another way: each bit can be set to 0 or 1. These bits are recorded in a number of ways: magnetic direction on a coil; magnetic dots on a tape; as a hole (or, in the absence of a hole — e.g. on a DVD surface), the electrical charge in a capacitor, and so on.

These bits do not mean much by themselves, so they are collected together to form binary numbers called bytes that represent letters, numbers, and other symbols, including delete, back space, and so on. For example, the ASCII code for the character 1 is 0110001 and 2 is 0110010. There is no ASCII code for 12 as that is two individual characters. If we are dealing with actual arithmetic then we would instead use the binary version of the numbers, where 1 = 0000001, 2 = 0000010, and 12 = 0001100. For negative numbers we can flip the lefthand bit.

[†] Probably.
[††] There are more options in Big O notation, but these will suffice for us here.

As a single byte has only eight characters it can only record an integer between 0 and 255 (2^8-1). 32- and 64-bit operating systems improve this by collecting four or eight bytes together respectively, giving considerably more flexibility and range in what we can store. This includes larger integers and floating point numbers, as well as multiple alphabets.

11.5.3 Quantum computers

Quantum computers also use arrays of bits, or rather qubits, but these process algorithms in a completely different way to classical computers, including having the qubits store multiple values simultaneously, thanks to quantum computing's strange properties[161]. In fact, everything involved with quantum mechanics is strange, so be prepared.

In a quantum computer we will use individual atomic particles, atoms or photons, to be our qubits. While a classical computer's bit might be set to 0 or 1 using an electrical charge or magnetic direction depending on the hardware, the qubit's value is determined by the spin direction of the particle[†]. Typically if we measure the spin to be 'up' then we count this as 0 and 'down' to be 1.

Note that we cannot measure what the spin direction actually is, only whether it is with or against our measurement method. For example, the direction might be relative to a magnetic field and the particles will be either aligned with the field or in the opposite direction. Or we might be using photons and testing to see whether they do or do not pass through a polariser[††]. Whatever method we use and whatever the particle's actual spin direction was before, after measurement it will be changed so that it is definitely up or down.

11.5.3.1 Superposition

At the quantum level things are both particles and waves, meaning that we can use either option to our advantage. One of these advantages is that of 'superposition', which means that we can store multiple values simultaneously. To explain, consider a single wave. It has a wavelength (the distance between its peaks) and an amplitude (the heights of its peaks) (Figure 11.1).

Figure 11.1: Single wave

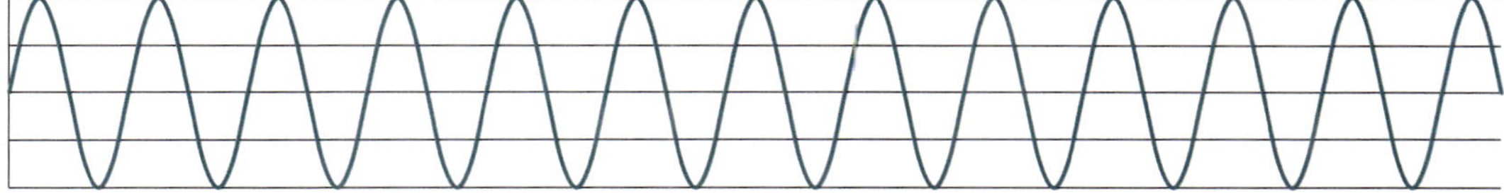

We can have waves of multiple frequencies existing simultaneously (Figure 11.2).

Figure 11.2: Simultaneous waves

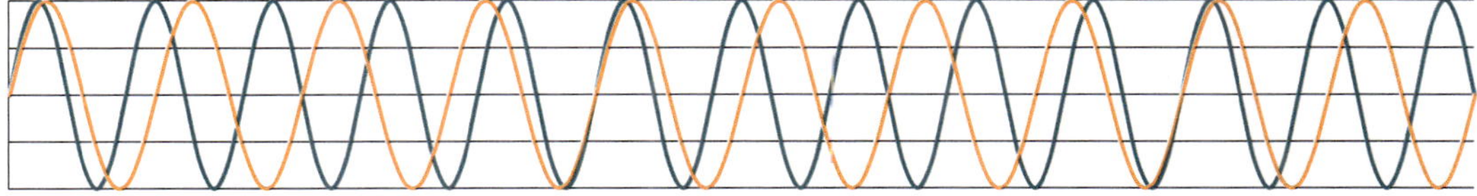

The result of which is that where the peaks coincide, we get a larger peak and where the peak of one meets the trough of the other, they cancel out, creating beats at a longer wavelength. The crucial thing to note here is that there is a superposition of the two waves, in that they are both still there, but they also combine to produce additional effects (Figure 11.3).

Similarly, if we have two waves at the same frequency and in phase with each other they will add together and will cancel out if they are completely out of phase.

[†] I will leave out discussion on whether the qubit is actually spinning or not; there are plenty of articles available that address this question. The probability is that after viewing or reading them your head will spin. It is a model that is useful.

[††] A photon's spin is the light wave's polarisation.

Figure 11.3: Wave interference

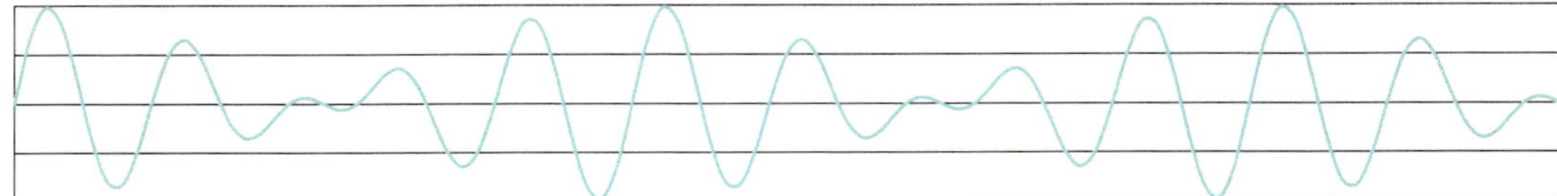

Because qubits can act like waves, we can use this to store multiple values simultaneously: a superposition of values. Likewise we can shift the phase and amplitude of these waves. The quantum algorithms can then take advantage of the constructive and deconstructive interference between the values, or use phase shifts to perform modulo maths (a 2π phase change is the same as no change at all).

Each qubit can thus hold the values of both 0 and 1, with some probability of giving either when measured. This means that a qubit register, a collection of qubits, can simultaneously hold larger numbers: a two-qubit register can hold the numbers 0–3 (00, 01, 10, 11), three qubits 0–7, and so on.

If we imagine the range of spin directions as a sphere (Figure 11.4), the axis of any one spin will point to one point on the sphere. If it points to the exact top, we can call this 0 or $|0\rangle$[†] in the Dirac or bra-ket notation used in quantum mechanics, and if to the very bottom then this is our 1 or $|1\rangle$. Similarly, the spin might point left $|-\rangle$ or right $|+\rangle$. The amount the spin points in the Z direction is the amplitude, and the rotation around Z is the phase.

Figure 11.4: Bloch sphere showing the qubit state space

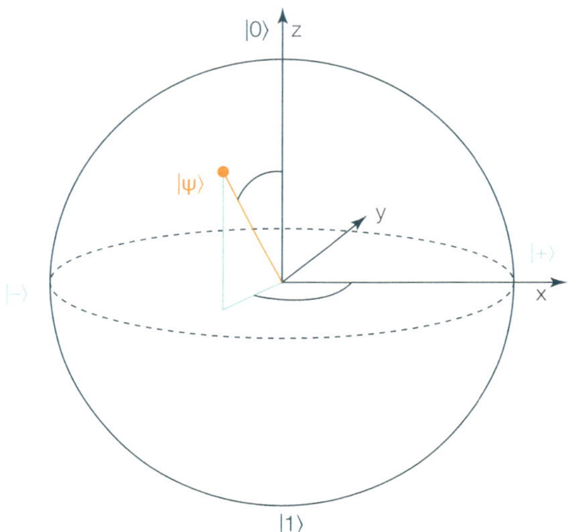

All quantum phenomena are probabilistic, in that we are uncertain what the answer is until we measure it, but the particle is also uncertain until that point[††]. This means that these directions on the Bloch sphere model probabilities not physical reality, but it can be useful to think of them in this way. Once we measure a spin it will retain that spin: in quantum terms its wave function has collapsed and now behaves in a way familiar to classical physics.

[†] We pronounce this "ket zero".

[††] The probabilistic nature of quantum objects was explored by physicist Erwin Schrödinger in his thought experiment: "Schrödinger's Cat". In it a cat is placed into a sealed box that also contains a quantum device that has a 50% chance of killing the cat. In classical physics the cat is either dead or alive, it is just that we do not know which, but with quantum mechanics the cat is in a superpositional state of both dead and alive at the same time until the box is opened. Schrödinger did not create the thought experiment to explain quantum mechanics but to ridicule it, and shortly after he changed fields to study Biology. Subsequent experiments (not including actual cats) have shown that the superposition of states is actually correct.
This superposition of possibilities is a natural state for cats: my cats manage to be simultaneously both innocent and guilty. Authors such as Ursula le Guin and Douglas Adams have proposed an additional state for Schrödinger's cat: when you open the box, the cat is now somewhere else. This is natural behaviour for cats.

If we create a superposition of both 0 and 1 on a qubit then we can describe that as $|\psi\rangle = \frac{1}{\sqrt{2}}|0\rangle + \frac{1}{\sqrt{2}}|1\rangle$, which means it has an equal chance of measuring as 0 or 1. The general case is $|\psi\rangle = \alpha|0\rangle + \beta|1\rangle$, where α and β represent complex numbers (combinations of real and imaginary[†] numbers) that are the amplitude (square root of the probability of getting 0 or 1) and $|\alpha|^2 + |\beta|^2 = 1$. When we measure this qubit it will be 0 or 1, with the probability of each based on α and β, but until that time it will simultaneously be both. Looking on the Bloch sphere representation of the qubit probabilities (Figure 11.4) we can think of the superposition as being a rotation of the spin direction about the Y axis, so that $|+\rangle = \frac{1}{\sqrt{2}}|0\rangle + \frac{1}{\sqrt{2}}|1\rangle$.

11.5.3.2 Quantum entanglement

Another of the key concepts in quantum computers is that of 'entanglement': two particles can be entangled so that if you measure the spin of one you will know the spin of the other. For example, if we entangle two particles in opposite directions they will always give opposite answers if we measure them in the same direction. If you measure the spin of the first entangled particle in the ↑↓ direction then it will be either ↑ or ↓; for argument's sake let us say that it was ↑. If you now measure the second particle using the same axis, then it will always be ↓[††].

If we have two entangled qubits then the pair may be $|00\rangle$, $|01\rangle$, $|10\rangle$ or $|11\rangle$, with a general description $|\psi\rangle = \alpha|00\rangle + \beta|01\rangle + \gamma|10\rangle + \delta|11\rangle$ where $|\alpha|^2 + |\beta|^2 + |\gamma|^2 + |\delta|^2 = 1$. While we might describe the state of a pair of classic bits using two pieces of information (each is 0 or 1), the state of an entangled qubit pair is described with four pieces of information: α, β, γ and δ. Or to put it another way, while a pair of classical bits holds two values, a pair of entangled qubits holds four. This means that quantum computers have the ability to hold larger amounts of information that a classical computer for a given number of bits.

We can describe these ket states as vectors (Appendix K) and so subject them to the usual rules for manipulating vectors, which offers one use for structural engineering: matrix maths. There are quantum algorithms for solving matrix simultaneous equations[231,232] and finding eigenvalues/eigenvectors[233,234]. Another application is likely to be Machine Learning and optimisation[235]: there are quantum algorithms for quantum neural networks[236], quantum annealing[237] to find minimums using a superposition of all possible answers such that the best answer is the one most likely to be returned, and so on.

The collapse of possibilities into the measurable result on the second particle happens instantaneously, which is a concept that famously unnerved Albert Einstein, whose theories of relativity say that nothing can travel faster than light. Calling it "spooky action at a distance", Einstein, along with his colleagues Podolsky and Rosen, proposed in 1935 that this entanglement was impossible and instead there must be some "hidden variables" that the particles store even though we cannot tell what they are[†††]. This would mean that the particles are not entangled but rather know in advance how they will behave[238]. In 1964 this idea was countered by John Stewart Bell[††††], who showed that entanglement and not hidden variables was the only explanation of experiments on quantum particles. Experiments based on Bell's paper have shown multiple times that quantum entanglement is real, even if we cannot understand it.

11.5.3.3 How quantum computers work

Quantum computers are effectively coprocessors for classical computers (Figure 11.5). Firstly, the classical computer takes the program then compiles it into a form that the quantum computer controller understands. The controller then sets up the quantum register by zeroing all the qubits, putting them into superposition and entangling them using various gates (Appendix K).

[†] The imaginary number $i = \sqrt{-1}$.

[††] You may be tempted to think that the qubits have definite values, it's just that we do not know them yet. Numerous experiments have shown that this is not the case and the probabilistic uncertainty is correct. For example, light tends to have polarisation in all directions, but a polariser filter, say your sunglasses, lets through exactly 50% of the light. Adding a second filter at 90° to this will result in all the light being stopped. The probabilistic nature of the quantum spin is shown by adding a third polariser, but at 45° to them both. This allows some light to pass through all three filters, but only if the third filter is in between the first two, not in front or behind.

[†††] The Einstein Podolsky Rosen paradox.

[††††] Bell's Theorem or Bell's Inequality. Explanation of Bell's Inequality is beyond the scope of this book.

Figure 11.5: Quantum computer architecture

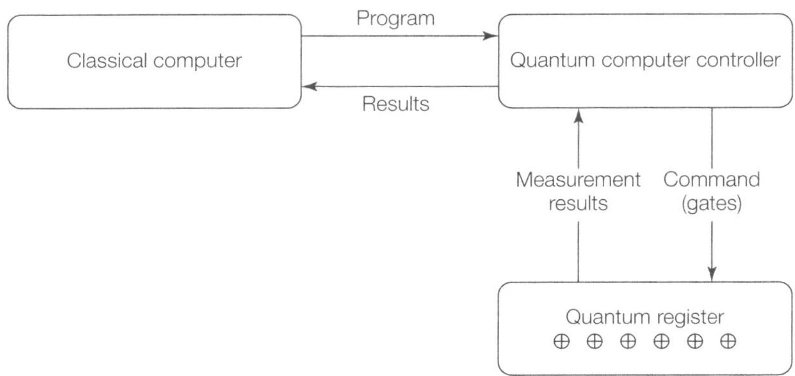

There are a number of ways of creating entanglement between particles[239]. The easiest to consider is the simultaneous creation of a pair of entangled photons: if they are created at the same time from the same atom then conservation of angular momentum rules that if one spins one way the other must spin in the opposite direction — they are polarised in orthogonal directions. If you can get these entangled photons to hit two separate atoms, these atoms also become entangled.

In quantum computers a common way to create entanglement is to place two atoms (our qubits) close together and excite them with a carefully controlled laser so they gain just the right amount of energy to trigger entanglement. This method also means that you can entangle a third atom with the first two, and so on until all the qubits are entangled. Entangling large numbers of qubits is difficult and is one of the main barriers to building usable quantum computers.

The quantum computer controller then performs various additional operations on the qubits before finally measuring the state of register. This measurement causes the qubits to collapse out of superposition and entanglement. The controller then passes the results to the classical computer for recording and post-processing.

As the state of a qubit is probabilistic, we may not always get the same answer every time that we run an algorithm. Instead we run the algorithms multiple times so that we can count the number of times each answer is given and thus we can calculate their probabilities: quantum computing uses Monte Carlo methods.

11.5.3.4 Will we be using quantum computers soon?
There is a lot of hype about quantum computers, some of it is correct, much is wrong, and how they will actually be used remains to be seen. What can we say about them with any certainty? Firstly, quantum computers are not fast. They are slow to program in comparison to classical computers, which means they are not going to be used for applications like Big Data in the short term. Instead they are best with problems where there are a small number of inputs but a large number of combinations.

Quantum computers are better than classical computers for problems where they can reduce the number of computing steps from an exponential number (k^N) to a polynomial number (N^k). This means that while the steps are slow, there are far fewer steps, so we get to the answer faster. At present there are only a few algorithms where this advantage applies, but more are likely to be invented. This means that quantum computers are not going to replace classical computers but instead will be used where there is a particular need.

Because the energy required to flip the electron spin is so low, any thermal or electromagnetic energy is likely to knock the spin off course. This means the atoms and equipment need to be chilled close to absolute zero and shielded from electromagnetic radiation, which is one reason you are unlikely to have a quantum smartphone anytime soon. Another is that it is very difficult to entangle and maintain large numbers of qubits.

If we even get usable quantum computers, it is likely that they will take a reasonable amount of time to become mainstream and for us to discover what we can do with them. Would Alan Turing or John Von Neumann, standing surrounded by their multi-ton creations, striving for calculations to help win WWII, have guessed all the things that we use computers for today? Who can predict what quantum computers, AI, Machine Learning, and all those other ideas not yet formed, will have on life, society and engineering in the next 70 years?

11.6 Optimisation for zero-carbon construction

Most civil engineers that I know are lovers of the natural environment. It is therefore ironic that we spend most of our careers either digging it up or concreting it over. Construction has a major impact on the environment: concrete and steel production generates global greenhouse emissions[129]; paving increases rain runoff and thus flooding; and dams can cause local earthquakes along with the inundation of valleys.

Globally, billions of tons of concrete are manufactured every year. Not only is cement manufacturing energy intensive, but it also needs billions of tons of sand, gravel and water. These all have to come from somewhere. For example, unauthorised over-quarrying of sand is a problem in India, whole beaches in Hungary and Jamaica were stolen in 2007 and 2008 respectively, and there are rumours of entire islands being taken. These thefts can cause erosion, increased flooding, and ingress of seawater inland which affects crops. Demand for sand drives up prices and profit, encouraging organised crime — you need to be organised to steal an entire beach — and this makes it extremely dangerous for those who try to intervene or investigate. Similarly, water is in such demand that nearly 200 violent conflicts have occurred around the world in the last 10 years, ranging from local murders to international military engagements, and they are becoming more frequent.

Steel is even more energy intensive, with smelting and fabrication producing $1,420kgCO_2$ and consuming 22GJ per ton. This is higher than concrete, but steel's high strength allows us to use far lighter sections. I quickly optimised both a steel and a concrete beam to carry the same load over the same distance. While the steel beam was 85% lighter than the concrete, it needed 2.5× the energy and released 1.5× the CO_2.

But concrete cannot rest on its laurels: I also checked a timber beam for the same span and loads. It required half the energy of the concrete section and a seventh of the CO_2 emissions — but timber does require preservatives and more fire-proofing, so there are still complications. Nevertheless, there is a lot of scope for creating environmentally friendly timber structures.

The most environmentally friendly approach would be to immediately stop all construction. Reuse and refurbishment of existing buildings is becoming more common, but population growth means that there is always demand for new construction. Instead, the building industry is aiming for zero-carbon construction[240]. This can only be achieved by considering the entire life of the structure across all disciplines: from construction, through service and renovation, to eventual demolition and recycling.

Structures will probably need to work harder — with the optimisation of topology, shapes or sections. Concrete structures are likely to have more void formers, cement replacement and efficient shaping of sections and members[241–243]. Large timber structures will become more common[244]. The high-strength materials currently in laboratories might also enable us to build with less structure.

Another consideration is avoiding manufacturing new structural elements, and instead reusing existing members recycled from demolished structures where possible. The circular economy, with its principles of refuse, reduce, reuse, repair, refurbish, remanufacture, repurpose, recycle and recover will be significant in helping us achieve zero-carbon construction[245].

One thing is clear: the future needs engineers, especially civil and structural engineers who can consider not just what is safe and cost-effective to build, but advocate the structure that has the lowest environmental impact from cradle to grave[246]. This is going to be a challenge, but that is the one thing engineers love.

11.7 What is the future role of the engineer?

In many respects the role of the engineer has not changed over the millennia. Technology means change, but the essential role of the engineer will be the same: interpret the client's and architect's requirements to produce a design that is safe, efficient and constructible. The tools and methods that we use have changed, and will change, but our role remains the same.

Civil engineers build the infrastructure necessary for human civilisation: we make useful and practical things that serve society. The root of the word 'engineer' is the same as 'ingenious'. We are to be clever and innovative. Engineers are usually thought of as problem solvers, but first we must be problem recognisers: we have to be aware of the problem before we can find a solution. We enable people to trade, to cross the river, to drink clean water and to dispose of the foul, and to seek shelter from the storm.

The first step in the engineering process is to understand the requirement or problem. I do a lot of engineering software support and I often get asked: *"How do I do X?"*. This is usually a detailed task that is either complex or convoluted; so often, my starting point is to understand why they want to undertake this task and thus whether it is the real goal or just what they think is the best way to get to something else. Finding the best answer to a question means that you first must determine exactly what the question is. AI is already good (and will get even better) at answering questions, but it will find it extremely hard to give an answer to those questions that have not been asked before. Working out what the real question is needs understanding and insight, which are not digital but human traits.

It is a safe bet that AI will change how we perform specific aspects of structural engineering. AI promises to be a digital engineering assistant, relieving us of many mundane tasks and automating many aspects of design and project work. Today you must specify beam locations and where the columns should be, tomorrow standard buildings might be automated, though an engineer will still be needed to check the result for quality and exceptions.

No matter how automatic the design, the engineer will still be responsible. With basic design being done with less human input, engineers will be freed up for the creative, the unusual and the special. The engineer of the future might be the high technologist, programming and controlling the digital tools, or the structural artist, closely aligned or even merged with the architects. It is true that we do already have technologists and structural artists today, but as William Gibson, the Cyberpunk author stated: *"The future has arrived — it's just not evenly distributed yet."*

Will AI and the Digital Revolution lead to job losses? It is more reasonable to say that it will lead to job changes. Originally our ancestors were all hunter-gatherers, then farming allowed them to grow in number at the expense of the old mobile way of life. Spare food allowed some to specialise in other jobs, such as potters and blacksmiths. The Agricultural Revolution allowed even more food to be grown by less people, displacing farm workers into the cities, where they then enabled the Industrial Revolution and factory production. The Digital Revolution has seen the invention of the computer, the internet and smartphones. Technological advances have always removed the need for some jobs while creating others. The computers have put the slide rule makers and log table printers out of business while creating jobs such as programmers. If the next half century advances as fast as the last, then who knows what lies ahead?

Engineering used to be all about doing the maths, but computers do that so much better and faster. They do not make mistakes if they are given the correct information and algorithms, they do not tire (though they may overheat) and will check tens of thousands of options without complaint. The software should be doing the hard work — letting engineers concentrate on what they do best: being creative, with purposeful insight and balanced judgement, dealing with information and new situations, and being human.

The future is in your hands and I look forward to seeing what you make of it.

Author's note

"If I have seen further it is by standing on the sholders [sic] of Giants."
Isaac Newton quoting Bernard of Chartres

While books like this are written by an author, that author is not in a vacuum. This particular book has been three years in the writing but over 30 years in preparation, aided by colleagues, lecturers, students, customers and project work. There are many people that I must thank for their assistance and encouragement.

A tonne of thanks must go to Lee Baldwin, my editor, for believing in me and being so flexible about deadlines. The book's Steering Group for guidance, encouragement and correction: David Brohn, Gareth Evans, Iain MacLeod, Paul Jeffries, Stephen Melville, and led by Jon Leach. Thanks also to the additional reviewers who were drafted in at various times: Paul Shepherd, Ramaseshan Kannan, Richard Feigin, Thomas Li and Will Wild (who also had to put up with random questions, terrible puns, and my borrowing his parametrics book — I do promise to return it soon). Without them there would be far more mistakes in the book; those that remain are mine.

I am very privileged to have three inspirational engineers supply forewords: Tristram Carfrae, Chris Wise and Jon Leach; all leaders in their field and experts in computational engineering.

All my colleagues in the Technical Software Group at Arup Digital Technology, or 'Oasys Software' as it is known externally, for putting up with all my daft questions and suggestions, and especially Stephen Hendry for his patient explanations, and Thomas Li and Ramaseshan Kannan for their suggestions and feedback on the manuscript.

The graduate structural engineers in Arup's Building Engineering group in Leeds, who suffer me telling them things when they would much rather be getting on with their work, or them asking me questions when I would rather get back to mine. Also, to Matt Lovell and Neil Hooton for some excellent ideas for the book.

The staff and students in the Civil Engineering department at Bradford University, especially Professor Ashraf Ashour and Dr Mick Honour for letting me loose on their students to teach them structural design and finite element analysis (I like to think I put the fun back into FEA). Also, to Vassili Toropov and Ozz Querin for letting me talk to the Leeds University students about optimisation: as you can see the lecture has grown.

There are many others who have contributed to this book, whether they realised it or not: The IStructE's Digital Workflows and Computational Design panel for ideas and inspiration; Marcin Karczmarczyk for both encouraging me and asking challenging questions; Saskia Lear and Andrew Tyas for real-life disasters and near misses; Tim Ibell and Chris Wise for inciteful conversations and inspiring presentations; Jude Widdowson for explaining how real-life genetic algorithms work, and a particular thanks to all of you who sent in models that were not behaving correctly: we all learned from those.

Posthumous thanks to J.E. Gordon, for showing me that engineering books do not have to be dull. Without you this book would have been quite different (and not about engineering, as I would have done something else with my life). If you have not yet read his books on structures and materials, wait no longer.

My family: Elle, Jon, Bea and Genni, who might not have followed me into an engineering career but still build my world. And of course, my wonderful and patient wife Lisa: forgive me for self-isolating while I was writing this.

Finally, apologies to all of you who do not like engineering books to have jokes. In my defence, they are better than most jokes in textbooks.

"If we shadows have offended,
think but this – and all is mended –
that you have but slumber'd here
while these visions did appear.
And this weak and idle theme,
no more yielding than a dream,
Gentles, do not reprehend;
if you pardon, we will mend."
William Shakespeare, A Midsummer Night's Dream

Appendix A: Section properties[26]

$$\frac{M}{I} = \frac{s}{y} = \frac{E}{R}$$

Where:

M = bending moment at the point of interest (Nm)
I = moment of inertia of section (m^4)
s = stress in material (N/m^2)
y = distance from neutral axis (m)
E = Young's modulus (N/m^2)
R = radius of curvature of beam (Nm)

The stress s at the extreme fibre of the section will be:

$$s = \frac{My}{I}$$

Radius of gyration i of section:

$$i = \sqrt{\frac{I}{A}}$$

Second moment of area I of section:

$$I = \sum_{bottom}^{top} ay^2$$

Figure A1: Typical section

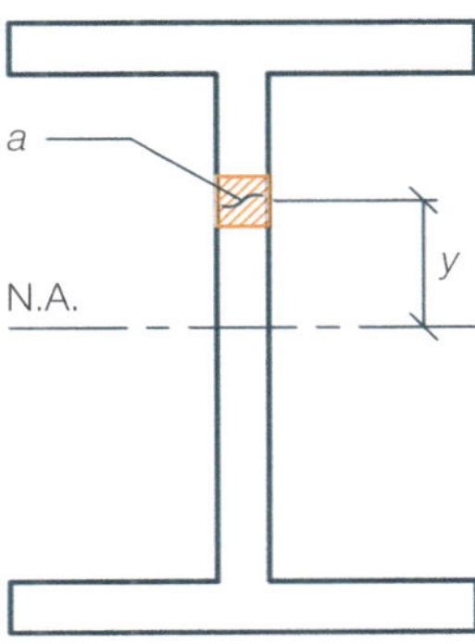

Where:

a = area of element of section (m^2)
y = distance to element from neutral axis (m)

Section	Area	Moment of inertia	Section moduli
Square	$A = bd$	$I_{xx} = \dfrac{bd^3}{12}, \; I_{yy} = \dfrac{db^3}{12}$	$Z_{xx} = \dfrac{bd^2}{6}, \; Z_{yy} = \dfrac{db^2}{6}$
Circle	$A = \pi r^2$	$I = \dfrac{\pi d^4}{64} = 0.7854 r^4$	$z = \dfrac{\pi d^3}{32} = 0.7854 r^3$

Parallel Axis theorem for second moment of area of section offset from axis (i.e. neutral axis of truss):

$$I_{x'} = I_x + Ad^2$$

Figure A2: Section offset from neutral axis

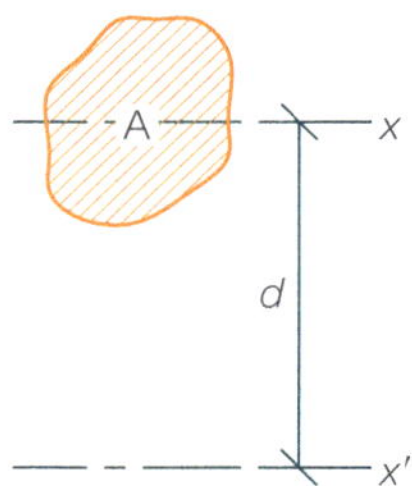

Where:

$I_{x'}$ = second moment of area about axis x'
I_x = second moment of area of section
A = area of section
d = distance from axis to section centroid

Appendix B: Strain Energy[247]

Strain Energy is a measure of the work done in a system or element. Work = Force × Displacement, while Strain Energy is calculated as Force × Strain. Elastic Strain Energy is also Potential Energy, so recoverable when the load is removed. If the material yields, then work has been done but only some of that energy is recoverable.

The Strain Energy U is:

$$U = \frac{1}{2}V\sigma\varepsilon = \frac{1}{2}VE\varepsilon^2 = \frac{1}{2}\frac{V}{E}\sigma^2$$

Where:

V = volume
σ = stress
ε = strain
E = Young's modulus

For a bar under axial load the Strain Energy is:

$$U = \frac{1}{2}P\Delta = \frac{P^2 L}{2EA}$$

Where:

P = axial load
Δ = deflection
L = length
A = cross-sectional area

For a beam under pure bending, with a curvature of θ:

$$U = \frac{1}{2}\theta M = \frac{M^2 L}{2EI}$$

Where

M = bending moment
I = moment of inertia

Strain Energy Density

The Strain Energy Density u is the Strain Energy in an element divided by its volume:

$$u = \frac{1}{2}\sigma\varepsilon = \frac{1}{2}E\varepsilon^2 = \frac{\sigma^2}{2E}$$

For a bar under axial load:

$$u = \frac{P^2}{2EA^2}$$

And for a beam in bending:

$$u = \frac{M^2}{2EIA}$$

Appendix C:
Finite Element Analysis (FEA)

The fundamental equation of the stiffness method for FEA is:

$$[K]\{u\} = \{f\}$$

Where:

$[K]$ = stiffness matrix
$\{u\}$ = vector of nodal displacements
$\{f\}$ = vector of external forces

We find the displacements and hence the element strains and subsequent results:

$$\{u\} = [K]^{-1}\{f\}$$

Matrices[248]

Example bar matrices

Bar element stiffness matrix in 1D and 2D[53].

Horizontal bar in 1D:

$$[K_e] = \frac{AE}{L} \begin{bmatrix} 1 & -1 \\ -1 & 1 \end{bmatrix}$$

Horizontal bar in 2D:

$$[K_e] = \frac{AE}{L} \begin{bmatrix} 1 & 0 & -1 & 0 \\ 0 & 0 & 0 & 0 \\ -1 & 0 & 1 & 0 \\ 0 & 0 & 0 & 0 \end{bmatrix}$$

Bar at angle α in 2D:

$$[K_e] = \frac{AE}{L} \begin{bmatrix} \cos^2\alpha & \cos\alpha\sin\alpha & -\cos^2\alpha & -\cos\alpha\sin\alpha \\ \cos\alpha\sin\alpha & \sin^2\alpha & -\cos\alpha\sin\alpha & -\sin^2\alpha \\ -\cos^2\alpha & -\cos\alpha\sin\alpha & \cos^2\alpha & \cos\alpha\sin\alpha \\ -\cos\alpha\sin\alpha & -\sin^2\alpha & \cos\alpha\sin\alpha & \sin^2\alpha \end{bmatrix}$$

Horizontal bar geometric stiffness matrix in 2D:

$$[K_g] = \frac{P}{L} \begin{bmatrix} 0 & 0 & 0 & 0 \\ 0 & 1 & 0 & -1 \\ 0 & 0 & 0 & 0 \\ 0 & -1 & 0 & 1 \end{bmatrix}$$

Bar mass matrix in 2D:

$$[M_e] = (\rho AL) \begin{bmatrix} \frac{1}{2} & 0 & 0 & 0 \\ 0 & \frac{1}{2} & 0 & 0 \\ 0 & 0 & \frac{1}{2} & 0 \\ 0 & 0 & 0 & \frac{1}{2} \end{bmatrix}$$

Matrix determinates

The determinate of a matrix is a function of that matrix that can give us useful information about it. If determinate is zero for example, then we know that it cannot be inverted.

If we have a 2×2 matrix:

$$A = \begin{bmatrix} a & b \\ c & d \end{bmatrix}$$

The determinate $|A| = ad - bc$.

If we look at a 3×3 matrix:

$$B = \begin{bmatrix} a & b & c \\ d & e & f \\ g & h & i \end{bmatrix}$$

The determinate is found by taking each of the terms in the first row and multiplying it by the determinate of the remaining sum matrix:

$$|B| = a \times \begin{vmatrix} e & f \\ h & i \end{vmatrix} + b \times \begin{vmatrix} f & d \\ i & g \end{vmatrix} + c \times \begin{vmatrix} d & e \\ g & h \end{vmatrix}$$

And so on recursively for larger matrices.

Matrix inversion

Square matrices with a non-zero determinate can be inverted.

For a 2×2 matrix:

$$A = \begin{bmatrix} a & b \\ c & d \end{bmatrix}$$

The inverse is:

$$A^{-1} = \frac{1}{|A|} \begin{bmatrix} d & -b \\ -c & a \end{bmatrix} = \frac{1}{ad - bc} \begin{bmatrix} d & -b \\ -c & a \end{bmatrix}$$

Note that the inverted matrix elements are all divided by the original matrix determinate.

There are several ways to invert a large matrix. A simple method is to use row operations to convert the original matrix into the identity matrix while doing the same operations on an identity matrix to produce the inverse.

$$\left(\begin{array}{ccc|ccc} a_{11} & \cdots & a_{1n} & 1 & \cdots & 0 \\ \vdots & \ddots & \vdots & \vdots & \ddots & \vdots \\ a_{n1} & \cdots & a_{nn} & 0 & \cdots & 1 \end{array} \right)$$

1. Divide all the items in the first row by a_{11}.
2. Subtract multiples of the first row from the second and subsequent rows so that the first column becomes 0.
   For example:

   $$a_{21} \rightarrow a_{21} - a_{11} \times a_{21}$$

   $$a_{2y} \rightarrow a_{2y} - a_{1y} \times a_{21}$$

   $$a_{xy} \rightarrow a_{xy} - a_{1y} \times a_{x1}$$

3. Divide the items in row two by a_{22}.
4. Subtract multiples of the second row from the subsequent rows to produce zeros in the second column below.
5. Repeat until all the diagonals on the left matrix are 1 and the values in the lower triangle are 0.
6. Repeat the process from step 1, but this time starting from the bottom row and working upwards. This will clear the upper triangle of the left matrix, leaving it as the identity matrix, and make the right matrix the inverse of the original.

$$\begin{pmatrix} 1 & \cdots & 0 & b_{11} & \cdots & b_{1n} \\ \vdots & \ddots & \vdots & \vdots & \ddots & \vdots \\ 0 & \cdots & 1 & b_{n1} & \cdots & b_{nn} \end{pmatrix}$$

Inversion example

Let us invert the 2×2 matrix:

$$\begin{bmatrix} a & b \\ c & d \end{bmatrix}$$

1. Starting point:

$$\begin{pmatrix} a & b & 1 & 0 \\ c & d & 0 & 1 \end{pmatrix}$$

2. Divide the first row by a:

$$\begin{pmatrix} 1 & \dfrac{b}{a} & \dfrac{1}{a} & 0 \\ c & d & 0 & 1 \end{pmatrix}$$

3. Take c multiplied by the first row away from the second row:

$$\begin{pmatrix} 1 & \dfrac{b}{a} & \dfrac{1}{a} & 0 \\ 0 & d - \dfrac{bc}{a} & \dfrac{-c}{a} & 1 \end{pmatrix}$$

4. Note $d - bc/a = (ad - bc)/a$:

$$\begin{pmatrix} 1 & \dfrac{b}{a} & \dfrac{1}{a} & 0 \\ 0 & \dfrac{(ad - bc)}{a} & \dfrac{-c}{a} & 1 \end{pmatrix}$$

5. Divide row 2 by $(ad - bc)/a$:

$$\begin{pmatrix} 1 & \dfrac{b}{a} & \dfrac{1}{a} & 0 \\ 0 & 1 & \dfrac{-c}{(ad - bc)} & \dfrac{a}{(ad - bc)} \end{pmatrix}$$

6. Take b/a times row 2 away from row 1:

$$\begin{pmatrix} 1 & 0 & \dfrac{d}{(ad - bc)} & \dfrac{-b}{(ad - bc)} \\ 0 & 1 & \dfrac{-c}{(ad - bc)} & \dfrac{a}{(ad - bc)} \end{pmatrix}$$

This gives us the standard inverted 2×2 matrix.

Finding eigenvectors and eigenvalues

We can think of a column matrix multiplied by a square matrix as a vector being transformed to a new vector by the square matrix. Usually the transformed vector will have a different direction to its original, but certain vectors retain the direction, changing only their length or magnitude. These vectors are particular to the matrix and are thus called eigenvectors; 'eigen' being the German word for 'particular'. The amount that an eigenvector is scaled by is called the eigenvalue, and is referred to by λ.

In matrix terms this equation is:

$Ax = \lambda x$ or $[A]\{x\} = \lambda\{x\} = \lambda[I]\{x\}$

Where I = the identity matrix sized to match.

We can rearrange this equation to:

$[\lambda[I] - [A]]\{x\} = 0$

Which means that:

$[\lambda[I] - [A]] = 0$

And the determinate of the resulting matrix is 0:

$|\lambda[I] - [A]| = 0$

Which we can solve by several methods, including the Jacobian method, which will find all the eigenvalues and eigenvectors, and subspace iteration, which finds the first few.

Generally, this solution will be a polynomial of the form:

$\lambda^n + c_1\lambda^{n-1} + \cdots + c_n = 0$

We then substitute each λ into $[\lambda[I] - [A]]\{x\} = 0$ to find each eigenvector $\{x\}$, which will require solving simultaneous equations of the form:

$ax + by + \cdots = 0$

These eigenvectors are then normalized to give us unique solutions.

Example eigensolution

Find the eigenvalues and eigenvectors of the following matrix:

$$[A] = \begin{bmatrix} 2 & 0 \\ 0 & 1 \end{bmatrix}$$

Which will double any x value while leaving the y unchanged.

$$|\lambda[I] - [A]| = 0$$

$$\left| \lambda \begin{bmatrix} 1 & 0 \\ 0 & 1 \end{bmatrix} - \begin{bmatrix} 2 & 0 \\ 0 & 1 \end{bmatrix} \right| = 0$$

$$\left| \begin{bmatrix} \lambda & 0 \\ 0 & \lambda \end{bmatrix} - \begin{bmatrix} 2 & 0 \\ 0 & 1 \end{bmatrix} \right| = 0$$

$$\begin{vmatrix} \lambda - 2 & 0 \\ 0 & \lambda - 1 \end{vmatrix} = 0$$

$$(\lambda - 2)(\lambda - 1) = 0$$

$$\lambda^2 - 3\lambda + 2 = 0$$

By inspection, the eigenvalues λ are 1 and 2. Substitute these into $[\lambda[I] - [A]]\{x\} = 0$

$$\begin{bmatrix} \lambda - 2 & 0 \\ 0 & \lambda - 1 \end{bmatrix} \begin{Bmatrix} x \\ y \end{Bmatrix} = \begin{Bmatrix} 0 \\ 0 \end{Bmatrix}$$

Where $\lambda = 1$

$$\begin{bmatrix} -1 & 0 \\ 0 & 0 \end{bmatrix} \begin{Bmatrix} x \\ y \end{Bmatrix} = \begin{Bmatrix} 0 \\ 0 \end{Bmatrix} \therefore x = 0, y = t$$

Where $\lambda = 2$

$$\begin{bmatrix} 0 & 0 \\ 0 & 1 \end{bmatrix} \begin{Bmatrix} x \\ y \end{Bmatrix} = \begin{Bmatrix} 0 \\ 0 \end{Bmatrix} \therefore x = t, y = 0$$

Where t can be any value.

So, the two pairs of eigenvalues and eigenvectors are:

$$\lambda = 1, x = \begin{Bmatrix} 0 \\ t \end{Bmatrix}$$

$$\lambda = 2, x = \begin{Bmatrix} t \\ 0 \end{Bmatrix}$$

Which shows that any vector that is only in the X direction is doubled in length, and any vector in the Y direction remains unchanged (multiplied by 1).

The eigenvectors can be of any initial length, but we require a single solution, so we choose an arbitrary length, which is normally 1 (normalized to 1). So, our final answers are:

$$\lambda = 1, x = \begin{Bmatrix} 0 \\ 1 \end{Bmatrix}$$

$$\lambda = 2, x = \begin{Bmatrix} 1 \\ 0 \end{Bmatrix}$$

Appendix D: Buckling

Euler buckling

The relationship between the length, material and cross-section of the column, and the load required to buckle it is:

$$P_{crit} = \frac{\pi^2 EI}{(KL)^2} = \frac{\pi^2 EI}{L_{eff}^2}$$

Where:

P_{crit} = critical axial load (N)
E = Young's modulus (N/m^2)
I = second moment of area (m^4)
L = length of column (m)
K = effective length factor based on end conditions (1 for pin-ended, 0.5 for fixed ends, etc.)
L_{eff} = effective length of member (m)

Buckling amplification[74,75]

Figure D1: Loaded strut

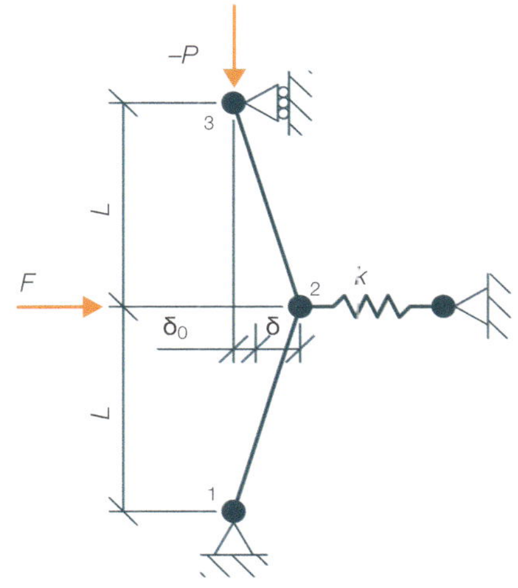

Where:

F = lateral force
k = system stiffness
δ = induced defection
δ_0 = initial imperfection
P = axial load
L = half the column height

When in equilibrium:

$$F = k\delta + \frac{2P}{L}(\delta + \delta_0)$$

Rearrange to give induced deflection:

$$F = k\delta + \frac{2P}{L}\delta + \frac{2P}{L}\delta_0$$

$$k\delta + \frac{2P}{L}\delta = F - \frac{2P}{L}\delta_0$$

$$\left(k + \frac{2P}{L}\right)\delta = F - \frac{2P}{L}\delta_0$$

$$\delta = \frac{F - (2P/L)\delta_0}{k + 2P/L}$$

Divide top and bottom by the stiffness (k):

$$\delta = \frac{F/k - (2P/kL)\delta_0}{1 + 2P/kL}$$

Substitute $\dfrac{-2P}{kL} \Rightarrow \dfrac{1}{\lambda_{crit}}$

$$\delta = \frac{F/k + (1/\lambda_{crit})\delta_0}{1 - 1/\lambda_{crit}}$$

$$\delta = \frac{F}{k(1 - 1/\lambda_{crit})} + \frac{\delta_0}{\lambda_{crit}(1 - 1/\lambda_{crit})}$$

$$\delta = \frac{\lambda_{crit}}{\lambda_{crit} - 1}\frac{F}{k} + \frac{1}{\lambda_{crit} - 1}\delta_0$$

This means that when the applied load approaches the critical buckling load ($\lambda_{crit} \Rightarrow 1$) the effects of destabilising loads and initial imperfections become increasingly significant.

Appendix E: Dynamics

Vibration of a string[249]

$$f = \sqrt{\frac{T}{4ml^2}}$$

Where:

f = frequency (Hz)
T = tension (N)
m = mass per unit length (kg/m)
l = length of string (m)

Vibration of a beam

Simple frequency check:

$$f = \frac{18}{\sqrt{\delta}}$$

Where:

f = natural frequency (Hz)
δ = unfactored load deflection (mm)

Detailed frequency check:

$$f = \frac{\pi}{2} \sqrt{\frac{EI}{wL^4}}$$

Where:

E = Young's modulus (N/m^2)
I = bending moment of area (m^4)
w = uniform load (kg)
L = length (m)

Vibration of a building

Simple frequency check:

$$f = \frac{10}{n}$$

Where n = number of storeys

Footfall/human-induced vibration

Footfall analysis is the assessment of floor vibrations caused by people walking and is not to be confused with the footfall analysis done by retailers measuring how many people pass or enter a shop[†]. Footfall analysis is not an examination into what will definitely happen, but a benchmark figure that is consistent between designs. This means that while we might gain only an approximate prediction of how the structure will actually behave, we do have an objective way to compare designs and establish definite behavioural requirements.

While the footfall analysis is intended to find the vibrations caused by people walking through the building, it is exceedingly difficult to model all the possible paths they might take through the building at all their speeds. A Monte Carlo analysis might plot lots of paths and therefore generate a reasonable average behaviour, but we can never be sure that we had found the worst case. In addition, a lot of the research into footfall response uses oscillating machines that vibrate a definite weight at a set frequency rather than actual walking people. While the analysis is based on excitation at fixed points, the formulas include adjustment factors to take this into account so that they are closer to the effect of moving people.

People come in all different shapes and sizes, and the weight of the person walking is a key factor in the footfall impact force. Unfortunately, while it is easy to research the public's walking step frequency from video feeds, it is some way from the anonymous data recovery to asking people how much they weigh, or even actually weighing them, and then correlating that information to their walking speed from the video. Thus, we take a single, 'average' weight.

We know the distribution of people's walking speeds, so we can determine the typical walking speeds. Someone might do a tap dance demonstration in the middle of the office, but it is not a design case unless the building is intended to also be a dance studio.

All structures vibrate when walked on, but there are a few factors to consider when you decide what the vibration limits are. For example, people on a footbridge are comfortable with some vibration as they know that they are on a slender structure that is clearly spanning a road or river. On the other hand, people naturally think that a building floor is solid all the way to the ground. This means that people on bridges are more tolerant of vibration than those in rooms.

What people are doing and whether they are standing, sitting or lying down changes how sensitive they are to vibration. People are less sensitive to vibration when standing and more when lying down. People are also more aware of horizontal vibrations than vertical, probably because lateral movements are more likely to destabilise us. If you imagine sitting on a tree branch, perhaps as one of our ancestors did regularly: a large bounce in the branch is not a problem but a sideways wobble has a good chance of throwing us off.

People's tasks also make them more sensitive to vibration. In the office, a computer screen bouncing when someone walks past is annoying but can be tolerable, but for surgeons, especially dealing with something as delicate as brain surgery, any noticeable vibration is unacceptable. This means that hospitals are very demanding for vibration control. Laboratories are even more demanding when it comes to vibration, whether for optical or electron microscopes, as these are sensitive to vibration below our human limit of perception.

We are not equally sensitive to all vibration frequencies. Studies have found that we are most sensitive in the 4–8Hz range, which happens to be the resonant frequency range of our internal organs. The problem for building designers is that 4–8Hz is a typical fundamental frequency for a floor.

People typically walk at frequencies between 1 and 2.5Hz, (typically at about 2 steps per second) or up to 4.5Hz on staircases. This means that we cannot allow the structure to have vertical natural frequencies in this range, otherwise the structure will bounce with every step, or if you prefer in technical terms, resonate with the first harmonic of the walking frequency.

[†]   Or football analysis, which is the regular suggestion from the spell checker.

Footbridges are also vulnerable to lateral footfall-induced vibration, as was demonstrated by London's Millennium Bridge in 2000. For lateral excitation we halve the walking frequency because the sideways force alternates between each leg: left for left and right for right. The added problem for footbridges is that, while we can walk out of step with a vertical bounce, we find it easier to walk in step with a lateral sway. This leads to a frequency lock-in, where people walk in time with the bridge, causing it to wobble even more.

The way to analyse for these problems is to first start with a modal analysis of the structure to find its natural frequencies. As mass resists accelerations ($F = ma$ or $a = F/m$, where the force F comes from the footsteps and the mass m from the structure, furniture and fittings plus its occupants) the worst case for vibration, in buildings at least, is when it is empty. But as we are not interested in the vibrations of empty buildings, design guides[70–72] usually recommend that we take 10% of the imposed load plus the minimum credible permanent load. Footbridges are different as the mass of a full crowd can lower the natural frequencies to a point where the bridge might resonate.

Next, we need to run a harmonic analysis on every node in the structure where people might walk, and calculate the resulting accelerations. A harmonic analysis is a sinusoidal vibration, which in this case is of equal weight to force induced by an average human tread. Harmonic vibrations are a little different to footsteps, but they are close enough for us to judge and compare design options[†].

The accelerations are then usually weighted by a baseline curve (Figure E1), giving us a 'Response factor' R. What R value is acceptable depends on what happens in that location. For example, hospital operating theatres need a minimum vibration so have a target R of 1, offices 4 or 8, and footbridges 32 or 64[70–72]. The important thing to note is that people are sensitive not to vibrations as such, but accelerations at certain frequencies. This means that finding and limiting the natural frequencies of the structure is only the first step, as it only indicates potential behaviour. The important thing is then to use the frequencies and modal masses to determine how the structure will actually respond to footsteps and other sources of excitation.

Figure E1: Frequency weighting curve (W_b) appropriate for vertical vibration[250]

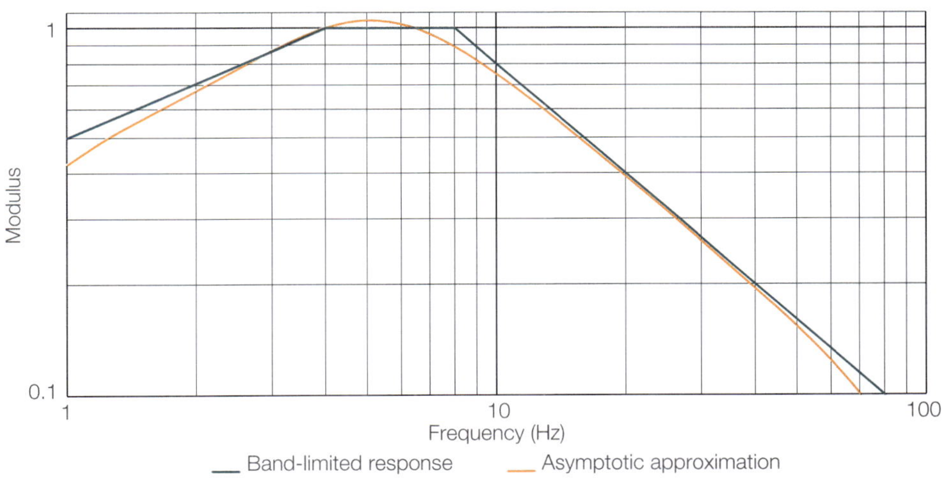

While we have ensured that the structure will not have a natural frequency that matches a walking frequency, that does not mean there will not be any resonance. Our footsteps will cause resonance by landing in time with every other bounce (2nd harmonic resonance), every third bounce (3rd harmonic), or fourth. The effect reduces for each harmonic, so we ignore harmonics above the 4th. Design guides do recommend that we check floor frequencies up to five or six times the maximum waking frequency though.

[†] Harmonic analysis is wrong but useful.

For structures with natural frequencies above about 10Hz, such as staircases or small floor bays, the damping in the structure can cause the vibration to fade away before the next step, so resonance is even more unlikely. Here, we are concerned with the impulse or transient accelerations on the structure, which is from the initial reaction to the footstep.

Dealing with a footfall vibration problem can be tricky. One way to reduce the vibration acceleration is by adding mass. This is ok in a new construction because you can increase the size of beams and columns for the added weight, but it is much harder in an existing structure. Adding mass reduces the accelerations but will also lower the natural frequencies, which might make the floor more likely to resonate, but it is likely that the floor is now stiffer.

I did hear of an alternative way to add mass to an existing structure from an Arup engineer: they added a large plant pot at the centre of resonance. Not only did this add mass at the most useful spot, it also prevented the occupants from walking in that location and thus inducing the worst vibrations. The appearance of the space might also have been improved by the presence of a large plant.

Adding stiffness (reducing it is not normally a possibility) will increase the natural frequencies, possibly reducing the resonance. But a stiffer beam or slab is likely to be heavier, which will again encourage the frequencies to reduce.

If the floor is resonating, then adding damping will help. Structures normally get this from friction between the various parts, such as bolted connections, raised floors and ceilings, partitions etc. The occupants themselves also add damping to the system. Adding mechanical damping such as tuned mass dampers and viscous dampers are options usually reserved for existing structures where other structural interventions are too difficult.

There are other ways of fine-tuning a model to get the results we need. Consider the size of the person and where their feet might land. A staircase might be most sensitive to vibrations on the edge, especially if torsional modes are significant. But people cannot walk on the very edge of a stair, rather at least 200–300mm in from the edge due to handrail and boundary effects (no one wants to accidentally kick a balustrade while running down a stair). Thus, it is conservative to assess the footfall vibration right at the stair or balcony edge and more realistic to start the checks a little way in. You might also ignore the response on the stair tread where the foot lands, and look only at the response values elsewhere on the stair as these will be the accelerations experienced by others.

It is also worth modelling the handrail on staircases and atrium edges; we may ignore their effect on the static load capacity, but they can add useful stiffening to the edge beams to help deal with the vibration.

Appendix F: Linear structures[60]

w   load per unit length
P   point load
L   length
E   Young's modulus of elasticity
I   Second moment of area of beam section

Table F1: Cantilevers

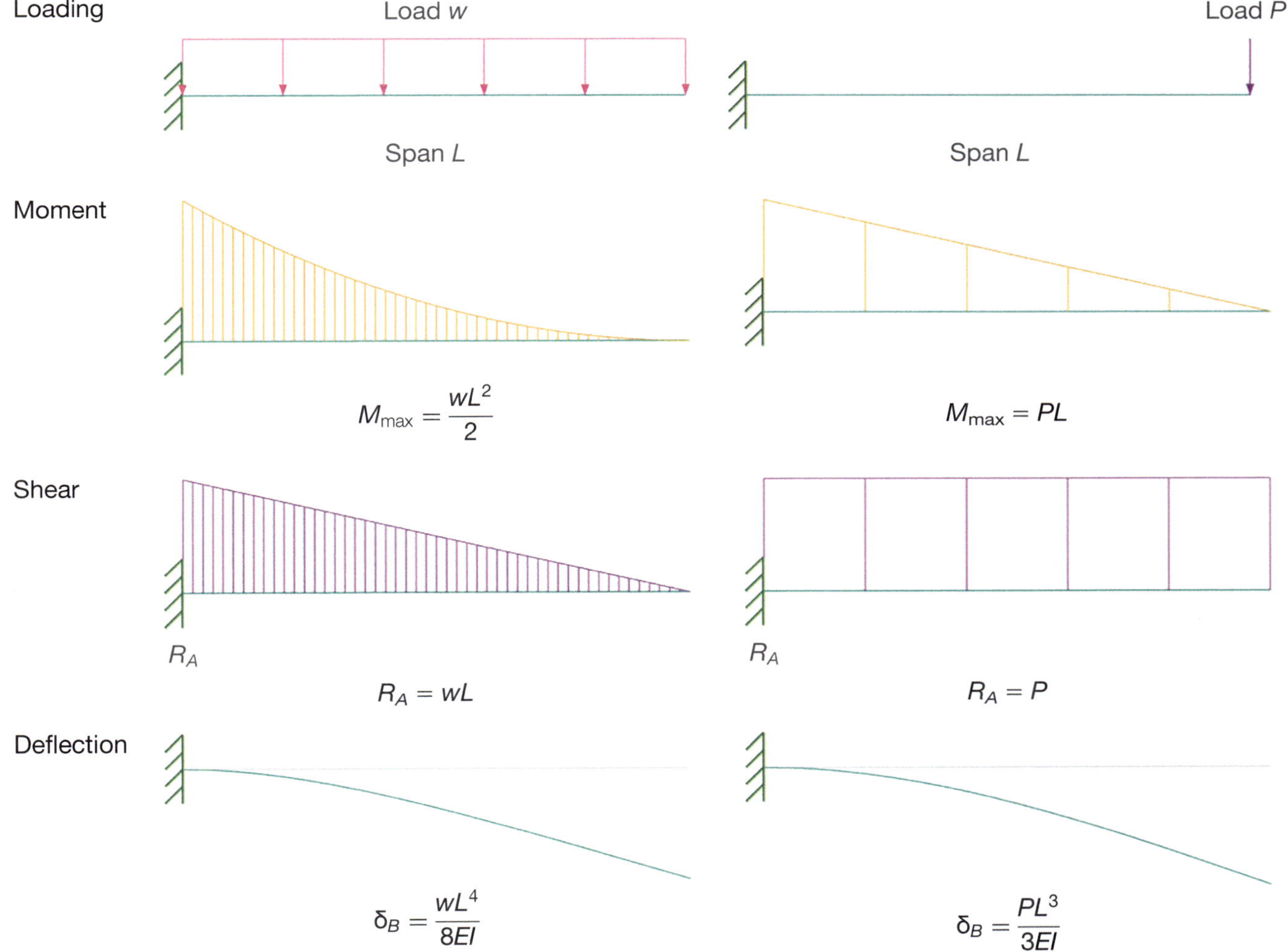

Table F2: Simply supported beams

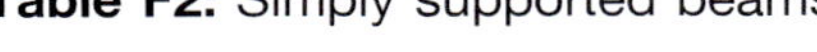

Loading	Load w	Load P

Span L | Span L

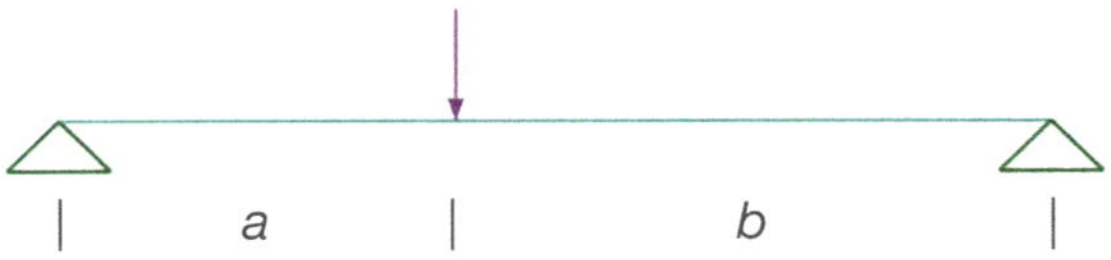

Moment

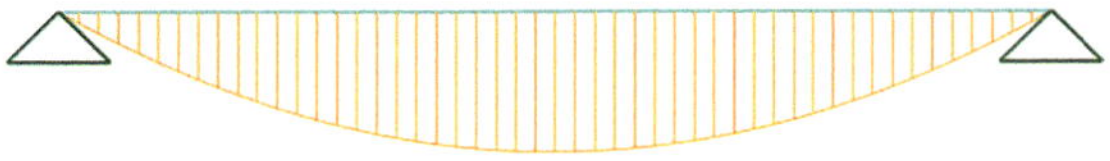

$$M_{max} = \frac{wL^2}{8}$$

$$M_{max} = \frac{Pab}{L} \quad \text{if } a = b: M_{max} = \frac{PL}{4}$$

Shear

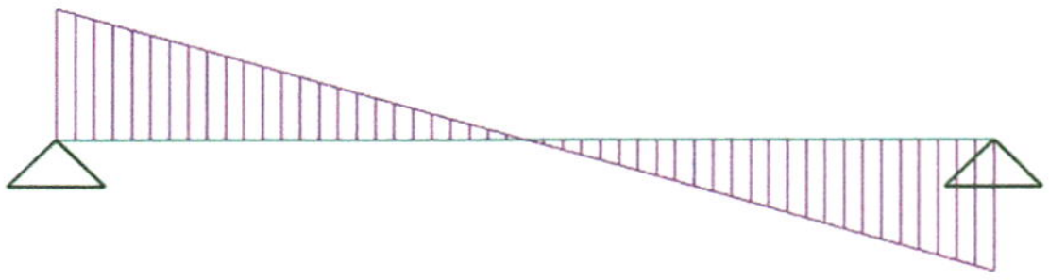

R_A | R_B | R_A | R_B

$$R_A = R_B = \frac{wL}{2}$$

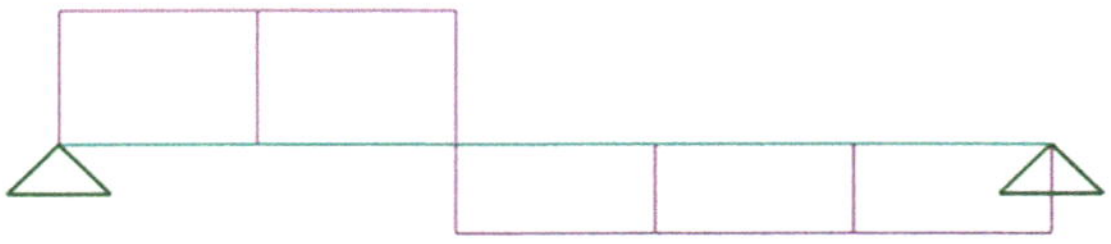

$$R_A = \frac{Pb}{L} \quad R_B = \frac{Pa}{L} \quad \text{if } a = b: R_A = R_B = \frac{P}{2}$$

Deflection

$$\delta_{max} = \frac{5wL^4}{384EI}$$

$$\delta_{centre} = \frac{PL^3}{48EI}\left[\frac{3a}{L} - 4\left(\frac{a}{L}\right)^3\right]$$

$$\text{if } a = b: \delta_{max} = \frac{PL^3}{48EI}$$

Table F3: Propped cantilevers

Loading	

Load w

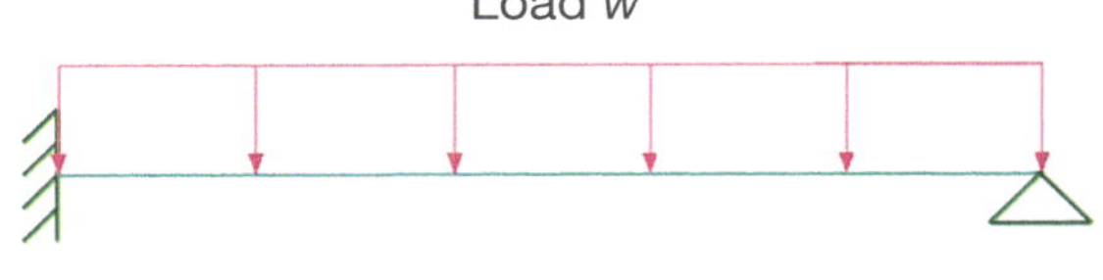

Span L

Load P

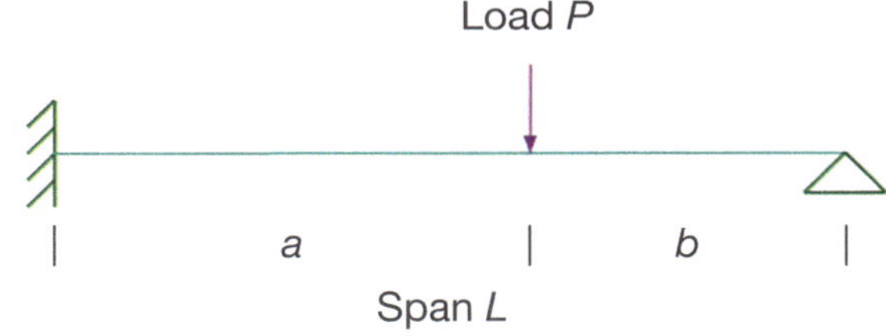

Span L

Moment

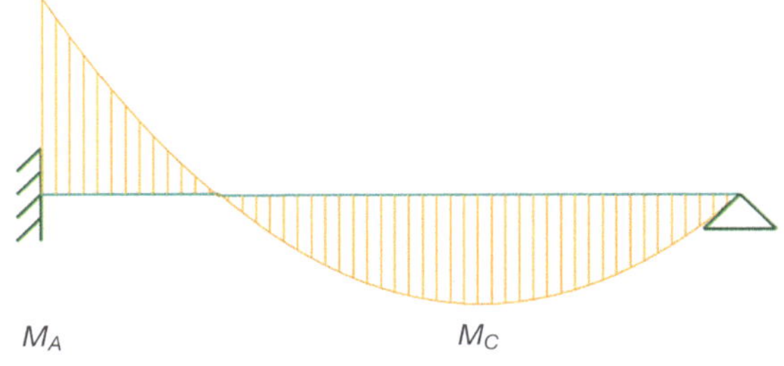

M_A M_C

$$M_A = -\frac{wL^2}{8} \quad M_C = \frac{9wL^2}{128}$$

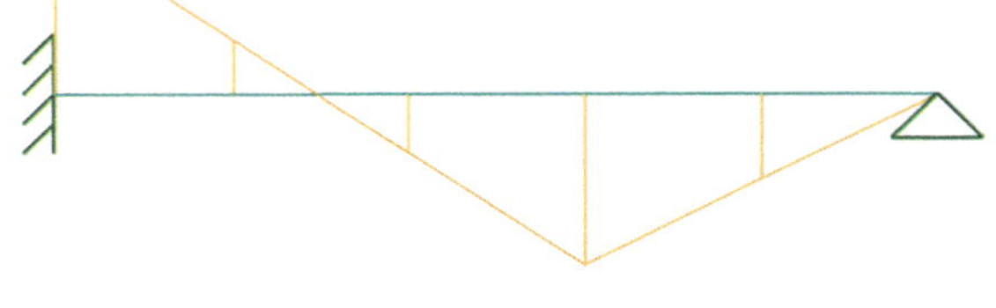

M_A M_C

$$M_A = -\frac{Pb(L^2 - b^2)}{2L^2}$$

$$\text{if } a = b: M_A = -\frac{3PL}{16}$$

$$M_C = \frac{Pb}{2}\left(2 - \frac{3b}{L} + \frac{b^3}{L^3}\right)$$

$$\text{if } a = b: M_C = \frac{5PL}{32}$$

Shear

R_A R_B

$$R_A = \frac{5wL}{8} \quad R_B = \frac{3wL}{8}$$

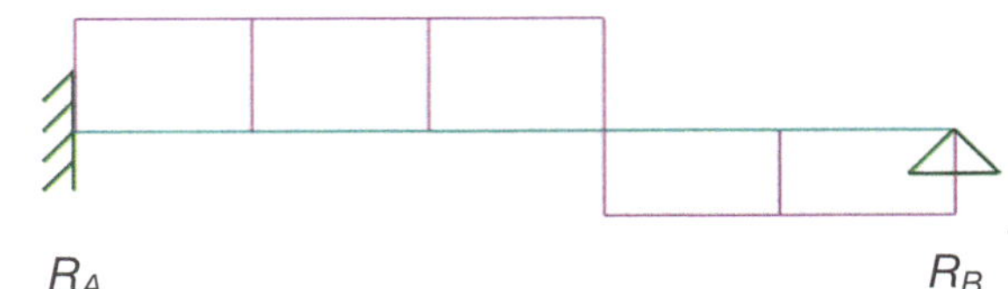

R_A R_B

$$R_B = \frac{Pa^2}{2L^3}(b + 2L)$$

$$R_A = P - R_B$$

$$\text{if } a = b: R_A = \frac{11P}{16} \quad R_B = \frac{5P}{16}$$

Deflection

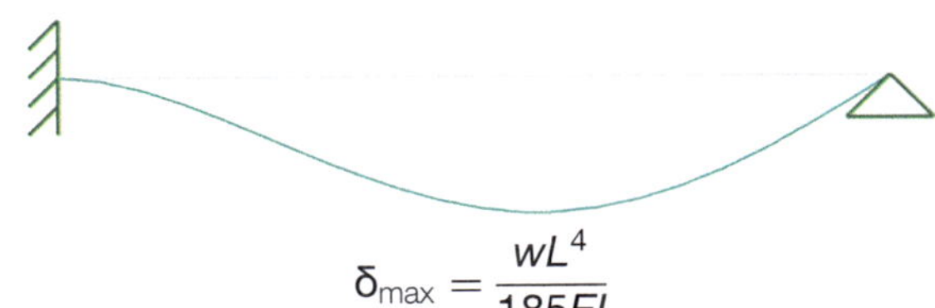

$$\delta_{max} = \frac{wL^4}{185EI}$$

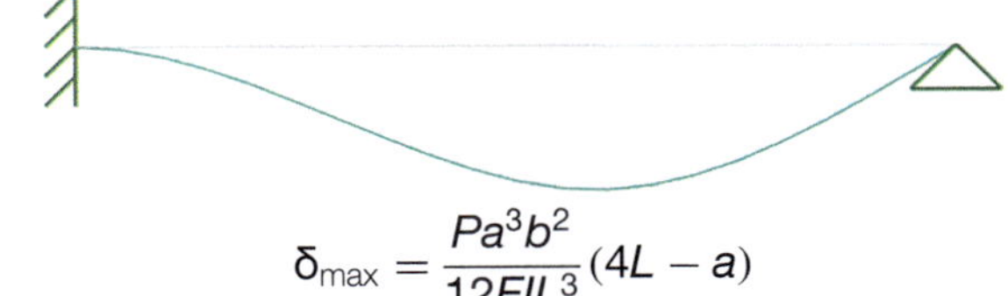

$$\delta_{max} = \frac{Pa^3b^2}{12EIL^3}(4L - a)$$

$$\text{if } a = b: \delta_{max} = 0.00932\frac{PL^3}{EI}$$

Table F4: Built-in beams

Loading

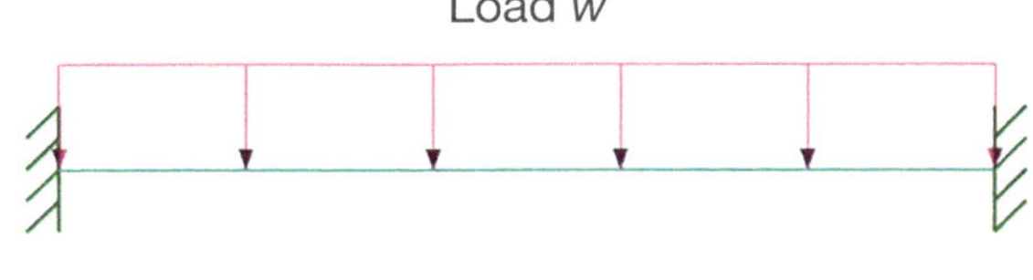

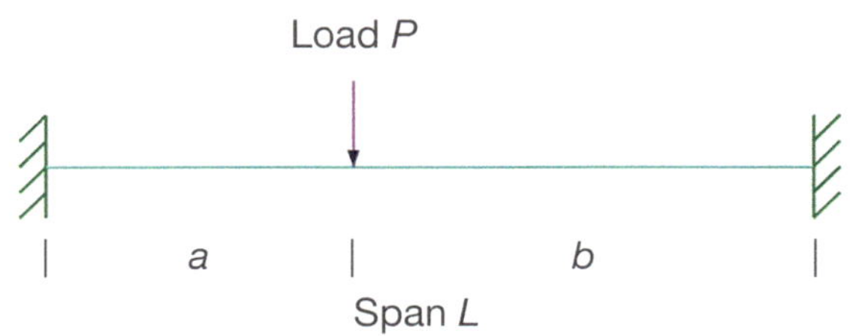

Moment

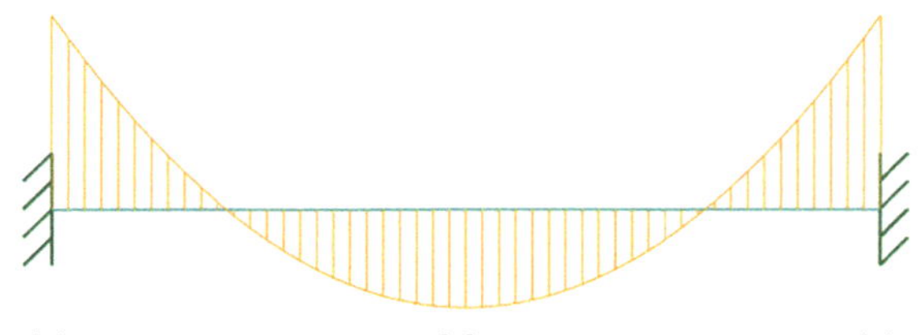

$$M_A = M_B = -\frac{wL^2}{12} \qquad M_C = \frac{wL^2}{24}$$

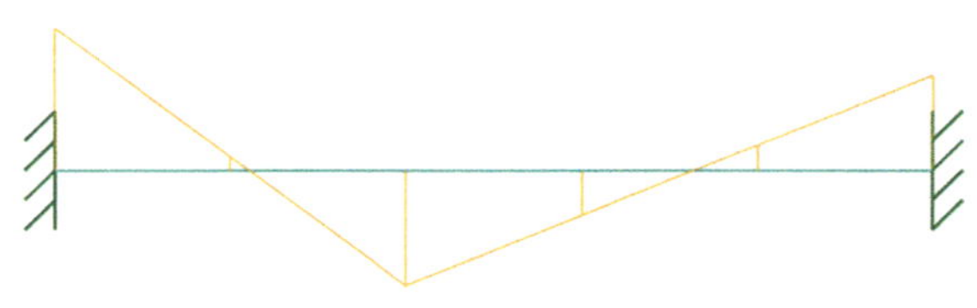

$$M_A = -\frac{Pab^2}{L^2}$$

$$M_B = -\frac{Pa^2b}{L^2}$$

$$M_C = \frac{2Pa^2b^2}{L^3}$$

$$\text{if } a = b: \ -M_A = -M_B = M_C = \frac{PL}{8}$$

Shear

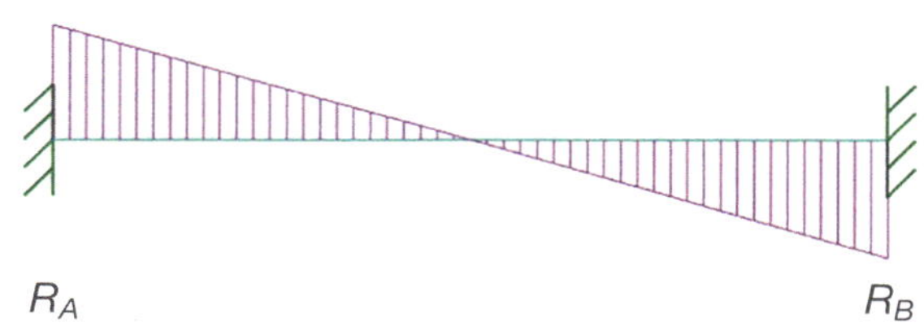

$$R_A = R_B = \frac{wL}{2}$$

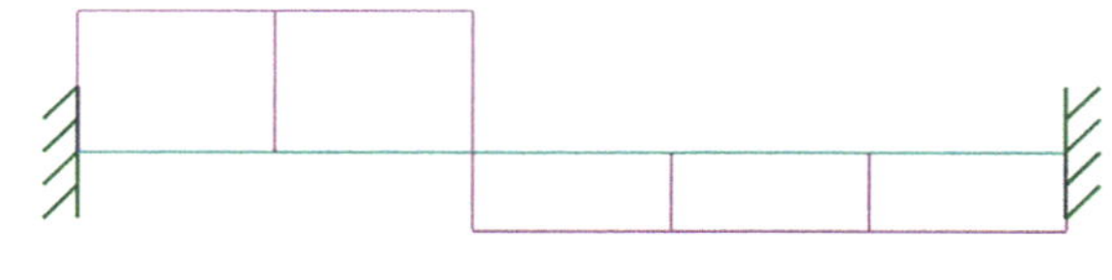

$$R_A = P\left(\frac{b}{L}\right)^2\left(1 + \frac{2a}{L}\right)$$

$$R_B = P\left(\frac{a}{L}\right)^2\left(1 + \frac{2b}{L}\right)$$

$$\text{if } a = b: R_A = R_B = \frac{P}{2}$$

Deflection

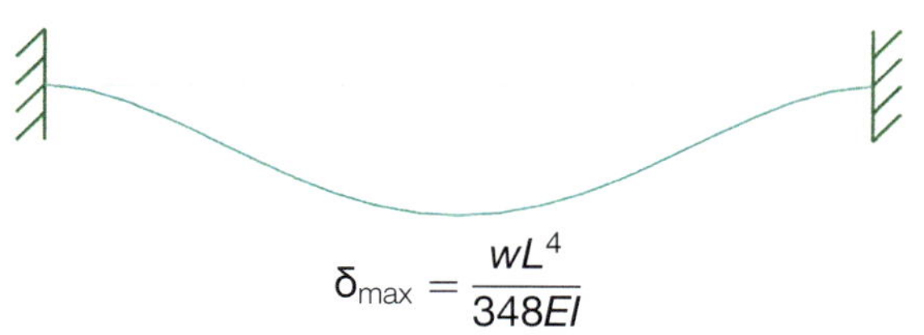

$$\delta_{max} = \frac{wL^4}{348EI}$$

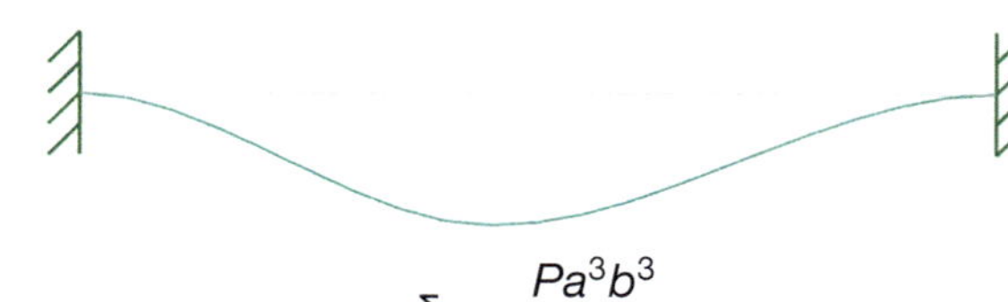

$$\delta_C = \frac{Pa^3b^3}{3EIL^3}$$

$$\text{if } a = b: \delta_{max} = \frac{PL^3}{192EI}$$

Appendix G: Cable structures

Figure G1: Loaded cable

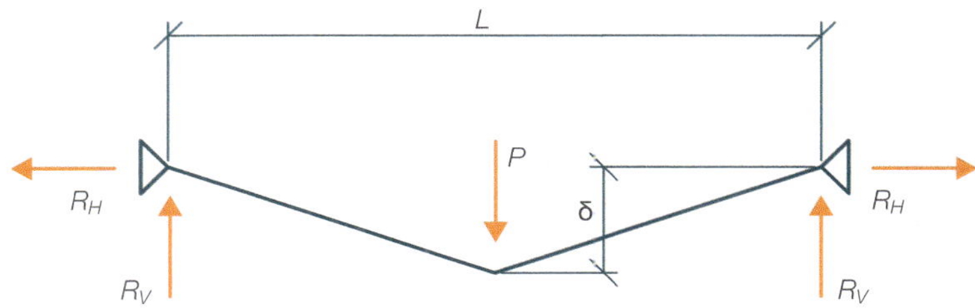

If the load is central:

$$R_v = \frac{P}{2}$$

$$R_h = \frac{PL}{4\delta}$$

$$T = \sqrt{\left(\frac{P}{2}\right)^2 + \left(\frac{PL}{2\delta}\right)^2}$$

$$\varepsilon = 2\sqrt{\frac{\delta^2 + (L/2)^2}{L}} = TAE$$

(if σ is still in the linear range).

Appendix H: Arches and domes

While individual masonry units might have a reasonable tensile capacity, especially natural stone, we have to avoid tension in masonry structures as a whole because they are generally held together with mortar, which is quite weak. The Ancient Greeks only knew about beams and so had to keep their temple columns close together to allow the stone to span. Other people in antiquity used similar methods, as well as the corbelled arch. It was the Romans who were the first to use the true arch; which enabled the spanning of large openings using only small pieces of stone or brick.

Corbelled arch

To stack bricks so that they have the largest cantilever and still remain self-stable, the top brick overlaps by 1/2 its length, the second by 1/4, the third by 1/8, and so on.

The plan opening of such an arch in regular UK bricks is $2 \times 215 \times (\frac{1}{2} + \frac{1}{4} + \frac{1}{8} + \ldots) + 10$ for the middle mortar joint. For n bricks this would be:

$$450 \sum_{i=1}^{n} \left(\frac{1}{2^n}\right) + 10$$

But this would not be a true arch as there will be no thrust at the top.

Parabola[251]

The form of a structure with a constant load on plan. For example, a suspension bridge considering only the deck weight:

$$y = \frac{x^2}{4a} + h$$

Where:

h = offset of parabola
a = a constant

Catenary[252]

The shape of a hanging chain: load is constant along its length. For example, the shape of the suspension bridge cable considering only the self-weight of the cable:

$$y = \frac{1}{2}a(e^{x/a} + e^{-x/a}) = a\cosh\left(\frac{x}{a}\right)$$

Where a is a parameter that determines how quickly the catenary 'opens up'.

Analysing arches

Figure H1: Arch

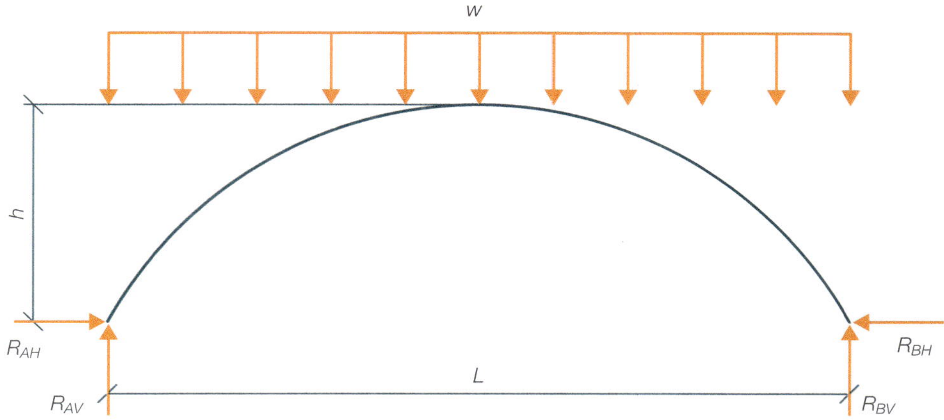

$$R_{AH} = R_{BH} = \text{compression in arch} = \frac{wL^2}{8h}$$

$$R_{AV} = R_{BV} = \frac{wL}{2}$$

Appendix I: Brohn diagrams

A key skill for those analysing structures is to have a reasonable idea of what results you expect from an analysis, both qualitative (shape) and quantitative (value). For these exercises you need to first sketch the deflected shape of the structure, then the bending moment diagram. You should check your results by building FEA models. If your diagrams and FEA results differ, work out why: was it a mistake in your sketch or model?

Things to note:

- Fixed restraints ⊣ mean that the beam cannot rotate at that point, so must remain straight locally
- Pinned restraints △ mean that the beam can rotate through the point but cannot translate
- Roller restraints △ are pinned restraints that allow the possibility of lateral movement
- Hinges —o— can transfer shear but not moment; the moments must be zero at these points

Tip: remember load superimposition

- Where there are multiple loads, determine the results for each and add them together
- Where a restraint makes the structure indeterminate, remove the restraint and replace with a force or moment that restores the structure to the original location

Figure I1: Understanding structural analysis examples[97]

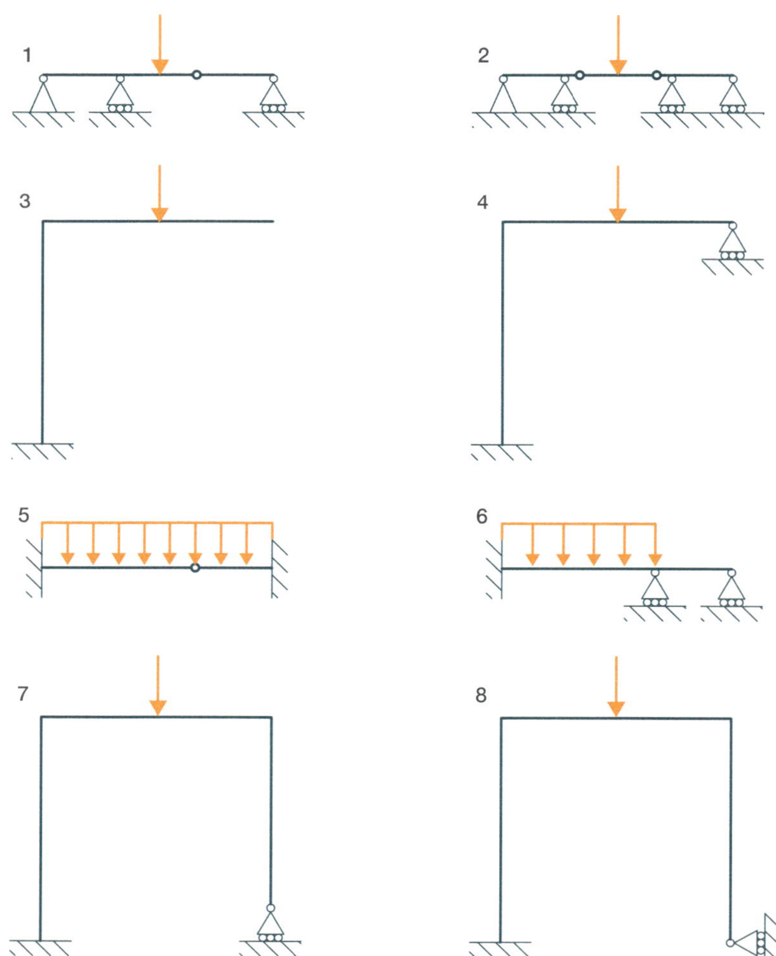

Appendix J: Braess' Paradox[103]

or why improving something can make it worse

On Earth Day in 1990 they closed New York's 42nd Street for the parade[253] and in 1999 one of the three main traffic tunnels in South Korea's capital city was shut down for maintenance[254]. Bizarrely, despite both routes being heavily used for traffic, the result was not the predicted chaos and jams; instead the traffic flows improved in both cases. Inspired by their experience, Seoul's city planners subsequently demolished a motorway leading into the heart of the city and experienced exactly the same strange result, with the added benefit of creating a 5-mile long, 1,000-acre park for the local inhabitants[255].

It is counter-intuitive that you can improve commuter travel times by reducing route options: planners normally want to improve things by adding routes. This paradox was first explored by Dietrich Braess[256] where he explored the maths behind how adding route choices to a network can sometimes make everyone's travel time worse.

Rather than looking at each journey individually, the mathematician John Forbes Nash proposed that a network is in equilibrium when there is no advantage for any individual driver to choose a different route. This means that in a Nash all routes have the same travel time and the network can be solved using simultaneous equations. To illustrate this lets looks at an example[257] (Figure J1).

Figure J1: Initial route network

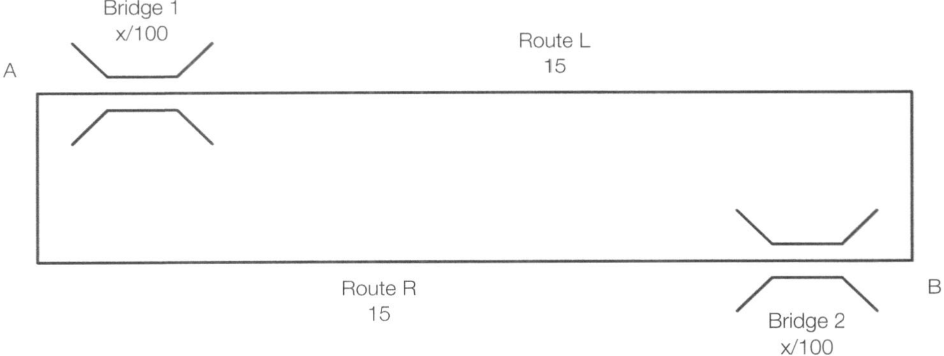

To get from A to B the travellers either take route L over Bridge 1 and then along a motorway, or route R along the other motorway and then over Bridge 2. The bridges are both subject to congestion such that the time (in minutes) to traverse them is the number of cars per hour divided by 100. The motorway sections always take 15 minutes. When the rush hour traffic has reached equilibrium, the time taken on both routes is the same. If we say that L_v is the traffic on route L and R_v the traffic on route R, then:

$$\text{Time } t = \frac{L_v}{100} + 15 = \frac{R_v}{100} + 15$$

If there are 1,000 cars going from A to B, then we know that:

$$L_v + R_v = 1{,}000$$

And we can see that:

$$L_v = R_v = 500$$

Substituting back into the first equation we can see that the trave time for all drivers (Lt and Rt) is 20 minutes.

The authorities want to improve things by adding a high-speed route, C, that only takes 7.5 minutes to travel, no matter what the level of traffic (Figure J2).

Figure J2: 'Improved' route network

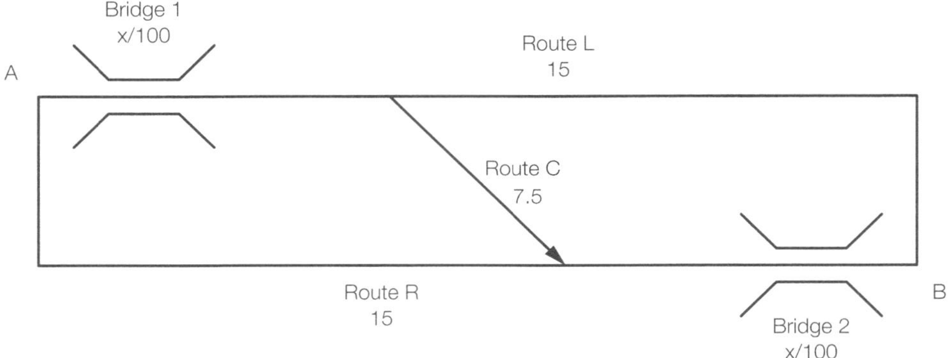

Let's refer to the traffic that passes over Bridge 1, then the new link, followed by Bridge 2 as C. The steady-state travel times for the three routes (Lt, Rt, Ct) are now:

$$L_t = R_t = C_t \Rightarrow \frac{L_v + C_v}{100} + 15 = \frac{L_v + C_v}{100} + 7.5 + \frac{R_v + C_v}{100} = \frac{R_v + C_v}{100} + 15$$

$$L_v + R_v + C_v = 1{,}000$$

From which we can calculate that $L_v = R_v = 250$ and $C_v = 500$. Thus, the travel time on each of the routes is now 22.5 minutes and again the building of the new road has made the daily commute worse not better.

The common cause in both these networks is that originally there was a balance of two moderate choices, but the additional route was so much better that it attracted an excessive amount of traffic to the approach routes and thus unbalanced the system.

The network does not have to be symmetrical to experience a Braess' Paradox. It might also manifest in other networks such as electronics or water supplies. It can certainly occur in structures (Figure J3)[258]. Consider this arrangement of springs: when the springs are connected together, they act in series and the overall deflection is large as they are both carrying the full load. Break the connection between them and they act in parallel, each carrying half the load, reducing the deflection.

Figure J3: Braess' paradox

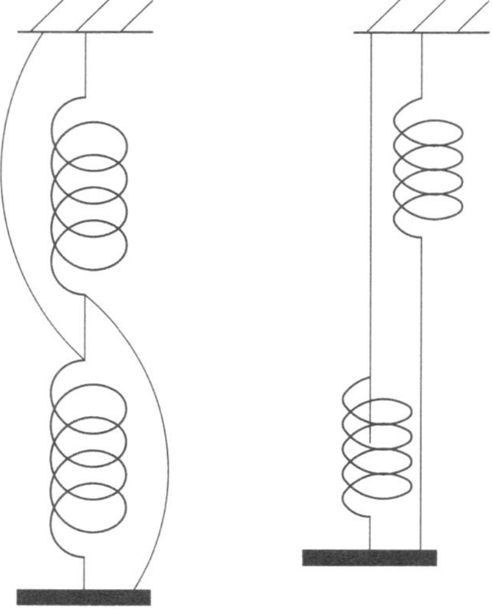

Does adding a route choice always result in more congestion? Or to put it another way, does road building generate traffic? The answer is: not necessarily. For example, one research project looked at routes through the city of Boston and found that of the 246 possible links on a journey between Harvard Square and Boston Common, closing one of six particular links did display the Braess' Paradox of improving traffic flow, but closing one of the other 240 did make things worse[259,260].

So how can additions induce the Braess' Paradox and how easy is it to spot one in action? Essentially, they occur when an improvement attracts a disproportionate load that the approaches cannot handle. If the induced congestion on these approaches affects other routes, then the entire system suffers as a result[261]. So, in conclusion, adding choices to your network can improve flow through it, or it can make things worse; you just have to check…

Appendix K: Quantum computing[230,262]

Quantum computing involves many things far beyond our experience here in the classical world, but the maths can be surprisingly straightforward.

Vector notation

We can represent quantum bits (qubits) with the Dirac vector notation:

$$0 = |0\rangle = \begin{pmatrix} 1 \\ 0 \end{pmatrix} \quad 1 = |1\rangle = \begin{pmatrix} 0 \\ 1 \end{pmatrix}$$

The values in Dirac vectors represent the amplitude (probability) of the qubit collapsing to that value when measured:

$$|\psi\rangle = \alpha|0\rangle + \beta|1\rangle = \begin{pmatrix} \alpha \\ \beta \end{pmatrix} = \begin{pmatrix} \sqrt{P(0)} \\ \sqrt{P(1)} \end{pmatrix}$$

Where α and β are complex numbers and $|\alpha|^2 + |\beta|^2 = 1$. The qubit will collapse to 0 with a probability of $|\alpha|^2$ and to 1 with a probability $|\beta|^2$. The qubit $\begin{pmatrix} 1 \\ 0 \end{pmatrix}$ has a 100% chance of collapsing to 0, and $\begin{pmatrix} 0 \\ 1 \end{pmatrix}$ has a 100% chance of collapsing to 1. A qubit in an equal superposition of 0 and 1 has the value $\begin{pmatrix} \frac{1}{\sqrt{2}} \\ \frac{1}{\sqrt{2}} \end{pmatrix}$, which shows that it has a $\left|\frac{1}{\sqrt{2}}\right|^2 = \frac{1}{2}$ chance of collapsing to 0 or 1.

Similarly, a two-qubit register's Dirac vector represents:

$$|\psi\rangle = \alpha|00\rangle + \beta|01\rangle + \gamma|10\rangle + \delta|11\rangle = \begin{pmatrix} \alpha \\ \beta \\ \gamma \\ \delta \end{pmatrix} = \begin{pmatrix} \sqrt{P(0)} \\ \sqrt{P(1)} \\ \sqrt{P(2)} \\ \sqrt{P(3)} \end{pmatrix} = \begin{pmatrix} \sqrt{P(00)} \\ \sqrt{P(01)} \\ \sqrt{P(10)} \\ \sqrt{P(11)} \end{pmatrix}$$

We derive these Dirac vectors by taking the tensor product of the individual qubit vectors, which we call the product state:

$$\begin{pmatrix} a \\ b \end{pmatrix} \otimes \begin{pmatrix} c \\ d \end{pmatrix} = \begin{pmatrix} a\begin{pmatrix} c \\ d \end{pmatrix} \\ b\begin{pmatrix} c \\ d \end{pmatrix} \end{pmatrix} = \begin{pmatrix} ac \\ ad \\ bc \\ bd \end{pmatrix}$$

where $|ac|^2 + |ad|^2 + |bc|^2 + |bd|^2 = 1$. We can factor the product state back to the individual state representation if the qubits are not entangled. The product state of n bits is a vector of size 2^n.

$$0 = |00\rangle = \begin{pmatrix} 1 \\ 0 \end{pmatrix} \otimes \begin{pmatrix} 1 \\ 0 \end{pmatrix} = \begin{pmatrix} 1 \\ 0 \\ 0 \\ 0 \end{pmatrix} \qquad 1 = |01\rangle = \begin{pmatrix} 1 \\ 0 \end{pmatrix} \otimes \begin{pmatrix} 0 \\ 1 \end{pmatrix} = \begin{pmatrix} 0 \\ 1 \\ 0 \\ 0 \end{pmatrix}$$

$$2 = |10\rangle = \begin{pmatrix} 0 \\ 1 \end{pmatrix} \otimes \begin{pmatrix} 1 \\ 0 \end{pmatrix} = \begin{pmatrix} 0 \\ 0 \\ 1 \\ 0 \end{pmatrix} \qquad 3 = |11\rangle = \begin{pmatrix} 0 \\ 1 \end{pmatrix} \otimes \begin{pmatrix} 0 \\ 1 \end{pmatrix} = \begin{pmatrix} 0 \\ 0 \\ 0 \\ 1 \end{pmatrix}$$

$$|6\rangle = |110\rangle = \begin{pmatrix} 0 \\ 1 \end{pmatrix} \otimes \begin{pmatrix} 0 \\ 1 \end{pmatrix} \otimes \begin{pmatrix} 1 \\ 0 \end{pmatrix} = \begin{pmatrix} 0 \\ 0 \\ 0 \\ 0 \\ 0 \\ 0 \\ 1 \\ 0 \end{pmatrix}$$

If we have two qubits in superposition we get:

$$\begin{pmatrix} \frac{1}{\sqrt{2}} \\ \frac{1}{\sqrt{2}} \end{pmatrix} \otimes \begin{pmatrix} \frac{1}{\sqrt{2}} \\ \frac{1}{\sqrt{2}} \end{pmatrix} = \begin{pmatrix} \frac{1}{2} \\ \frac{1}{2} \\ \frac{1}{2} \\ \frac{1}{2} \end{pmatrix}$$

Note that $\left|\frac{1}{2}\right|^2 = \frac{1}{4}$, so there is a $\frac{1}{4}$ chance of collapsing to $|00\rangle$, $|01\rangle$, $|10\rangle$, or $|11\rangle$.

While measuring a qubit will collapse it, we can manipulate the qubits without measuring them and thus leave them in superposition. We mathematically manipulate the qubit vectors using matrix operations.

Logic gates
Classical computing uses a number of logic gates.

NOT	
Input	Output
0	1
1	0

OR		
Input 1	Input 2	Output
0	0	0
0	1	1
1	0	1
1	1	1

AND		
Input 1	Input 2	Output
0	0	0
0	1	0
1	0	0
1	1	1

XOR		
Input 1	Input 2	Output
0	0	0
0	1	1
1	0	1
1	1	0

Quantum computing also uses logic gates:

CNOT gate

The Controlled NOT or CNOT gate has a control bit that remains unchanged, but the target bit is flipped if the control qubit is equal to 1.

CNOT			
Input		Output	
Control	Target	Control	Target
0	0	0	0
0	1	0	1
1	0	1	1
1	1	1	0

The matrix form of the CNOT gate is:

$$c = \begin{pmatrix} 1 & 0 & 0 & 0 \\ 0 & 1 & 0 & 0 \\ 0 & 0 & 0 & 1 \\ 0 & 0 & 1 & 0 \end{pmatrix}$$

$$c|00\rangle = c\left(\begin{pmatrix}1\\0\end{pmatrix} \otimes \begin{pmatrix}1\\0\end{pmatrix}\right) = \begin{pmatrix} 1 & 0 & 0 & 0 \\ 0 & 1 & 0 & 0 \\ 0 & 0 & 0 & 1 \\ 0 & 0 & 1 & 0 \end{pmatrix}\begin{pmatrix}1\\0\\0\\0\end{pmatrix} = \begin{pmatrix}1\\0\\0\\0\end{pmatrix} = \begin{pmatrix}1\\0\end{pmatrix} \otimes \begin{pmatrix}1\\0\end{pmatrix} = |00\rangle$$

$$c|01\rangle = c\left(\begin{pmatrix}1\\0\end{pmatrix} \otimes \begin{pmatrix}0\\1\end{pmatrix}\right) = \begin{pmatrix} 1 & 0 & 0 & 0 \\ 0 & 1 & 0 & 0 \\ 0 & 0 & 0 & 1 \\ 0 & 0 & 1 & 0 \end{pmatrix}\begin{pmatrix}0\\1\\0\\0\end{pmatrix} = \begin{pmatrix}0\\1\\0\\0\end{pmatrix} = \begin{pmatrix}1\\0\end{pmatrix} \otimes \begin{pmatrix}1\\0\end{pmatrix} = |01\rangle$$

$$c|10\rangle = c\left(\begin{pmatrix}0\\1\end{pmatrix} \otimes \begin{pmatrix}1\\0\end{pmatrix}\right) = \begin{pmatrix} 1 & 0 & 0 & 0 \\ 0 & 1 & 0 & 0 \\ 0 & 0 & 0 & 1 \\ 0 & 0 & 1 & 0 \end{pmatrix}\begin{pmatrix}0\\0\\1\\0\end{pmatrix} = \begin{pmatrix}0\\0\\0\\1\end{pmatrix} = \begin{pmatrix}0\\1\end{pmatrix} \otimes \begin{pmatrix}0\\1\end{pmatrix} = |11\rangle$$

$$c|11\rangle = c\left(\begin{pmatrix}0\\1\end{pmatrix} \otimes \begin{pmatrix}0\\1\end{pmatrix}\right) = \begin{pmatrix} 1 & 0 & 0 & 0 \\ 0 & 1 & 0 & 0 \\ 0 & 0 & 0 & 1 \\ 0 & 0 & 1 & 0 \end{pmatrix}\begin{pmatrix}0\\0\\0\\1\end{pmatrix} = \begin{pmatrix}0\\0\\1\\0\end{pmatrix} = \begin{pmatrix}0\\1\end{pmatrix} \otimes \begin{pmatrix}1\\0\end{pmatrix} = |10\rangle$$

CCNOT or Toffoli gate

The Toffoli or controlled-controlled-not gate is similar to the CNOT gate but works on three qubits. It inverts the third bit only if both the first bits are set to 1.

CCNOT					
Input A	Input B	Input C	Output A	Output B	Output C
0	0	0	0	0	0
0	0	1	0	0	1
0	1	0	0	1	0
0	1	1	0	1	1
1	0	0	1	0	0
1	0	1	1	0	1
1	1	0	1	1	1
1	1	1	1	1	0

The matrix form is:

$$CC = \begin{pmatrix} 1 & 0 & 0 & 0 & 0 & 0 & 0 & 0 \\ 0 & 1 & 0 & 0 & 0 & 0 & 0 & 0 \\ 0 & 0 & 1 & 0 & 0 & 0 & 0 & 0 \\ 0 & 0 & 0 & 1 & 0 & 0 & 0 & 0 \\ 0 & 0 & 0 & 0 & 1 & 0 & 0 & 0 \\ 0 & 0 & 0 & 0 & 0 & 1 & 0 & 0 \\ 0 & 0 & 0 & 0 & 0 & 0 & 0 & 1 \\ 0 & 0 & 0 & 0 & 0 & 0 & 1 & 0 \end{pmatrix}$$

Hadamard gate

The Hadamard gate takes a 0- or 1-bit and puts it into equal superposition:

$$H|0\rangle = \begin{pmatrix} \frac{1}{\sqrt{2}} & \frac{1}{\sqrt{2}} \\ \frac{1}{\sqrt{2}} & \frac{-1}{\sqrt{2}} \end{pmatrix} \begin{pmatrix} 1 \\ 0 \end{pmatrix} = \begin{pmatrix} \frac{1}{\sqrt{2}} \\ \frac{1}{\sqrt{2}} \end{pmatrix}$$

$$H|1\rangle = \begin{pmatrix} \frac{1}{\sqrt{2}} & \frac{1}{\sqrt{2}} \\ \frac{1}{\sqrt{2}} & \frac{-1}{\sqrt{2}} \end{pmatrix} \begin{pmatrix} 0 \\ 1 \end{pmatrix} = \begin{pmatrix} \frac{1}{\sqrt{2}} \\ \frac{-1}{\sqrt{2}} \end{pmatrix}$$

The Hadamard gate is reversible, so if we reapplied it to the superimposed qubits they would revert to their original state: they would transition out of superposition without measurement and thus the quantum calculation can be deterministic.

Entanglement

We can entangle two qubits by using a Hadamard gate on one, then using it as the control of a CNOT gate with the other (Figure K1). First initialise both qubits to $|0\rangle$, then pass one through the Hadamard gate to put it into superposition, before using the CNOT gate to entangle them.

Figure K1: Entanglement of two qubits

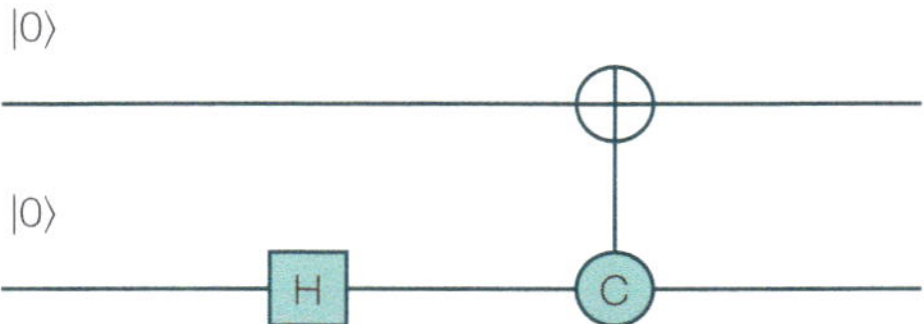

$$CH_1\left(\begin{pmatrix}1\\0\end{pmatrix}\otimes\begin{pmatrix}1\\0\end{pmatrix}\right) = C\left(\begin{pmatrix}\frac{1}{\sqrt{2}}\\\frac{1}{\sqrt{2}}\end{pmatrix}\otimes\begin{pmatrix}1\\0\end{pmatrix}\right) = \begin{pmatrix}1&0&0&0\\0&1&0&0\\0&0&0&1\\0&0&1&0\end{pmatrix}\begin{pmatrix}\frac{1}{\sqrt{2}}\\0\\\frac{1}{\sqrt{2}}\\0\end{pmatrix} = \begin{pmatrix}\frac{1}{\sqrt{2}}\\0\\0\\\frac{1}{\sqrt{2}}\end{pmatrix}$$

If the product state of two qubits cannot be factored back to the individual qubits, they are said to be entangled.

$$\begin{pmatrix}\frac{1}{\sqrt{2}}\\0\\0\\\frac{1}{\sqrt{2}}\end{pmatrix} = \begin{pmatrix}a\\b\end{pmatrix}\otimes\begin{pmatrix}c\\d\end{pmatrix} \qquad \begin{aligned} ac &= \tfrac{1}{\sqrt{2}}\\ ad &= 0\\ bc &= 0\\ bd &= \tfrac{1}{\sqrt{2}} \end{aligned}$$

This has no solution but does have a 50% chance of collapsing to $|00\rangle$ and 50% to $|11\rangle$.

If a qubit had been initialised to $|1\rangle$ the entangled pair would collapse to $|01\rangle$ or $|10\rangle$, which I will leave to you to prove.

For further reading, two important quantum algorithms are Grover's algorithm[263], for fast searching of unordered databases, and Shor's algorithm[264], which factors exceptionally large numbers exponentially faster than a classical computer.

References

1   Whitelaw, J. 'Profile: Chris Wise'. *The Structural Engineer*. 94(8), August 2016, pp.46–48
2   Hamming, R.W. *Numerical Methods for Scientists and Engineers*. New York: McGraw-Hill, 1962
3   Turing, A.M. 'Computing Machinery and Intelligence'. *MIND*. LIX(236), October 1950, pp.433–460.
     Available at: https://academic.oup.com/mind/article/LIX/236/433/986238 [Accessed: September 2020]
4   Strohmeyer, R. 'The 7 Worst Tech Predictions of All Time'. *PC World*. 31 December 2008.
     Available at: www.pcworld.com/article/155984/worst_tech_predictions.html [Accessed: September 2020]
5   Wise, C. *et al*. 'An amphitheatre for cycling: the design, analysis and construction of the London 2012
     Velodrome'. *The Structural Engineer*. 90(6), June 2012, pp.13–25
6   Tedeschi, A. (Ed.) *AAD Algorithms-aided design: parametric strategies using Grasshopper*.
     Brienza: Le Penseur, 2014
7   Kasparov, G. *Deep Thinking: Where Machine Intelligence Ends and Human Creativity Begins*.
     London: John Murray, 2017
8   Agarwal, A. 'Explained Simply: How an AI program mastered the ancient game of Go'. *freeCodeCamp.org*.
     10 March 2018. Available at: www.freecodecamp.org/news/explained-simply-how-an-ai-program-mastered-the-
     ancient-game-of-go-62b8940a9080/ [Accessed: September 2020]
9   Babbage, C. *Passages from the Life of a Philosopher*. London: Longman, Green, Longman, Roberts & Green,
     1864
10   Verulam. 'An even more quirky structure!'. *The Structural Engineer*. 70(18), September 1992, pp.327–328
11   Verulam. 'Quirky structures'. *The Structural Engineer*. 70(22), November 1992, pp.402–404
12   Gates, B. *The Road Ahead* (2nd edition). London: Penguin, 1996
13   Handsford, M. 'ICE at odds with members on importance of A level maths'. *New Civil Engineer*. 16 January 2003.
     Available at:
     www.newcivilengineer.com/archive/ice-at-odds-with-members-on-importance-of-a-level-maths-16-01-2003/
     [Accessed: September 2020]
14   Ibell, T. 'Virtual by design'. *The Structural Engineer*. 94(3), March 2016, pp.88–89
15   Arnold, W. 'The structural engineer's responsibility in this climate emergency'. *The Structural Engineer*. 98(6),
     June 2020, pp.10–11
16   Ibn-Mohammad, T. *et al*. 'Operational vs. embodied emissions in buildings — A review of current trends'.
     Energy and Buildings. 66, November 2013, pp.232–245
17   Thirion, C. 'Putting the material in the right place: Investigations into the sustainable use of structural materials to
     reduce the initial embodied environmental impact of building structures' [Masters thesis]. UCL, 2013.
     Available at: https://discovery.ucl.ac.uk/id/eprint/1396779/ [Accessed: September 2020]
18   Orr, J. *et al*. 'Minimising energy in construction: practitioners' views on material efficiency'. *Resources,
     Conservation and Recycling*. 140, January 2019, pp.125–136
19   Moynihan, M. and Allwood, J. 'Utilization of structural steel in buildings'. *Proceedings of the Royal Society A*.
     8 August 2014. Available at: https://royalsocietypublishing.org/doi/10.1098/rspa.2014.0170
     [Accessed: September 2020]
20   Petroski, H. *To Engineer is Human: The Role of Failure in Successful Design*. New York: Vintage Books, 1992
21   McCullagh, P. and Nelder, J.A. *Generalized Linear Models*. Berlin: Springer, 1983
22   Box, G.E.P. 'Science and statistics'. *Journal of the American Statistical Association*. 71(356), 1 May 1976,
     pp.791–799
23   Arup, O. 'The world of the structural engineer'. *The Structural Engineer*. 47(1), January 1969, pp.3–12
24   Everything is a Remix (2020). Available at: www.everythingisaremix.info/ [Accessed: September 2020]
25   Campbell, J. *The Hero with a Thousand Faces* (3rd edition). Novato, California: New World Library, 2012
26   Gordon, J.E. *Structures: Or Why Things Don't Fall Down*. London: Penguin, 1978
27   Moore, R. 'Roma Agrawal: "Structural engineers are unsung heroes" '. *The Guardian*, 11 February 2018.
     Available at: www.theguardian.com/artanddesign/2018/feb/11/roma-agrawal-structural-engineers-are-unsung-
     heroes-interview-shard [Accessed: September 2020]
28   Mandelbrot, B. *Les Objets fractals: Forme, Hasard et Dimension* [in French]. Paris: Flammarion, 1975
29   Hegarty, J. *Hegarty on Creativity: There are No Rules*. London: Thames and Hudson, 2014
30   Gell-Mann, M. *The Quark And The Jaguar: Adventures in the Simple and the Complex*. London: Abacus, 1995

31   Billington, D.P. *The Tower and the Bridge: The New Art of Structural Engineering*. Princeton, New Jersey: Princeton University Press, 1985

32   Tusa, J. *On Creativity*. London: Methuen Publishing Ltd, 2004

33   Design Council. *What is the framework for innovation? Design Council's evolved Double Diamond*. Available at: www.designcouncil.org.uk/news-opinion/what-framework-innovation-design-councils-evolved-double-diamond [Accessed: September 2020]

34   Simon, H.A. *The Sciences of the Artificial* (3rd edition). Cambridge, Massachusetts: MIT Press, 1996

35   Belbin. *The Nine Belbin Team Roles*. Available at: www.belbin.com/about/belbin-team-roles/ [Accessed: May 2020]

36   The Myers & Briggs Foundation. *MBTI Basics*. Available at: www.myersbriggs.org/my-mbti-personality-type/mbti-basics/ [Accessed: May 2020]

37   Shirky, C. *Here Comes Everybody: The Power of Organizing Without Organizations*. London: Allen Lane, 2008

38   Arup. *Completing La Sagrada Familia: a collaboration in digital and stone*. Available at: www.arup.com/projects/sagrada-familia [Accessed: September 2020]

39   Sweigart, A. *Automate the Boring Stuff with Python* (2nd Edition). San Francisco, California: No Starch Press, 2019

40   Menabrea, L.F. [translated with notes by Ada Augusta, Countess of Lovelace] *Sketch of The Analytical Engine Invented by Charles Babbage*. Available at: www.fourmilab.ch/babbage/sketch.html [Accessed: September 2020]

41   Leavitt, D. *The Man Who Knew Too Much: Alan Turing and the invention of computers*. London: W&N, 2007

42   Hodges, A. *Alan Turing: The Enigma*. New York: Vintage Books, 2014

43   Turing, A.M. *On Computable Numbers, with an Application to the Entscheidungsproblem*. Available at: https://londmathsoc.onlinelibrary.wiley.com/doi/abs/10.1112/plms/s2-42.1.230 [Accessed: September 2020]

44   Trautman, P.S. 'A Computer Pioneer Rediscovered, 50 Years On'. *New York Times*, 20 April 1994. Available at: https://web.archive.org/web/20161104051054/http:/www.nytimes.com/1994/04/20/news/20iht-zuse.html [Accessed: September 2020]

45   Dyson, G. *Turing's Cathedral: The Origins of the Digital Universe*. New York: Pantheon Books, 2012

46   von Neumann, J. *The Computer & the Brain*. London: Yale University Press, 2012

47   Reis, R. 'How to think like a programmer — lessons in problem solving.' *freeCodeCamp.org*. 10 April 2018 Available at: www.freecodecamp.org/news/how-to-think-like-a-programmer-lessons-in-problem-solving-d1d8bf1de7d2/ [Accessed: September 2020]

48   Debney, P. 'CAD – today is only just the beginning'. *The Structural Engineer*. 77(3), February 1999, pp.16–20

49   Structural-Safety. *Report 614: Columns missing due to 3D modelling*. Available at: www.structural-safety.org/publications/view-report/?report=8837 [Accessed: September 2020]

50   BuildingSMART. *Industry Foundation Classes (IFC)*. Available at: www.buildingsmart.org/standards/bsi-standards/industry-foundation-classes/ [Accessed: September 2020]

51   Carroll, L. *Sylvie and Bruno Concluded*. New York: MacMillan Publishers, 1893

52   Young, T. *A Course of Lectures on Natural Philosophy and the Mechanical Arts, Volume 1*. London: Joseph Johnson, 1807

53   Hellen, T.K. and Becker, A.A. *Finite Element Analysis for Engineers — A Primer*. Glasgow: NAFEMS, 2013

54   Read, C. *Logic: Deductive and Instructive*. London: John Murray, 1915 (first published in 1898)

55   Kannan, R., Hendry, S. and Kaethner, C. 'What is your structural model not telling you?'. *The Structural Engineer*. 94(3), March 2016, pp.44–47

56   Adams, D. *The Hitchhiker's Guide to the Galaxy*. London: Pan Books, 1979

57   Fioretti, G. 'Agent-Based Simulation Models in Organization Science'. *Organizational Research Methods*. 16(2), December 2012, pp.227–242

58   de Saint-Exupéry, A. *Terre Des Hommes* (in French). Paris: Éditions Gallimard, 1939

59   Einstein, A. *Selected Writings*. New Dehli: LeftWord Books, 2003

60   Davison, B. and Owens, G.O. *Steel Designers' Manual* (7th edition). Hoboken, New Jersey: Wiley-Blackwell, 2016

61   Barre de Saint-Venant, A.J.C. *Memoire sur la torsion des prismes* [in French, 1855]. Oxford: Wentworth Press/Blackwell's, 2018

62   Love, A.E.H. *A Treatise on The Mathematical Theory of Elasticity*. Cambridge: Cambridge University Press, 2013

63   Levy, M. and Salvadori, M. *Why Buildings Fall Down: How Structures Fail* (2nd edition). New York: W.W. Norton & Co., 2002

64   Mohammad, A. *Investigation of March 15, 2018 Pedestrian Bridge Collapse at Florida International University, Miami, FL*. Available at: www.osha.gov/doc/engineering/pdf/2019_r_03.pdf [Accessed: September 2020]

65 *BS EN 1993-1-1:2005+A1:2014 Eurocode 3. Design of steel structures. General rules and rules for buildings.*
London: BSI, 2005

66 Brooker, O. 'How to design reinforced concrete flat slabs using Finite Element Analysis'. *The Concrete Centre.*
Available at: www.concretecentre.com/Publications-Software/Publications/How-to-design-reinforced-concrete-flat-slabs-using.aspx [Accessed: September 2020]

67 Arnott, K. 'Shear wall analysis — new modelling, same answers'. *The Structural Engineer.* 83(3), February 2005, pp.20–22

68 Heyman, J. *The Stone Skeleton: Structural Engineering of Masonry Architecture* (1st edition).
Cambridge: Cambridge University Press, 1997

69 Debney, P. 'Footfall vibration and finite element analysis'. *The Structural Engineer.* 86(3), February 2008, pp.37–40

70 Wilford, M. and Young, P. *A Design Guide for Footfall Induced Vibration of Structures.*
Camberley: MPA The Concrete Centre, 2006

71 Murray, T.M. *et al. Design Guide 11: Vibrations of Steel-Framed Structural Systems Due to Human Activity* (Second edition). Chicago: American Institute of Steel Construction, 2016

72 Smith, A.L., Hicks, S.J., and Devine, P.J. *Design of Floors for Vibration: A New Approach* (2nd edition).
Ascot: Steel Construction Institute, 2009

73 *The Matrix.* Motion picture. Village Roadshow Pictures/Groucho II Film Partnership/Silver Pictures,
United States/Australia, 1999

74 Dallard, P. *Buckling: An Approach Based on Geometric Stiffness and Eigen Analysis.*
London: Ove Arup & Partners, 1998

75 McKechnie, S. *Buckling Analysis Using the Dallard Method in GSA.* London: Ove Arup & Partners, 2002

76 Felippa, C. 'Introduction to Finite Element Analysis (Ch.31: Lumped and Consistent Mass Matrices)'.
University of Colorado. Available at: http://kis.tu.kielce.pl/mo/COLORADO_FEM/colorado/IFEM.Ch31.pdf
[Accessed: September 2020]

77 Vibrationdata. *El Centro Earthquake.* Available at: www.vibrationdata.com/elcentro.htm [Accessed: September 2020]

78 *BS EN 1998-1:2004 + A1:2013: Eurocode 8: Design of structures for earthquake resistance. General rules, seismic actions and rules for buildings.* London: BSI, 2005

79 Gleick, J. *Chaos: Making A New Science.* London: Cardinal, 1991

80 Arup. *Nuclear Transport Risk Assessment.* Available at: www.arup.com/projects/nuclear-transport-ra
[Accessed: September 2020]

81 YouTube. *Train Test Crash 1984 — Nuclear Flask Test.* Available at: www.youtube.com/watch?v=ZY446h4pZdc
[Accessed: September 2020]

82 Carfrae, T. and Michael, D. 'Alistair Day and the origins of dynamic relaxation. Part 1: Bringing an idea to life'.
The Structural Engineer. 97(6), June 2019, pp.20–26

83 Carfrae, T. and Michael, D. 'Alistair Day and the origins of dynamic relaxation. Part 2: A practice viewed through projects'. *The Structural Engineer.* 97(7), July 2019, pp.12–18

84 ASCE Technical Council on Forensic Engineering (TCFE). *Failure Case Studies: Hartford Civic Center.*
Available at: https://eng-resources.uncc.edu/failurecasestudies/building-failure-cases/hartford-civic-center/
[Accessed: September 2020]

85 Martin, R. and Delatte, N.J. 'Another Look at Hartford Civic Center Coliseum Collapse'. *Journal of Performance of Constructed Facilities.* 15(1), February 2001, pp.31–36

86 Selby, R.G., Vecchio, F.J. and Collins, M.P. 'The Failure of an Offshore Platform'. *Concrete International.* 19(8), January 1997, pp.28–35

87 Jakobsen, B. and Rosendahl, F. 'The Sleipner Platform Accident'. *Structural Engineering International.* 94(3), 1994, pp.190–193

88 Schlaich, J. and Reineck, K-H. 'Die Ursache Für Den Totalverlust Der Betonplattform Sleipner A' [in German].
Beton- und Stahlbetonbau. 88(1), January 1993, pp.1–4

89 Canterbury Earthquakes Royal Commission. *Final Report: Volume 6: Canterbury Television Building (CTV).*
Available at: https://canterbury.royalcommission.govt.nz/Final-Report-Volume-Six-Contents
[Accessed: September 2020]

90 Structural-Safety. *Report 441: Computer analysis and slab design twisting moments.* October 2014.
Available at: www.structural-safety.org/publications/view-report/?report=4544 [Accessed: October 2020]

91 Structural-Safety. *Report 10: Computer aided design.* November 2005.
Available at: www.structural-safety.org/publications/view-report/?report=2430 [Accessed: October 2020]

92 Kinsell, R.H. 'Raising the Roof — A Picture Story of the Fabrication and Erection of the Hartford Civic Center Coliseum Space Frame Roof Structure'. ASCE Section Website Program. Available at: http://sections.asce.org/connecticut/sites/sections.asce.org.connecticut/files/1975_03_HartfordCenter-compressed.pdf [Accessed: October 2020]

93 *ISO 16730-1:2015: Fire safety engineering — Procedures and requirements for verification and validation of calculation methods — Part 1: General*. Geneva: ISO, 2015

94 Structural-Safety. *Report 372: Understanding the difference between analysis and design*. April 2015. Available at: www.structural-safety.org/publications/view-report/?report=4821 [Accessed: October 2020]

95 New Zealand History. *Christchurch earthquake kills 185*. Available at: https://nzhistory.govt.nz/page/christchurch-earthquake-kills-185 [Accessed: October 2020]

96 Hope, T. *Personal correspondence*

97 Brohn, D. *Understanding Structural Analysis* (3rd edition). Kingsbridge: New Paradigm Solutions Ltd, 2005

98 Adams, D. *Mostly Harmless*. London: Pan Books. 2009

99 Gibbons, O.P. and Orr, J.J. *How to calculate embodied carbon*. The Institution of Structural Engineers. Available at: www.istructe.org/resources/guidance/how-to-calculate-embodied-carbon/ [Accessed: October 2020]

100 SteelConstruction.info. *Cost of structural steelwork*. Available at: www.steelconstruction.info/Cost_of_structural_steelwork [Accessed: October 2020]

101 Nolan, J. 'Cost versus Value — The role of the Consulting Structural Engineer'. *The Structural Engineer*. 90(2), February 2012, pp.13–22

102 Mitchell, M. *Complexity: A Guided Tour*. New York: Oxford University Press, 2011

103 Debney, P. *Braess' Paradox – or Why improving something can make it worse*. May 2012. Available at: www.oasys-software.com/news/braess-paradox-or-why-improving-something-can-make-it-worse/ [Accessed: October 2020]

104 Carfrae, T. *Designing with computers — 2014 Gold Medal Address*. Available at: www.istructe.org/resources/career-profiles/lecture-gold-medal-2014-tristram-carfrae/ [Accessed: October 2020]

105 MX3D (2019). *MX3D Bridge*. Available at: https://mx3d.com/projects/mx3d-bridge/ [Accessed: October 2020]

106 Fairclough, H. *et al*. 'Optimisation-driven conceptual design: case study of a large transfer truss'. *The Structural Engineer*. 97(10), October 2019, pp.20–26

107 Maxwell, J.C. 'On reciprocal figures and diagrams of forces'. *The London, Edinburgh and Dublin Philosophical Magazine and Journal of Science*. 27(4), April 1864, pp.250–261. Available at: www.biodiversitylibrary.org/item/121147#page/264/mode/2up [Accessed: October 2020]

108 Michell, A.G.M. 'The limits of economy of material in frame-structures'. *The London, Edinburgh and Dublin Philosophical Magazine and Journal of Science*. 8(47), 1904, pp.589–597. Available at: doi:10.1080/14786440409463229 [Accessed: October 2020]

109 Hegemier, G.A. and Prager, W. 'On Michell trusses'. *International Journal of Mechanical Sciences*. 11(2), February 1969, pp.209–215

110 Prager, W. 'A note on discretized Michell structures'. *Computer Methods in Applied Mechanics and Engineering*. 3(3), May 1974, pp.349–355

111 Prager, W. 'Optimal layout of cantilever trusses'. *Journal of Optimization Theory and Applications*. 23(1), September 1977, pp.111–117

112 Gibson, T. *et al*. *P150: Design for Manufacture Guidelines*. Available at: www.steelconstruction.info/images/c/c0/SCI_P150.pdf [Accessed: October 2020]

113 Debney, P. *Evolutionary Topology Optimisation with GSA*. Available at: www.oasys-software.com/news/evolutionary-topology-optimisation-gsa/ [Accessed: October 2020]

114 Fairclough, H.E., Pritchard, T.J., He, L., Gilbert, M. (2020) *LayOpt: A truss layout optimization web application*, LimitState Ltd, Sheffield [online] Available at: https://www.layopt.com [Accessed: October 2020]

115 Block, P. *et al*. 'Redefining structural art: strategies, necessities and opportunities'. *The Structural Engineer*. 98(1), January 2020, pp.66–72

116 Allen, E. and Zalewski, W. *Form and Forces: Designing Efficient, Expressive Structures*. Hoboken, New Jersey: John Wiley & Sons, 2009

117 Akbarzadeh, M., van Mele, T. and Block, P. 'Compression-only form finding through finite subdivision of the force polygon'. *Proc. IASS-SLTE 2014 Symp.: Shells, Membranes and Spatial Structures: Footprints*. Available at: www.block.arch.ethz.ch/brg/files/akbarzadeh-2014-iass-compression-form-finding-finite-subdivision-force-polygon_1410358989.pdf [Accessed: October 2020]

118 YouTube. *Reimagining Shell Structures — Philippe Block*.
Available at: www.youtube.com/watch?v=vAavRx7uoeA&t=2607s [Accessed: October 2020]

119 Block, P. 'Equilibrium systems: studies in masonry structure'. *MIT Libraries*.
Available at: http://hdl.handle.net/1721.1/32096 [Accessed: October 2020]

120 Block, P. 'Thrust Network Analysis: exploring three-dimensional equilibrium'. *MIT Libraries*.
Available at: http://hdl.handle.net/1721.1/49539 [Accessed: October 2020]

121 Davis, L. *et al*. 'Innovative funicular tile vaulting: A prototype vault in Switzerland'. *The Structural Engineer*.
90(11), November 2012, pp.46–56

122 Youtube. *William F. Baker — On The Harmony Of Theory And Practice In The Design Of Tall Buildings*.
Available at: www.youtube.com/watch?v=_XrvfBlcWcs [Accessed: July 2020]

123 Baker, W.F. 'Structural Innovation: Combining Classic Theories with New Technologies'. *Engineering Journal*.
52(3), 2015, pp.203–217

124 Baker, W.F. *et al*. 'Maxwell's reciprocal diagrams and discrete Michell frames'. *Structural and Multidisciplinary Optimization*. 48(2), March 2013, pp.267–277

125 Beghini, L.L. *et al*. 'Structural optimization using graphic statics'. *Structural and Multidisciplinary Optimization* 49(3), March 2014, pp.351–366

126 Hartz, C. *et al*. 'The Application of 2D and 3D Graphic Statics in Design'. *Journal of the International Association for Shell and Spatial Structures*. 59(4), December 2018, pp.235–242

127 Debney, P. *'Form-Finding Asymmetric Bridges'*.
Available at: www.oasys-software.com/news/form-finding-asymmetric-bridges/ [Accessed: October 2020]

128 YouTube. *Peter Debney — Form-Finding: Fabrics, Cables, & Shells*.
Available at: www.youtube.com/watch?v=32b4-t8q5x0 [Accessed: October 2020]

129 Lehne, J. and Preston, F. *'Making Concrete Change: Innovation in Low-Carbon Cement and Concrete'*.
Available at: www.chathamhouse.org/sites/default/files/publications/research/2018-06-13-making-concrete-change-cement-lehne-preston.pdf [Accessed: October 2020]

130 Meyer, C. *Concrete Materials and Sustainable Development in the United States*.
Available at: www.columbia.edu/cu/civileng/meyer/publications/publications/83%20Concrete%20Materials%20and%20Sustainable%20Development.pdf [Accessed: October 2020]

131 wbcsd. 2020. *'Low Carbon Technology Partnerships initiative: Cement'*.
Available at: http://docs.wbcsd.org/2015/11/LCTPi-Cement-Report.pdf [Accessed: October 2020]

132 Schmit, L.A. 'Structural Design by Systematic Synthesis'. *Proc. 2nd ASCE Conf. on Electronic Computation*,
1960, pp.105–122

133 Belegundu, A.D. and Chandrupatla, T.R. *Optimization Concepts and Applications in Engineering* (2nd edition).
Cambridge: Cambridge University Press, 2011

134 Solnit, R. *A Field Guide to Getting Lost*. Edinburgh: Canongate Books, 2006

135 British Steel (2020). *Sections (Price extras pounds sterling)*.
Available at: https://britishsteel.co.uk/media/323461/sections-price-list.pdf [Accessed: October 2020]

136 AECOM. *Spon's Architects' and Builders' Price Book 2020*. Boca Raton, Florida: CRC Press, 2019

137 The Institution of Structural Engineers. *Manual for the design of steelwork building structures* (3rd edition).
London: IStructE Ltd., 2008

138 Approved Index (2020). *Steel Building Prices*.
Available at: www.approvedindex.co.uk/steel-buildings/steel-building-prices [Accessed: October 2020]

139 Weisstein, E.W. *Newton's Method*. Available at: http://mathworld.wolfram.com/NewtonsMethod.html
[Accessed: October 2020]

140 Rutten, D. *Evolutionary Principles applied to Problem Solving*.
Available at: www.grasshopper3d.com/profiles/blogs/evolutionary-principles [Accessed: October 2020]

141 Mitchell, M. *An Introduction to Genetic Algorithms (Complex Adaptive Systems)*. Cambridge,
Massachusetts: MIT Press, 1998

142 Richards, M., Smith, N. and Dobbyn C. *Natural and Artificial Intelligence: Evolutionary Computing* (2nd edition).
London: Open University Press, 2009

143 Dias, W.P.S. 'Engineering as cyclic problem solving — some insights from Karl Popper'. *The Structural Engineer*.
85(2), January 2007, pp.32–37

144 Jenkins, W.M. 'Structural optimisation with the genetic algorithm'. *The Structural Engineer*. 69(24),
December 1991, pp.418–422

145 Britannica.com (2020). *Lamarckism*. Available at: www.britannica.com/science/Lamarckism
[Accessed: October 2020]

146 Dobbyn, C. *et al*. *Natural and Artificial Intelligence: Natural Intelligence* (2nd edition).
London: Open University Press, 2008

147 Li, L.J. *et al*. 'A heuristic particle swarm optimizer for optimization of pin connected structures'.
Computers & Structures. 85(7–8), April 2007, pp.340–349

148 Dorigo, M. and Stützle, T. *Ant Colony Optimization*. Cambridge, Massachusetts: MIT Press, 2004

149 Dorigo, M., Birattari, M. and Stützle, T. 'Ant colony optimization'. *IEEE Computational Intelligence Magazine*. 1(4),
November 2006, pp.28–39

150 Trudeau, R.J. *Introduction to Graph Theory*. New York: Dover Publications, 2003

151 Oasys Software (2019). *Optimisation of High-Rise Structures*.
Available at: www.oasys-software.com/case-studies/optimisation-of-high-rise-structures/ [Accessed: October 2020]

152 Hill, T. 'Knowing When To Stop: How to gamble if you must — the mathematics of optimal stopping'. *American Scientist*. 97(2), March/April 2009, pp.126–133

153 Wolpert, D.H. and Macready, W.G. 'No free lunch theorems for optimization'. *IEEE Transactions on Evolutionary Computation*. 1(1), April 1997, pp.67–82

154 Descartes, R. *Principles of philosophy* [1644]. SMK Books, 2009

155 McCarthy, J. *et al*. *A proposal for the Dartmouth Summer Research Project on Artificial Intelligence*.
Available at: http://jmc.stanford.edu/articles/dartmouth/dartmouth.pdf [Accessed: October 2020]

156 Wojcik, S. *5 things to know about bots on Twitter*.
Available at: www.pewresearch.org/fact-tank/2018/04/09/5-things-to-know-about-bots-on-twitter
[Accessed: October 2020]

157 Licklider, J.C.R. 'Man-Computer Symbiosis'. *IRE Transactions on Human Factors in Electronics*. 1(1),
March 1960, pp.4–11

158 Kurzweil, R. *The Singularity is Near*. London: Duckworth Books, 2006

159 Deeplearning.ai (2019). *AI For Everyone*. Available at: www.deeplearning.ai/ai-for-everyone/
[Accessed: October 2020]

160 Searle, J.R. 'Minds, Brains, and Programs'. *Behavioral and Brain Sciences*. 3(3), September 1980, pp.417–424

161 Penrose, R. *The Emperor's New Mind* (2nd edition). Oxford: Oxford University Press, 1999

162 *Rain Man*. Motion picture. United Artists, United States, 1988

163 IBM.com (2020). *IBM 100: Icons of progress — Deep Blue*.
Available at: www.ibm.com/ibm/history/ibm100/us/en/icons/deepblue/ [Accessed: October 2020]

164 Silver, D. *et al*. 'Mastering the game of Go with deep neural networks and tree search'. *Nature*. 529(7587),
January 2016, pp.484–489

165 Leibniz, G.W. 'On Universal Synthesis and Analysis, or the Art of Discovery and Judgment'. *Philosophical Papers and Letters*. 1679, pp.229–234

166 IBM.com (2020). *Watson Anywhere*. Available at: www.ibm.com/uk-en/watson [Accessed: October 2020]

167 IBM.com (2009). *News release: IBM Developing Computing System to Challenge Humans on America's Favorite Quiz Show, Jeopardy!* Available at: www-03.ibm.com/press/us/en/pressrelease/27324.wss
[Accessed: October 2020]

168 Davis, E., Morgenstern, L. and Ortiz, C. *The Winograd Schema Challenge*.
Available at: https://cs.nyu.edu/faculty/davise/papers/WinogradSchemas/WS.html [Accessed: October 2020]

169 Susskind, R. and Susskind, D. *The Future of The Professions*. Oxford: Oxford University Press, 2015

170 Autodesk (2020). *Revit Products: Automatic checks*. Available at: https://knowledge.autodesk.com/support/revit-products/learn-explore/caas/CloudHelp/cloudhelp/2014/ENU/Revit/files/GUID-2EBFF6A4-0BA9-486B-9CA2-A53F4DE36003-htm.html [Accessed: October 2020]

171 Autodesk (2019). *Autodesk Model Checker for Revit*.
Available at: www.biminteroperabilitytools.com/modelchecker.php [Accessed: October 2020]

172 Mitchell, M. *Artificial Intelligence: A Guide for Thinking Humans*. London: Pelican Books, 2019

173 TrainingZone (2017). *Curse of Knowledge: why experts struggle to explain their know-how*.
Available at: www.trainingzone.co.uk/develop/talent/curse-of-knowledge-why-experts-struggle-to-explain-their-know-how [Accessed: October 2020]

174 Lidwell, W., Butler, J. and Holden, K. *The Pocket Universal Principles of Design*.
Beverly, Massachusetts: Rockport Publishing, 2015

175 Smith, N. *Natural and Artificial Intelligence: Symbolic Intelligence* (Second edition).
Milton Keynes: Open University Press, 2009

176 Norman, D. *The Design of Everyday Things*. Cambridge, Massachusetts: MIT Press, 2013

177 Rota, G-C. 'In Memoriam of Stan Ulam — The Barrier Of Meaning'. *Physica D: Nonlinear Phenomena*. 22(1–3), Oct–Nov 1986, pp.1–3

178 Lowe, D. '"Artificial Intelligence" or statistics?'. *Significance*. 16(4), August 2019, p.7

179 Kubat, M. *An Introduction to Machine Learning*. New York: Springer Publishing, 2015

180 Medium (2018). *What is underfitting and overfitting in Machine Learning and how to deal with it*. Available at: https://medium.com/greyatom/what-is-underfitting-and-overfitting-in-machine-learning-and-how-to-deal-with-it-6803a989c76 [Accessed: October 2020]

181 Surowiecki, J. *The Wisdom of Crowds: Why the Many Are Smarter Than the Few*. London: Abacus, 2005

182 BBC (2014). *Wisdom of the Crowd: The Myths and Realities*. Available at: www.bbc.com/future/article/20140708-when-crowd-wisdom-goes-wrong [Accessed: October 2020]

183 Fry, H. *Hello World: How to be Human in the Age of the Machine*. London: Transworld Publishers, 2018

184 Hawkins, J. and Blakeslee, S. *On Intelligence: How a New Understanding of the Brain Will Lead to the Creation of Truly Intelligent Machines*. New York: St Martin's Press, 2004

185 Luger, G.F. and Stubblefield, W.A. *Artificial Intelligence* (Third edition). Harlow: Addison-Wesley Longman, Inc., 1997

186 The Human Memory (2020). *Brain Neurons and Synapses*. Available at: https://human-memory.net/brain-neurons-synapses/ [Accessed: October 2020]

187 Rose, S. *The Britannica Guide to the Brain*. London: Robinson Publishing, 2008

188 Minsky, M. and Papert, S. *Perceptrons: An Introduction to Computational Geometry*. Cambridge, Massachusetts: MIT Press, 1969

189 Engineering.com (2018). *Artificial Intelligence Detects Construction Problems in Hours, Not Weeks*. Available at: www.engineering.com/BIM/ArticleID/16674/Artificial-Intelligence-Detects-Construction-Problems-in-Hours-Not-Weeks.aspx [Accessed: October 2020]

190 David Cope. *Experiments in Musical Intelligence*. Available at: http://artsites.ucsc.edu/faculty/cope/experiments.htm [Accessed: October 2020]

191 Boole, G. *An investigation of the Laws of Thought, on which are founded the Mathematical Theories of Logic and Probabilities* [1854]. Cambridge: Cambridge University Press, 2009

192 Medium (2018). *Genetic Algorithms + Neural Networks=Best of Both Worlds*. Available at: https://towardsdatascience.com/gas-and-nns-6a41f1e8146d [Accessed: October 2020]

193 Einstein, A. *Sidelights on Relativity* [1922]. New York: Dover Publications, 2010

194 Tegmark, M. *Life 3.0: Being Human in the Age of Artificial Intelligence*. London: Allen Lane, 2017

195 Shelley, M. *Frankenstein* [1818]. London: Alma Classics, 2017

196 *R.U.R. (Rossumovi Univerzální Roboti): A Fantastic Melodrama in Three Acts and an Epilogue*. Play. Doubleday, Page and Company, 1923

197 *The Terminator*. Motion picture. Orion Pictures, United States, 1984

198 tylervigen.com (2020). *Spurious correlations*. Available at: www.tylervigen.com/spurious-correlations [Accessed: October 2020]

199 KD nuggets (2019). "Why ice cream is linked to shark attacks — correlation/causation smackdown". Available at: www.kdnuggets.com/2019/01/dr-data-ice-cream-linked-shark-attacks.html [Accessed: October 2020]

200 McKinsey & Company (2019). *Tackling bias in artificial intelligence (and humans)*. Available at: www.mckinsey.com/featured-insights/artificial-intelligence/tackling-bias-in-artificial-intelligence-and-in-humans# [Accessed: October 2020]

201 Berkeley DeepDrive (2020). *Robust Visual Understanding in Adversarial Environments*. Available at: https://deepdrive.berkeley.edu/project/robust-visual-understanding-adversarial-environments [Accessed: October 2020]

202 Wald, A. *A method of estimating plane vulnerability based on damage of survivors*. Available at: https://apps.dtic.mil/docs/citations/ADA091073 [Accessed: October 2020]

203 Bongard, M. *Pattern Recognition*. New York: Macmillan Publishing, 1970

204 Stabinger, S., Rodríguez-Sánchez, A. and Piater, J. '25 years of CNNs: Can we compare to human abstraction capabilities?'. *Proc. Int. Conf. on Artificial Neural Networks (ICANN 2016)*, pp.380–387

205 Li, L. *et al*. 'Improved genetic algorithm and its application to determination of critical slip surface with arbitrary shape in soil slope'. *Frontiers of Architecture and Civil Engineering in China*. 2, May 2008, pp.145–150

206 Reale, C. *et al*. 'Automatic classification of fine-grained soils using CPT measurements and Artificial Neural Networks'. *Advanced Engineering Informatics*. 36, April 2018, pp.207–215

207 Arup.com (2019). *Water Research Review*.
Available at: www.arup.com/perspectives/publications/research/section/water-research-review
[Accessed: October 2020]

208 Arup.com (2019). *Machine learning graduates from Facebook to the built environment*.
Available at: www.arup.com/perspectives/machine-learning-graduates-from-facebook-to-the-built-environment
[Accessed: October 2020]

209 DeVries, P.M.R. *et al.* 'Deep learning of aftershock patterns following large earthquakes'. *Nature*. 560,
August 2018, pp.632–634

210 Casas, J.R. and Moughty, J.J. 'Bridge damage detection based on vibration data: past and new developments'.
Frontiers in Built Environment. 3(4), February 2017. Available at: doi.org/10.3389/fbuil.2017.00004
[Accessed: October 2020]

211 Doxel Inc. (2020). *Artificial Intelligence for Construction Productivity*. Available at: www.doxel.ai/
[Accessed: October 2020]

212 YouTube (2018). *Spot Robot Testing at Construction Sites*.
Available at: www.youtube.com/watch?v=wND9goxDVrY [Accessed: October 2020]

213 YouTube (2020). *Structural steel: Robot technology for structural steel automation*.
Available at: www.youtube.com/playlist?list=PLfAPt5-t5BSH__kEvMXr4bdExxVvD6_7U [Accessed: October 2020]

214 Redshift by Autodesk (2019). *This bricklaying robot is changing the future of construction*.
Available at: www.autodesk.com/redshift/bricklaying-robot/ [Accessed: October 2020]

215 The Construction Index (2018). *Brick-laying robot reaches the UK*.
Available at: www.theconstructionindex.co.uk/news/view/brick-laying-robot-reaches-the-uk
[Accessed: October 2020]

216 Frey, C.B. and Osborne, M.A. Working paper: 'The future of employment: How susceptible are jobs to
computerisation?'. *Oxford Martin Programme on the Impacts of Future Technology* (Oxford Martin School,
University of Oxford). Available at: www.oxfordmartin.ox.ac.uk/downloads/academic/future-of-employment.pdf
[Accessed: October 2020]

217 Medium (2018). *How 22 Years of AI Superiority Changed Chess*.
Available at: https://towardsdatascience.com/how-22-years-of-ai-superiority-changed-chess-76eddd061cb0
[Accessed: October 2020]

218 Deepmind.com (2017). *Exploring the mysteries of Go with Alphago and China's top players*.
Available at: https://deepmind.com/blog/article/exploring-mysteries-alphago [Accessed: October 2020]

219 de Saint-Exupéry, A. *Citadelle (The Wisdom of the Sands)* [1948]. New York: Amereon Ltd., 2006

220 Betancourt, M. *Automated Labor: The 'New Aesthetic' And Immaterial Physicality*.
Available at: https://journals.uvic.ca/index.php/ctheory/article/view/14934/5827 [Accessed: October 2020]

221 YouTube (2018). *The Robot Revolution: The New Age of Manufacturing: Moving Upstream*.
Available at: www.youtube.com/watch?v=HX6M4QunVmA [Accessed: October 2020]

222 YouTube (2018). *The Construction Robots Are Coming*. Available at: www.youtube.com/watch?v=nKGGHdl3NyQ
[Accessed: October 2020]

223 Health and Safety Executive (2003). *Causal factors in construction accidents*.
Available at: www.hse.gov.uk/research/rrpdf/rr156.pdf [Accessed: October 2020]

224 Amazon.com (2019). *Amazon Prime Air*.
Available at: www.amazon.com/Amazon-Prime-Air/b?ie=UTF8&node=8037720011 [Accessed: October 2020]

225 IceFormwork.com (2020). *A model of lean digital fabrication for precast concrete industry*.
Available at: https://iceformwork.com/ [Accessed: October 2020]

226 Foster + Partners (2020). *Lunar Habitation*. Available at: www.fosterandpartners.com/projects/lunar-habitation/
[Accessed: October 2020]

227 NASA (2020). *STMD: Centennial Challenges*.
Available at: www.nasa.gov/directorates/spacetech/centennial_challenges/3DPHab/index.html
[Accessed: October 2020]

228 The New York Times (2011). *The iPad in Your Hand: As Fast as a Supercomputer of Yore*. Available at:
https://bits.blogs.nytimes.com/2011/05/09/the-ipad-in-your-hand-as-fast-as-a-supercomputer-of-yore/
?searchResultPosition=1 [Accessed: October 2020]

229 Guinness World Records (2015). *1973: First Mobile Phone Call*.
Available at: https://www.guinnessworldrecords.com/news/60at60/2015/8/1973-first-mobile-phone-call-392969
[Accessed: October 2020]

230 FutureLearn (2020). *Understanding Quantum Computers*.
Available at: www.futurelearn.com/courses/intro-to-quantum-computing [Accessed: October 2020]

231 Barz, S. *et al*. 'A two-qubit photonic quantum processor and its application to solving systems of linear equations'. *Scientific Reports*. 4, August 2014. Available at: doi:10.1038/srep06115 [Accessed: October 2020]

232 Schleich, P. 'How to solve a linear system of equations using a quantum computer'. [Seminar report]
Available at: www.mathcces.rwth-aachen.de/_media/3teaching/00projects/schleich.pdf [Accessed: October 2020]

233 Shao, C. 'Computing Eigenvalues of Matrices in a Quantum Computer'.
Available at: https://arxiv.org/abs/1912.08015 [Accessed: October 2020]

234 Zhou, X-Q. *et al*. 'Calculating unknown Eigenvalues with a Quantum Algorithm'. *Nature Photonics*. 7, February 2013, pp.223–228

235 Avanade (2019). *Quantum Computing: An optimization example*.
Available at: www.avanade.com/nl-nl/blogs/be-orange/technology/quantum-computing-an-optimization-example [Accessed: October 2020]

236 Beer, K. *et al*. 'Training deep quantum neural networks'. *Nature Communications*. 11(808), February 2020.
Available at: doi:10.1038/s41467-020-14454-2 [Accessed: October 2020]

237 D-Wave (2020). *What is Quantum Annealing?* Available at: https://docs.dwavesys.com/docs/latest/c_gs_2.html [Accessed: October 2020]

238 Einstein, A., Podolsky, B. and Rosen, N. 'Can Quantum-Mechanical Description of Physical Reality Be Considered Complete?'. *Physical Review*. 47, May 1935, pp.777–780

239 Forbes (2017). *How Do You Create Quantum Entanglement?*
Available at: www.forbes.com/sites/chadorzel/2017/02/28/how-do-you-create-quantum-entanglement/ [Accessed: October 2020]

240 UKGBC.org (2019). *Net Zero Carbon Buildings: A Framework Definition*. Available at:
www.ukgbc.org/wp-content/uploads/2019/04/Net-Zero-Carbon-Buildings-A-framework-definition.pdf [Accessed: October 2020]

241 Orr, J.J. *et al*. 'Concrete structures using fabric formwork'. *The Structural Engineer*. 89(8), April 2011, pp.20–26

242 Digital Building Technologies (2018). *The Smart Slab*. Available at: http://dbt.arch.ethz.ch/project/smart-slab/ [Accessed: October 2020]

243 Institution of Structural Engineers (2018). *Knitting bespoke reinforcement for new concrete structures*.
Available at: www.istructe.org/resources/training/lecture-knitting-bespoke-reinforcement-concrete/ [Accessed: October 2020]

244 Wood Design & Building (2018). *Computational Design with Timber*.
Available at: www.wooddesignandbuilding.com/computational-design-timber/ [Accessed: October 2020]

245 Fivet, C. and Brutting, J. 'Nothing is lost, nothing is created, everything is reused: structural design for a circular economy'. *The Structural Engineer*. 98(1), January 2020, pp.74–81

246 The Institution of Structural Engineers (2020). *Climate emergency*.
Available at: www.istructe.org/resources/climate-emergency/ [Accessed: October 2020]

247 Timoshenko, S.P. and Young, D.H. *Theory of Structures* (2nd edition). New York: McGraw-Hill, 1965

248 Howard, A. and Rorres, C. *Elementary Linear Algebra with Supplemental Applications* (11th edition).
New York: John Wiley & Sons Inc., 2014

249 Parker, J. 'Milne Medal 2018: Unexpected education'. *The Structural Engineer*. 97(6), June 2019, pp.12–19

250 *BS 6841:1987: Guide to measurement and evaluation of human exposure to whole-body mechanical vibration and repeated shock*. London: BSI, 1987

251 Wolfram MathWorld (2020). *Parabola*. Available at: https://mathworld.wolfram.com/Parabola.html [Accessed: October 2020]

252 Wolfram MathWorld (2020). *Catenary*. Available at: https://mathworld.wolfram.com/Catenary.html [Accessed: October 2020]

253 The New York Times (1990). *What if They Closed 42nd Street and Nobody Noticed?*
Available at: http://www.nytimes.com/1990/12/25/health/what-if-they-closed-42d-street-and-nobody-noticed.html [Accessed: October 2020]

254 The Guardian (2006). *Heart and soul of the city*.
Available at: www.guardian.co.uk/environment/2006/nov/01/society.travelsenvironmentalimpact [Accessed: October 2020]

255 Scientific America (2009). *Removing Roads And Traffic Lights Speeds Urban Travel*.
Available at: www.scientificamerican.com/article.cfm?id=removing-roads-and-traffic-lights [Accessed: October 2020]

256 Braess, D., Nagurney, A. and Wakolbinger, T. 'On a Paradox of Traffic Planning'. *Transportation Science*. 39(4), November 2005, pp.446–450

257 Virtual Cell Program (2019). *Braess-Paradox*. Available at: http://vcp.med.harvard.edu/braess-paradox.html [Accessed: October 2020]

258 YouTube (2011). *Braess' Paradox: A Physical Demonstration*. Available at: www.youtube.com/watch?v=nMrYlspifuo [Accessed: October 2020]

259 The Economist (2008). *Queuing conundrums*. Available at: www.economist.com/node/12202559 [Accessed: October 2020]

260 Youn, H., Gastner, M.T. and Jeong, H. 'The Price of Anarchy in Transportation Networks: Efficiency and Optimality Control'. *Physical Review Letters*. 101, October 2008.

261 Hagstrom, J.N. and Abrams, R.A. 'Characterizing Braess's Paradox for Traffic Networks'. *Proc. IEEE ITSC 2001*. pp.836–841

262 Microsoft (2018). *Quantum Computing for Computer Scientists*. Available at: www.microsoft.com/en-us/research/video/quantum-computing-computer-scientists/ [Accessed: October 2020]

263 Grover, L.K. 'A fast quantum mechanical algorithm for database search'. *Proc. STOC*. pp.212–219

264 Shor, P.W. 'Polynomial-Time Algorithms for Prime Factorization and Discrete Logarithms on a Quantum Computer'. *SIAM Journal on Computing*, 26(5), 1997, pp.1484–1509